Plastic Remediation

Scrivener Publishing
100 Cummings Center, Suite 541J
Beverly, MA 01915-6106

Publishers at Scrivener
Martin Scrivener (martin@scrivenerpublishing.com)
Phillip Carmical (pcarmical@scrivenerpublishing.com)

Plastic Remediation

Transformative Pathways for a Rejuvenated Sustainable Environment

Edited by

Neha Agarwal
Department of Chemistry, Navyug Kanya Mahavidyalaya, Affiliated with the University of Lucknow, Lucknow, Uttar Pradesh, India

Brijesh Pare
Department of Chemistry, Government Madhav Science Post Graduate College, Ujjain, Madhya Pradesh, India

Vijendra Singh Solanki
Department of Chemistry, Institute of Science and Research, IPS Academy, Indore, Madhya Pradesh, India

and

Sankara Rao Miditana
Department of Chemistry, Government Degree College, Puttur, Andhra Pradesh, India

Scrivener Publishing

WILEY

Preface

Plastic pollution has emerged as one of the most challenging environmental crises of the 21st century, threatening ecosystem stability, biodiversity, and long-term sustainability. As global plastic production continues to rise, so does the urgency to develop innovative, scalable, and ecosystem-friendly strategies for managing and remediating plastic waste. Plastic Remediation: Transformative Pathways for a Rejuvenated Sustainable Environment brings together diverse scientific, technological, and community-driven perspectives to illuminate the multifaceted nature of this challenge and the opportunities for impactful solutions.

The book is organized into three comprehensive parts. Part I lays the scientific foundation by examining the sources, types, and persistence of plastic waste, while unveiling its profound impacts on marine, terrestrial, and ecological systems. Part II highlights various cutting-edge innovations from nanotechnology and nano-biomaterials to biological degradation, catalytic depolymerization, and waste-to-energy technologies, showcasing many promising pathways toward sustainable remediation. Part III shifts the focus to policy frameworks, circular economy strategies, and community-led interventions that are essential for translating scientific advancements into real-world outcomes.

By integrating fundamental knowledge with pioneering advancements and forward-looking policy insights, this volume aspires to guide researchers, policymakers, and practitioners in reimagining plastic remediation. It emphasizes that a truly sustainable future requires collaborative action, technological innovation, and a shared responsibility toward environmental stewardship. This book stands as a collective effort to inspire transformative progress toward a cleaner, healthier, and more resilient planet.

The Editors

Acknowledgements

We extend our deepest gratitude to all the contributors whose dedication, expertise, and scholarly commitment made Plastic Remediation: Transformative Pathways for a Rejuvenated Sustainable Environment possible. Each chapter in this volume reflects the hard work, scientific rigor, and forward-thinking vision of researchers, academicians, and practitioners from diverse disciplines, all united by a shared goal of addressing the global plastic crisis.

Our sincere thanks go to the editorial and review team for their meticulous evaluation, constructive feedback, and unwavering support throughout the development of this book. Their insights have significantly enriched the clarity, depth, and coherence of the chapters.

We are immensely grateful to the institutions, research laboratories, and organizations that provided intellectual, infrastructural, and administrative support to the authors and editors. Their encouragement has been instrumental in fostering an environment conducive to interdisciplinary collaboration and innovation.

We also acknowledge the global community of environmental scientists, policymakers, and activists whose ongoing efforts to combat plastic pollution continue to inspire meaningful research and action. Their work reinforces the importance of sustainable pathways addressed within this book.

Furthermore, we express heartfelt appreciation to our families and colleagues for their patience, encouragement, and constant motivation throughout this journey. Without their support, the completion of this volume would not have been possible. Finally, we thank Martin Scrivener and the team at Scrivener Publishing for their efforts and guidance in bringing this work to publication.

This book is dedicated to all who strive for a cleaner, healthier, and more sustainable planet.

Part I

FOUNDATIONAL INSIGHTS: DECODING PLASTIC WASTE AND UNRAVELING ITS FAR-REACHING IMPACTS

1.1 Introduction

The Ancient Greek plastikos, which means "able to be shaped or molded," is from where the word plastic originates. Plastics are such polymers that come under the primary component of a wide range of synthetic or semi-synthetic materials. Owing to its fluidity, it can be extruded, molded, or pressed into solid stuff of different shapes. Plastics are widely used around the world due to their exceptional properties, including strength, flexibility, non-toxicity, affordability, and lightweight nature, which make them highly adaptable for various applications. The production of plastics primarily relies on petroleum and natural gas as key raw materials. Renewable resources like polylactic acid share little roles in the manufacturing of plastic. It is estimated that 9.2 billion metric tons of plastic were produced between 1950 and 2017, with almost half of that amount occurring since 2004. According to preliminary data, global plastic production exceeded 400 million metric tons in 2023 alone. If current trends in plastic demand persist, global plastic production is projected to exceed 1.3 billion metric tons annually by 2060. Approximately 40% of plastic consumption is attributed to packaging, while around 20% is used in the building and construction sector, making these the two largest applications of plastic [1, 2].

Speculations about the probable threats to flora and fauna point to the widespread presence of plastic pollution. This pollution results from human activities as well as natural occurrences. According to research studies, when plastic waste enters the environment, it takes nearly 100 to 1000 years or longer to disintegrate, depending on the conditions. Plastic pollution can break down into smaller particles ranging in size from five millimeters and one nanometer, known as MPs and nanoplastics. MPs and nanoplastics are found in every ecosystem on Earth, from tropical coral reefs to the Antarctic tundra. This chapter discusses various aspects of plastic pollution, including its types and sources.

1.2 Impacts, Sources, Types and Persistence of Plastics

1.2.1 Environmental Impacts of Plastics and Sources

The aquatic environment is under threat due to plastic degradation, as aquatic creatures may ingest plastic and suffer from suffocation. They often become entangled in plastic waste. According to research reports, more than 1500 species living in marine and terrestrial habitats have been found to consume plastic. OECD (Organization for Economic Cooperation and Development) estimates that in 2019, 90% of the greenhouse gas emissions from plastic products—accounting for 3.4% of global emissions—originated from the production and conversion of fossil fuels into new plastic products over the course of their life cycles. If human behavior regarding the use of plastic does not change, greenhouse gas (CO_2, CH_4, etc.) emissions associated with the life cycle of plastic products are expected to double by 2060, according to the OECD. According to the predictions by the World Economic Forum, the global plastics industry could be responsible for up to 15% of carbon emissions and 20% of total oil consumption by 2050, in the absence of effective measures to prevent greenhouse gas emissions [2]. According to Kim *et al.* [3], the development of biodegradable plastics is utmost necessary in order to curb the harmful effects of waste plastics on the environment. One hopeful solution to the problem of plastic waste is the use of biodegradable plastics, rather than allowing plastic to accumulate in the soil and oceans. Microbes break down biodegradable plastics into smaller molecules, such as H_2O, CO_2, and CH_4 [3].

In India, cows often consume plastic waste commonly found in garbage dumps, which can lead to intestinal ailments. Traces of plastic have also been discovered in the stomachs of endangered California condors and various land birds [4]. Even though plastics decompose in the environment, as stated by James *et al.* [5], the degradation periods of plastic products have been analyzed in recent studies [6–14]. These estimates, which can range from months to decades or more, are highly variable [8]. Plastics are broken up, broken down, transformed, assimilated, and mineralized by both, biotic and abiotic processes [6, 13, 14]. Environmental factors, the kind of plastic, and the product's geometry and functionality all influence the effectiveness and selectivity of these processes [6]. Therefore, it is important to consider environmental degradation when designing plastic products. Regardless of waste management practices and end-of-life policies, certain plastic items will inevitably pollute the environment. Thus, addressing the persistence of these materials is essential [15–18].

Environmental persistence is a fundamental principle in legal frameworks, as well as in green chemistry and engineering [19–21]. It provides valuable information; the more persistent a plastic material is, the greater its potential to harm ecosystems and human health [16, 17, 22]. Thus, there is a chance to reduce threats to ecosystems and human health by taking persistence into account when designing products and choosing materials that decompose rapidly when released into the environment [5]. Reduce, Reuse,

Recycle (also known as the 3R policy) is a waste management strategy that promotes the recycling and responsible use of materials such as plastics [23]. In waste management, the three "R's" stand for Reduce (the most preferred option), Reuse, and Recycle (the least preferred choice) [24, 25]. All the components corresponding to 3R policy are interconnected (Table 1.1) [23], and the concept was developed to maximize resource utilization before disposal. There are two approaches to implementing the 3R policy: (i) on-site promotion, which involves treatment at the source of generation, prior to discharge by the individual, and (ii) off-site promotion, which involves handling services provided by private enterprise and/or the government agencies after waste collection (Table 1.1). The transition must be addressed through a value chain that includes (i) production, (ii) product design, (iii) usage and reuse, and (iv) end-of-life, in order to establish a model that supports a circular economy in plastic waste management.

Initiatives supporting the development of more environmentally friendly plastics are being incorporated into the product design. For example, EU Directive 2019/904 of June 5, 2019, aims to reduce the environmental impact of specific plastic products by establishing design and performance requirements [26]. To minimize environmental damage, manufacturers also have an obligation to design and produce plastics that are more easily retained within the circular value chain. The EU Commission Regulation 10/2011 of January 14, 2011, is another regulation that focuses on the product design of plastic materials intended for contact with food stuffs. It specifies guidelines for the chemical composition of plastic products that come into contact with food [23].

According to the United Nations Environment Program, MPs have also been found in human livers, kidneys, and placentas. Furthermore, carcinogenic chemicals from plastic products can leach into tap water, potentially causing immunological, neurological, reproductive, and developmental disorders, according to the International Union for Conservation of Nature. Similar concerns regarding endocrine-disrupting effects have been raised in certain animal studies. However, further research is needed to fully understand the potential impacts of MPs on human health [2].

1.2.2 Effects of Plastic Pollution on Human Health

MPs have been detected in human blood and placentas, as well as in commonly consumed food and beverages such as beer, salt, and tap water. Many of the chemicals used in plastic production are known to be carcinogenic and can adversely affect the immune, reproductive, developmental,

Table 1.1 Different ways to promote the 3R policy [23].

S. no.	Element	Approaches	Detailed methods
1	Reduce	Government sets some guidelines and a legal framework	• Polluter-Pays standard • Long-drawn-out producer responsibility • Encouragement of the production of environmentally friendly goods • Campaign for reduced packaging and green purchasing schemes (e.g., eco-label systems) • Buy less and use fewer materials
		Raise public awareness	• Purchase environmentally friendly products "My bag promotion" (Carrying reusable bags) • Reduce waste discharge through the reuse and recycling of waste at its source
2	Reuse	Government initiates plan of action and a officially permitted system	• Promotion of dump system for reusable/recyclable goods • Encourage the use of returnable bottles and packaging
		Develop public awareness	• Purchase goods with returnable containers and packaging • Categorize reusable goods at the source of generation
3	Recycle	Stuff recycling	• Reusable/recyclable wastes • Composting, sorting
		Thermal recycling	• Refused-derived fuel (RDF) manufacture • Incineration for production of power and heat supply, production of biogas

and nervous systems [27]. By 2018, MPs had been detected in over 114 aquatic species throughout freshwater and marine environments. It has been observed that MPs are trapped in the tissues and digestive system of many invertebrate marine creatures, especially crustaceans like crabs. This occurs because fish and birds often ingest food in the form of plastic fragments. They consume MPs floating on the surface of the water, mistaking them for food. After ingesting MPs, aquatic organisms may experience neurological and reproductive issues, along with reduced food intake. As a result, they lack sufficient energy to perform their normal activities. The presence of MPs in organisms ranging from zooplankton and small fish to large marine predators suggests that they have become an integral part of the marine food chain [28].

Terrestrial, freshwater, and marine ecosystems are severely affected by plastic contamination. Preserving these natural ecosystems is essential, as they provide a wide range of services that are fundamental to the economy and human well-being. For example, wetlands are essential for the availability of freshwater whereas healthy mangroves offer coastline protection services. The harmful effects of plastic waste are similar among aquatic and airborne creatures, often resulting in asphyxiation, ingestion, and entanglement. Wildlife such as turtles, fish, whales, and birds often mistake plastic waste for food. As a result, their intestines become filled with indigestible plastic, leading to starvation. It also causes both external and internal injuries, impairing the swimming and flying abilities of affected animals. Domesticated farm animals are similarly vulnerable to the harmful effects of plastic waste. The spread of invasive alien species *via* floating plastics debris is considered a major factor contributing to extinction of species and the loss of biodiversity. Plastic materials and fireproof coatings often contain carcinogenic compounds that can leach into the soil, leading to environmental contamination. These harmful substances may enter groundwater and nearby rivers, exposing both humans and ecosystems to significant health and ecological risks. The conversion of extracted oil and gas into plastics has a far-reaching impact on the climate. The combustion of plastic waste releases greenhouse gases such as carbon dioxide, methane, dioxins, and other pollutants. These harmful emissions contribute to environmental pollution and exacerbate climate change [27].

1.2.3 Types of Plastics

Nowadays, we use the word "plastic" is commonly used as a catch-all term to describe a wide range of shapes and forms that this material can take [29].

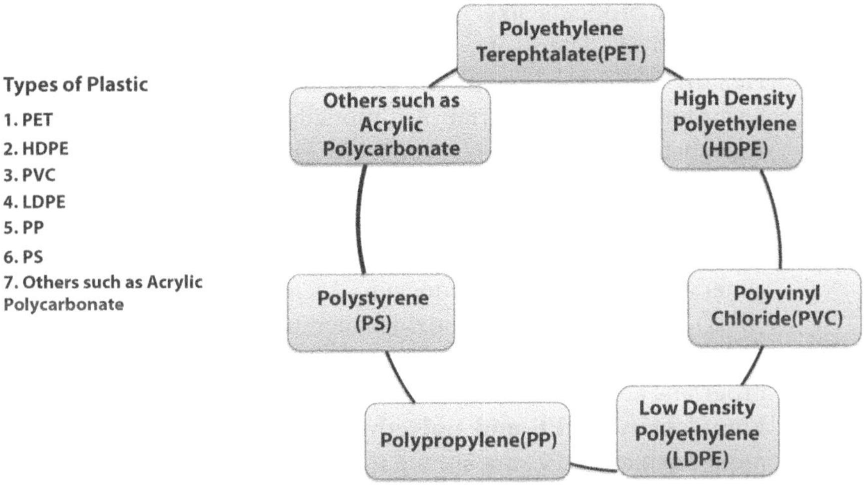

Figure 1.1 Types of plastics.

There are actually seven different kinds of plastic, each with a unique chemical makeup, function, recyclable nature, and level of hazard (Figure 1.1) [29].

(1) **PET or polyethylene terephthalate:** PET plastics are present in every aisle of the grocery store, and we come encounter them on a daily basis. They are commonly used as containers for water, soft drinks, medicine bottles, household cleaning products, and other items. PET plastics are widely used because they can prevent oxygen from entering the container and spoiling the perishable contents. This type of plastic is also among the most widely recycled plastics [29].

(2) **HDPE or high-density polyethylene:** HDPE is an opaque plastic that is typically used in containers for engine oil, milk, shampoos, conditioners, and other personal hygiene products. Compared to PET plastics, HDPE is tougher, more stable, and often recyclable. However, HDPE plastics can only be reused for food and drink purposes if they were originally used for the same application. Otherwise, they may pose a risk by interfering with human hormones.

(3) **Polyvinyl chloride (PVC):** PVC is a durable plastic primarily used in plumbing applications, such as fittings and pipes as well as products like titles and flooring. It is also

found in products like inflatable mattresses, cooking oil bottles, toys, detergents, and many others. According to Healthy Child Healthy World, PVC is considered the most hazardous plastic material. Additionally, the International Agency for Research on Cancer (IARC) of the World Health Organization has classified vinyl chloride (the chemical used to produce PVC) as a known carcinogen. The use of this type of plastic can release several harmful contaminants that are generally considered carcinogenic, including bisphenol A (BPA), phthalates, lead, dioxins, mercury, and cadmium. Moreover, this type of plastic is rarely accepted by recycling programs [29].

(4) **Low-density LDPE and polyethylene:** Low-density polymers, such as LDPE, are among the most flexible and durable plastics. They are also the easiest and least expensive to process. LDPE is commonly used in products like squeezable bottles, cling wrap, shopping bags, and other everyday items. While LDPE is well-suited for convenient human use, it is difficult to recycle and often ends up in landfills [29].

(5) **Polypropylene:** Due to its high melting point, polypropylene is suitable for storing microwaveable meals and hot foods. Its durability also makes it useful in products such as prescription bottles, syrup containers, and yogurt cups. Despite its many advantageous properties, such as heat resistance and strength, polypropylene is not easily recyclable and is considered less suitable for prolonged human use in certain applications [29].

(6) **PS/polystyrene:** Polystyrene (PS) is a thermoplastic that is naturally translucent and can appear either as rigid foam or as a solid plastic. It is a versatile material used in a wide range of consumer products. In its solid form, polystyrene is a hard plastic commonly used in items that require transparency, such as takeout containers, food packaging, and laboratory equipment. When combined with various colorants, additives, or other polymers, polystyrene is also used to manufacture appliances, electronics, automotive parts, toys, plant pots, and more. However, these plastics are rarely recycled and never fully decompose. If not properly collected and managed, they can persist in the environment for hundreds of years [29].

(7) **Others:** The seventh category includes all plastics that do not fall into the previously mentioned types. Because these plastics are difficult to identify and separate, most recycling programs do not accept them. Polycarbonate, the most commonly used plastic in this category, has been associated with the harmful chemical bisphenol A (BPA). It is used in products such as compact discs, baby bottles, and medical storage containers. This category also includes mixed materials like bioplastics blended with conventional plastics.

It is of utmost importance to have an analytical discussion about the detrimental effects of plastic on the environment. Plastics come in various types, each with its own unique set of characteristics. While some plastics are more flexible, others are more durable and resilient. However, plastic of any kind will always have an adverse impact on the ecosystem. Petroleum, a non-renewable resource, is used to make plastic. It means that there are no more plastics left after the consumption of total petroleum stocked on the earth. This implies that once all of the Earth's petroleum reserves are consumed, no more plastic can be produced [30]. PET is one of the most widely used plastics. Water bottles, food containers, and packaging for several other products are also made with it. Because PET is strong and difficult to decompose, it can persist in the environment for a very long time [30].

Polyvinyl chloride is another kind of plastic. PVC is used to make flooring, pipelines, and some food product packaging. Additionally, certain toys and clothing contain it. Hazardous compounds found in PVC have the potential to leak into the environment and threaten both human and animal health. PET and PVC are not renewable resources because these are derived from petroleum. Because they are difficult to decompose, these plastics can persist in the environment for a long time and cause various environmental issues. However, some types of plastic are made from renewable resources, such as corn and sugarcane, offering a more sustainable alternative. These plastics are known as bioplastics. They are considered more environmentally friendly than petroleum-based polymers because they are made from renewable plant resources [30]. Since bioplastics are still relatively new, their long-term durability is unknown. Additionally, because they are plant-based, bioplastics may compete with food crops for water and land resources. Therefore, despite being more environmentally friendly than petroleum-based polymers, bioplastics have certain drawbacks. Completely avoiding plastic is the most effective way to reduce its harmful impact on the environment. When plastic use is necessary, we should opt for types derived from renewable resources, such as corn or sugarcane. Additionally, consistently recycling plastic containers and bottles is essential to help mitigate the detrimental effects of plastic on the environment [30].

1.2.4 Potential Sources of Plastic Waste

Invention of Bakelite in 1907 revolutionized the international market by introducing completely synthetic plastic resins. By the end of the 20th century, plastics were regarded as one of the chronic pollutants affecting various ecological niches, ranging from the sea floor to Mount Everest. Plastic pollution has received considerable attention as a widespread environmental concern. Animals often mistake plastics for food, leading to ingestion and harm. It clogs drainage systems and contributes to flooding in low-lying areas. Additionally, it causes serious aesthetic degradation [4]. Major land-based sources of plastic pollution include urban and stormwater runoff, littering, industrial processes, tire abrasion, construction activities, and agriculture. In marine environments, plastic pollution often originates from land drainage, abandoned fishing gear, shipping paint, and other maritime sources. Various sources responsible for plastic pollution are depicted in Figure 1.2.

Solar radiation, wind, currents, and other natural factors break down plastic into MP particles (less than 5 mm) and nanoplastic particles (less than 100 nm). Tires and synthetic textiles release "primary" MP particles through abrasion. Nanoplastics are small enough to pass through cell membranes and penetrate living organisms [27].

1.2.5 Persistence of MPs and Plastic Trash in the Environment

Walker and others have long studied the effects of environmental exposure to a wide range of toxins on human health. We know that lung and cardiovascular diseases have been linked to exposure to air pollution particles, Walker explains. Although MPs have not been directly implicated, they are likely to play a role. Nanoplastics, which are even smaller fragments of plastic, are a subject of current research. These particles are too small to be seen with the naked eye. Due to their ability to enter the body's cells and organs, nanoplastics may pose more serious health risks to humans. Nanoplastics are particularly difficult to detect and study due to their microscopic size. MPs, which are tiny pieces of plastic, are increasingly associated with health concerns. Generally, MPs are smaller than sesame

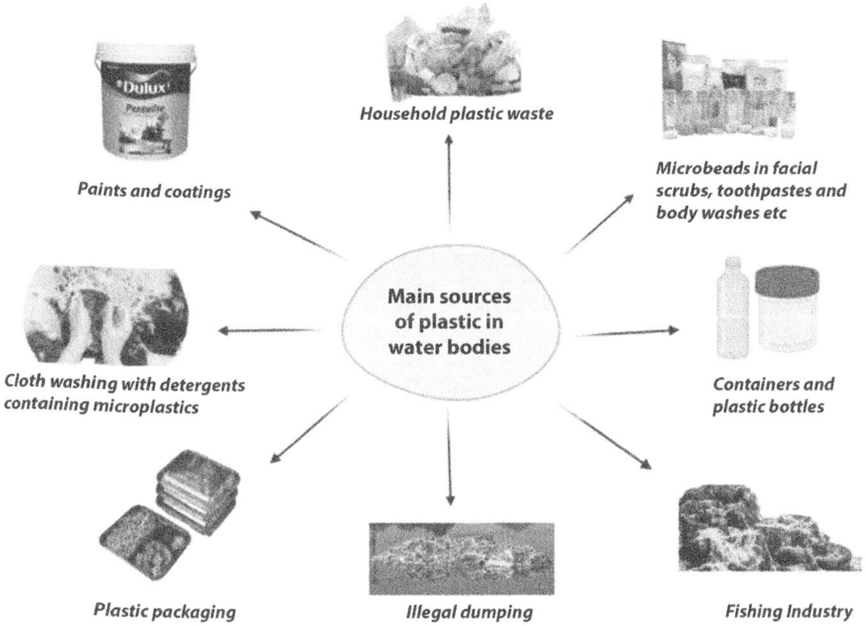

Figure 1.2 Sources of plastic pollution [31].

seeds. They can enter the human body through the food we eat and the air we breathe. Researchers have discovered MPs in human feces, stomach, lungs, blood, and also in breast milk and the placenta. MPs have been detected in the human body, but it remains unclear whether, and to what extent, these particles affect human health [32]. According to Pilapitiya *et al.* [23], the main reasons behind the persistence of plastic for centuries in the environment is its resistance to biodegradation, intricate three-dimensional structure, stability, high molecular weight, durability, and hydrophobic nature [33, 34]. In the environment, plastics undergo various weathering processes, including thermal oxidation, hydrolysis, photodegradation, biodegradation, and fragmentation [35, 36]. Plastics are found in a wide range of environments, including mountaintops, agricultural fields, deserts, and oceans [37]. Significant sources of plastic waste are agricultural plastics, including packing materials, wrappings, and polytunnel plastic films. These plastics break down into MPs, which can persist in the soil for up to 15 years. Furthermore, improper disposal and direct dumping contribute to the accumulation of plastic waste, leading to its buildup on open land [38, 39]. Terrestrial environments serve as sinks for plastic pollution,

5. James, B.D., Ward, C.P., Hahn, M.E., Thorpe, S.J., Reddy, C.M., Minimising of environmental impacts of plastic pollution through ecodesign of products with low environmental persistence. *ACS Sustain. Chem. Eng.*, 12, 1185–1194, 2024.
6. Chamas, A., Moon, H., Zheng, J., Qiu, Y., Tabassum, T., Jang, J.H., Abu-Omar, M., Scott, S.L., Suh, S., Degradation rates of plastics in the environment. *ACS Sustain. Chem. Eng.*, 8, 3494–3511, 2020.
7. Mazzotta, M.G., Reddy, C.M., Ward, C.P., Rapid degradation of cellulose diacetate by marine microbes. *Environ. Sci. Technol. Lett.*, 9, 37–41, 2022.
8. Ward, C.P. and Reddy, C.M., We need better data about the environmental persistence of plastic goods. *Proc. Natl. Acad. Sci. U.S.A.*, 117, 14618–14621, 2020.
9. Min, K., Cuiffi, J.D., Mathers, R.T., Ranking environmental degradation trends of plastic marine debris based on physical properties and molecular structure. *Nat. Commun.*, 11, 727, 2020.
10. Dilkes-Hoffman, L.S., Lant, P.A., Laycock, B., Pratt, S., The rate of biodegradation of PHA bioplastics in the marine environment: A meta-study. *Mar. Pollut. Bull.*, 142, 15–24, 2019.
11. Sang, T., Wallis, C.J., Hill, G., Britovsek, G.J.P., Polyethylene terephthalate degradation under natural and accelerated weathering conditions. *Eur. Polym. J.*, 136, 109873, 2020.
12. Zhu, L., Zhao, S., Bittar, T.B., Stubbins, A., Li, D., Photochemical dissolution of buoyant MPs to dissolved organic carbon: Rates and microbial impacts. *J. Hazard. Mater.*, 383, 121065, 2020.
13. Gewert, B., Plassmann, M.M., MacLeod, M., Pathways for degradation of plastic polymers floating in the marine environment. *Environ. Sci. Process Impacts*, 17, 1513–1521, 2015.
14. Andrady, A.L., Barnes, P.W., Bornman, J.F., Gouin, T., Madronich, S., White, C.C., Zepp, R.G., Jansen, M.A.K., Oxidation and fragmentation of plastics in a changing environment: from UV-radiation to biological degradation. *Sci. Tot. Environ.*, 851, 158022, 2022.
15. Law, K.L. and Narayan, R., Reducing environmental plastic pollution by designing polymer materials for managed end-of-life. *Nat. Rev. Mater.*, 7, 104–116, 2022.
16. Degli Innocenti, F. and Breton, T., Intrinsic biodegradability of plastics and ecological risk in the case of leakage. *ACS Sustain. Chem. Eng.*, 8, 9239–9249, 2020.
17. Worm, B., Lotze, H.K., Jubinville, I., Wilcox, C., Jambeck, J., Plastic as a persistent marine pollutant. *Annu. Rev. Environ. Resour.*, 42, 1–26, 2017.
18. Ghosh, K. and Jones, B.H., Roadmap to biodegradable plastics—Current state and research needs. *ACS Sustain. Chem. Eng.*, 9, 6170–6187, 2021.
19. Anastas, P.T. and Zimmerman, J.B., Design through the 12 principles of green engineering. *Environ. Sci. Technol.*, 37, 94A–101A, 2003.

20. Registration, Evaluation, Authorization and Restriction of Chemicals (REACH), Regulation (EC) No. 1907/2006 of the European Parliament and of the Council. *Off. J. Eur. Commun.*, 396, 1–849, 2006.
21. Anastas, P.T. and Breen, J.J., Design for the environment and green chemistry: The heart and soul of industrial ecology. *J. Clean. Prod.*, 5, 97–102, 1997.
22. Rochman, C.M., Browne, M.A., Halpern, B.S., Hentschel, B.T., Hoh, E., Karapanagioti, H.K., Rios-Mendoza, L.M., Takada, H., The, S., Thompson, R.C., Classify plastic waste as hazardous. *Nature*, 494, 169–171, 2013.
23. Pilapitiya, P.G.C.N.T. and Ratnayake, A.S., The world of plastic waste: A review. *Cleaner Mater.*, 11, 100220, 2024.
24. Sakai, S.-I., Yoshida, H., Hirai, Y., Asari, M., Takigami, H., Takahashi, S., Tomoda, K., Peeler, M.V., Wejchert, J., Schmid-Unterseh, T., Douvan, A.R., Hathaway, R., Hylander, L.D., Fischer, C., Oh, G.J., Jinhui, L., Chi, N.K., International comparative study of 3R and waste management policy developments. *J. Mater. Cycles Waste Manage.*, 13, 86–102, 2011.
25. Babaremu, K.O., Okoya, S.A., Hughes, E., Tijani, B., Teidi, D., Akpan, A., Igwe, J., Karera, S., Oyinlola, M., Akinlabi, E.T., Sustainable plastic waste management in a circular economy. *Heliyon*, 8, e09984, 2022.
26. Roy, A., Naskar, A., Ghosh, A., Adhikari, J., Saha, P., Ghosh, M., Hybrid plastics and natural materials. *Encycl. Mater. Plast. Polym.*, 3, 288–305, 2022.
27. IUCN (International Union for Conservation of Nature and Natural Resources), Plastic pollution, 2024, https://iucn.org/resources/issues-brief/plastic-pollution.
28. Rogers, K., MPs (plastic particulate), 2025, https://www.britannica.com/technology/MP.
29. Umapathy, A., 7 types of plastic: How do they impact our environment?, 2022, https://repurpose.global/blog/post/7-types-of-plastic.
30. The Vanella Group, The negative effects of plastic on the environment, 2022, https://www.vanellagroupmn.com/the-negative-effects-of-plastic-on-the-environment.
31. Yasin, R., Tehseen, P., Shakir, T., Mumtaz, A., Subtain, M., Abbas, A., Naeem, M., Zaman, M., Manzoor, M., Analyzing the impacts of plastic wastes in water bodies and suitable methodology to prevent it. *Indones. J. Agric. Environ. Anal.*, 3, 53–64, 2024.
32. NIH (News in Health), The persistence of plastics: Can tiny plastic pieces affect our health?, 2024, https://newsinhealth.nih.gov/2024/03/persistence-plastics.
33. Wang, C., Zhao, L., Lim, M.K., Chen, W.-Q., Sutherland, J.W., Structure of the global plastic waste trade network and the impact of China's import ban. *Resour. Conserv. Recycl.*, 153, 104591, 2020.

34. Huang, S., Wang, H., Ahmad, W., Ahmad, A., Vatin, N.I., Mohamed, A.M., Deifalla, A.F., Mehmood, I., Plastic waste management strategies and their environmental aspects: A scientometric analysis and comprehensive review. *Int. J. Environ. Res. Public Health*, 19, 4556, 2022.
35. Wang, Y., Huang, J., Zhu, F., Zhou, S., Airborne MPs: A review on the occurrence, migration and risks to humans. *Bull. Environ. Contam. Toxicol.*, 107, 657–664, 2021b.
36. Wang, X., Li, P., Su, M., Zou, X., Duan, L., Zhang, H., Characteristics of plastic pollution in the environment: A review. *Bull. Environ. Contam. Toxicol.*, 107, 577–584, 2021a.
37. Cózar, A., Echevarría, F., González-Gordillo, J.I., Irigoien, X., Úbeda, B., Hernández-León, S., Palma, AT., Navarro, S., García-de-Lomas, J., Ruiz, A., Fernández-de-Puelles, M.L., Duarte, C.M., Plastic debris in the open ocean. *Proc. Natl. Acad. Sci. U.S.A.*, 111, 10239–10244, 2014.
38. Gabbott, S., Key, S., Russell, C., Yonan, Y., Zalasiewicz, J., The geography and geology of plastics. *Plast. Waste Recycl.*, 3, 33–63, 2020.
39. Hurley, R., Horton, A., Lusher, A., Nizzetto, L., Plastic waste in the terrestrial environment. *Plast. Waste Recycl.*, 7, 163–193, 2020.
40. Verma, R., Vinoda, K.S., Papireddy, M., Gowda, A.N.S., Toxic pollutants from plastic waste – A review. *Pro. Environ. Sci.*, 35, 701–708, 2016.

2

Plastic Contamination in Marine and Soil Ecosystems: Innovative Strategies for Clean-Up and Mitigating Impact

Pooja Talukdar[1,2], Bornali Bora[1,2] and Sachin Rameshrao Geed[1,2]*

[1]Biological Sciences and Technology Division, CSIR–North East Institute of Science and Technology (CSIR-NEIST), Jorhat, Assam, India
[2]Academy of Scientific and Innovative Research (AcSIR), Ghaziabad, Uttar Pradesh, India

Abstract
Plastic pollution in marine and terrestrial ecosystems has emerged as a critical global environmental concern, with serious implications for biodiversity, ecosystem integrity, and human livelihoods. This chapter explores the multifaceted aspects of plastic pollution, beginning with a review of its sources, including primary plastics from direct consumption and secondary plastics generated through environmental degradation. It examines the pathways through which plastics are transported within marine and terrestrial ecosystems and highlights the extent of their widespread presence.

The chapter also discusses the ecological and toxicological impacts of plastic litter, detailing its consequences for organisms and habitats. To address these challenges, various remediation strategies are explored, including innovative clean-up methods and advancements in microbiological and enzymatic degradation technologies, illustrated with examples of successful ocean and soil clean-up initiatives. Mitigation approaches are further considered, with a focus on the development and use of sustainable materials. Emphasis is placed on the importance of community-driven initiatives and public awareness campaigns in reducing plastic waste. Finally, the chapter identifies key research gaps and technological barriers and proposes interdisciplinary approaches to achieve long-term solutions. By integrating innovative strategies with policy measures and community efforts, the

**Corresponding author*: sachinelite.geed@gmail.com

Neha Agarwal, Brijesh Pare, Vijendra Singh Solanki and Sankara Rao Miditana (eds.) Plastic Remediation: Transformative Pathways for a Rejuvenated Sustainable Environment, (19–54)

chapter envisions a sustainable future for marine and terrestrial ecosystems free from plastic contamination.

***Keywords*:** Plastic contamination, marine ecosystems, ecosystem health, strategies for ocean clean-up, soil remediation, sustainable materials

2.1 Introduction

Plastics, derived from the Greek word *plastikos* (moldable), are synthetic organic polymers renowned for their durability, affordability, and versatility. Primarily sourced from byproducts of petroleum, coal, and natural gas, plastics are composed of long-chain polymer molecules. Their lightweight nature and adaptability have made them indispensable across diverse industries, including packaging, construction, healthcare, electronics, agriculture, and aeronautics [1].

Despite these advantages, plastics pose significant environmental challenges. Their inherent durability, while beneficial during use, results in persistent pollution post-disposal. Once discarded, plastics degrade slowly, fragmenting into microplastics (MPs) (≤5 mm) and nanoplastics (NPs) (≤1 μm) through processes such as photodegradation and mechanical degradation. Photodegradation, primarily induced by solar ultraviolet (UV) radiation, initiates polymer breakdown and enhances susceptibility to further degradation. Mechanical degradation, driven by physical forces, temperature fluctuations, and moisture, is generally less impactful on its own but can accelerate photodegradation [2]. The fragmentation rate of plastics depends on their type and morphology, with thinner plastics degrading more rapidly. In sediment or aquatic environments, plastics degrade at an average rate of less than 1–3% of their weight per year. The maritime industry including fishing, aquaculture, and shipping is a major contributor to plastic leakage into marine ecosystems [3].

Global plastic production has increased dramatically, rising from 234 million tonnes (Mt) in 2000 to 460 Mt in 2019. Over the same period, plastic waste has been doubled from 156 Mt to 353 Mt. However, only 9% of this waste was recycled, 19% was incinerated, and approximately 50% was landfilled. The remaining 22% was disposed of in uncontrolled environments, such as open dumps or through open burning. In addition to physical pollution, plastics can release toxic substances, including monomers, oligomers, and various additives, thereby compounding their environmental impact [4]. Plastics are classified based on factors such as chemical composition, polymerization method, thermal behavior (thermoplastics vs. thermosets), and mechanical properties (e.g., rigidity or flexibility). Common types include polyvinyl chloride (PVC), low-density polyethylene (LDPE), high-density polyethylene (HDPE), polypropylene (PP), polystyrene (PS), polyethylene terephthalate (PET), polyurethane (PU), (PA), and polyester (PES). They tend to contain additives such as flame retardants, stabilizers, pigments, fillers, and plasticizers that contribute to their enhanced performance and make them alternatives to conventional materials such as metals [5]. The manufacturing process for plastics includes extraction of the petroleum-based raw materials followed by refining and chemical processing. Additives are added to enhance durability, after which the material is processed through monomer preparation and forming into consumer and industrial goods like packaging, electronics, and autocomponents. Inappropriate disposal of these materials is a leading cause of environmental pollution, emphasizing the compelling need for effective waste management measures [6].

Though plastics have revolutionized contemporary industries, their persistence in the environment poses severe ecological threats. Poor management of plastic waste widens the pollution gap and imperils terrestrial and aquatic ecosystems. Overcoming these challenges requires global efforts aimed at sustainable production, responsible consumption, and proper treatment of wastes.

This chapter addresses the increasing issue of plastic contamination in marine and terrestrial ecosystem environments. Its strength is its integrated approach to innovative remediation techniques, such as enzymatic and microbial degradation, and current examples of effective removal of plastics. Through analyses of the environmental effects and actionable responses, the chapter promotes emergency, collaborative, science-driven, and community-led measures to abate plastic contamination.

2.2 Sources and Pathways of Plastic Pollution

Plastic pollution arises from diverse sources and follows multiple pathways to enter environmental systems, including aquatic, terrestrial, and atmospheric environments. Major contributors to plastic debris in water bodies include wastewater treatment plants (WWTPs), domestic sewage, urban runoff, industrial activities, and stormwater discharge. WWTPs are particularly significant, discharging both manufactured and degraded plastic particles, with concentrations ranging from thousands to millions of particles per cubic meter [7].

Plastics can move freely among water, air, and soil *via* the so-called plastic cycle. Lightweight plastics, in particular, are highly susceptible to wind and water transport, enabling their dispersion even into remote or pristine regions. Human exposure to plastics occurs mainly through ingestion of contaminated food and beverages (e.g., drinking water, salt, bivalves, and fish), inhalation of airborne plastic particles, and in the case of NP dermal absorption [4]. Plastics are broadly categorized into primary and secondary forms. Primary MPs are intentionally manufactured for use in products such as cosmetics and 3D printing materials. In contrast, secondary MPs and ultimately NPs are formed through the environmental breakdown of larger plastic items, primarily *via* UV radiation and microbial degradation.

In terrestrial ecosystems, approximately 32% of all produced plastics remain within continental environments, originating from both intentional and unintentional sources. Unintentional contamination includes agricultural practices such as the application of compost or sewage sludge and irrigation with polluted water. Intentional sources include littering, plastic mulching, the use of untreated wastewater, and the application of plastic-based fertilizers. Additional entry routes include flooding, airborne deposition, tire abrasion, and leakage from landfills [8]. Plastic pollution is classified into micro-, macro-, and mega-plastics, with larger debris commonly found in urban areas and along riverbanks, especially in the Northern Hemisphere. These often originate from discarded household items like packaging and footwear and are carried by ocean currents to island coastlines. By source, plastic waste includes industrial (clean, recyclable, but growing), commercial (often contaminated polyethylene), agricultural (e.g., packaging, irrigation pipes), and municipal waste (widespread in Asian cities, from homes and public spaces) [9].

Plastic pollution, originating from multiple sources and traversing varied environmental pathways, disperses widely across water, land, and air. Understanding these origins and movements is essential to mitigate the environmental and health impacts of plastic contamination. The classification of plastic waste into primary and secondary forms, along with its movement through marine and terrestrial ecosystems, reveals the mechanisms by which plastics enter and persist in critical environmental compartments (see Table 2.1).

2.2.1 Primary Plastics

Virgin plastics, also known as primary plastic polymers (PPPs), are synthetic or semi-synthetic materials used for the first time in the manufacture of plastic products. These include thermoplastics, thermosets, elastomers,

Table 2.1 Overview of common plastic polymers, associated health risks, environmental impacts, and mitigation strategies.

Polymer	Place of study	Health issues	Environmental impact	Solutions	Reference
Bisphenol A (BPA), phthalates, antiminitroxide, brominated flame retardants, PFAS	Bangladesh	Eye irritation, respiratory problems, liver dysfunction, cancer, reproductive issues, genotoxicity, gastrointestinal issues	Soil, water, and air pollution	Enforce proper production and usage regulations	[10]
Microplastics (MPs), nanoplastics (NPs), polystyrene (PS), polyvinyl chloride (PVC), PAHs, PCBs	Global	Respiratory, immune, reproductive, and digestive system effects; genotoxicity, cytotoxicity, oxidative stress, metabolism disruption, DNA damage, inflammation, immunological responses; potential carcinogenicity	Ubiquitous presence in ecosystems and food chains	Highlight exposure pathways, recommend further investigations	[11]

(Continued)

Table 2.1 Overview of common plastic polymers, associated health risks, environmental impacts, and mitigation strategies. (*Continued*)

Polymer	Place of study	Health issues	Environmental impact	Solutions	Reference
Mulch film (secondary microplastics from degradation)	Europe	Potential particulate matter related health damages	Air and soil pollution due to photooxidative and mechanical degradation of mulch films into secondary microplastic	Integrate degradation and emissions modules in LCAs; fill data gaps; regulate littering	[12]
Nanoplastics (NPs), microplastics (MPs)	Global	Potential harmful consequences *via* ingestion, inhalation, or absorption through skin; biological stress, physical damage	Atmospheric fallout, contamination of various environmental compartments (water, air, soil), sorption of organic pollutants and toxic metals	Develop better techniques to identify and quantify nanoplastics, further research on environmental fate and human exposure	[13]

(*Continued*)

Table 2.1 Overview of common plastic polymers, associated health risks, environmental impacts, and mitigation strategies. (*Continued*)

Polymer	Place of study	Health issues	Environmental impact	Solutions	Reference
Nanoplastics (NPs)	Global	Bioaccumulation, metabolic disruption, organ dysfunction, altered physiology and behavior, cytotoxicity	Environmental accumulation, interference with ecosystems	Focused studies on human health impacts, risk assessment, management strategies, biodegradation	[14]
Microplastics	Global	Potential cellular and molecular toxicity, health effects depending on size, type, and concentration	Widespread contamination in oceans, remote islands, polar regions	Standardized monitoring methods, development of substitutes, further toxicological studies	[15]

(*Continued*)

urban and stormwater runoff, sewer overflows, industrial discharges, tire wear, construction activities, and illegal dumping [19].

Each year, an estimated 14 million tons of plastic enter the ocean, accounting for approximately 80% of marine debris found across aquatic environments, from surface waters to deep-sea sediments. Plastic pollution poses a serious threat to marine life, causing suffocation, entanglement, lacerations, infections, and internal damage. Due to their resistance to biodegradation, plastics persist in the environment and cause long-term ecological disruptions. The scale of global marine plastic pollution is alarming, with billions of pounds accumulating in oceanic gyres and covering nearly 40% of the Earth's ocean surface [22]. Figure 2.1 depicts the pathways through which plastic waste enters aquatic environments and its subsequent ecological effects.

Marine plastic litter is generally categorized into two groups: macroplastics (particles larger than 5 mm) and MPs (ranging from 1 μm to 5 mm). MPs originate from various polymers and sources, including industrial abrasives, cleaning products, and the fragmentation of larger plastic debris. Their small size, low density, and high durability enable them to disperse widely through ocean currents. Due to their hydrophobic and

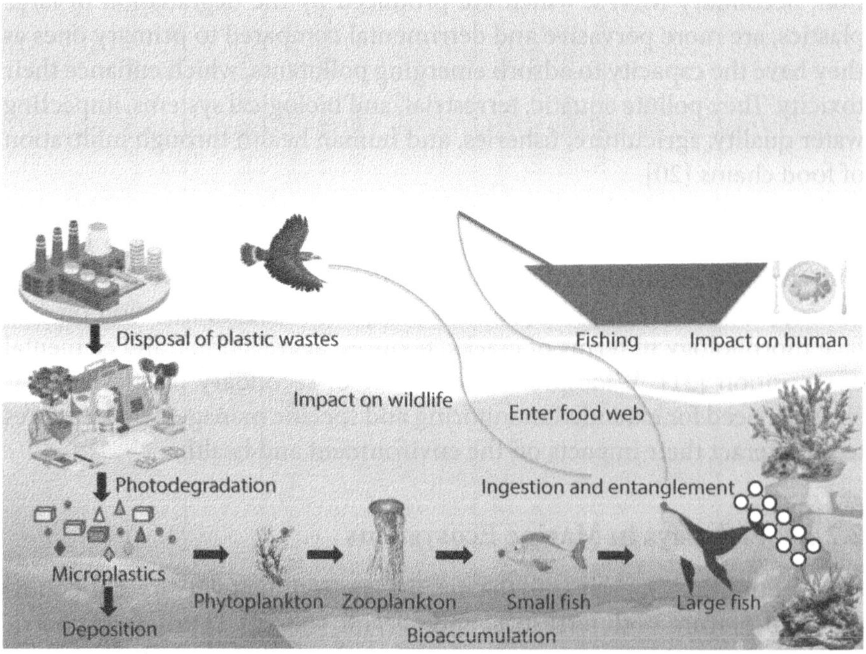

Figure 2.1 Pathways and ecological impact of plastic waste in aquatic ecosystems.

high-molecular-weight properties, MPs resist degradation and are increasingly recognized as emerging environmental pollutants [23].

Floating MPs may decrease penetration of light into water and hence have an impact on photosynthetic organisms. After being consumed and egested by marine species, the particles have the potential to aggregate with organic matter to form "marine snow," which precipitates to the seafloor. This converts the deep sea into a possible sink for MPs, where their deposition can hinder gas exchange and oxygen diffusion, leading to marine organism hypoxia [24]. In addition, MPs amplify climate change and eutrophication by promoting microbial processes like mineralization, nitrification, and denitrification that release greenhouse gases such as CO_2, CH_4, and N_2O. Resuspension of MPs also promotes algal blooms, thus augmenting nutrient pollution and adding other toxins to aquatic food webs.

At a larger ecological level, MNPs obstruct marine ecosystems by interfering with phytoplankton photosynthesis, inhibiting zooplankton growth, and altering the oceanic carbon cycle. These interferences undermine the ocean's ability to store CO_2 and transfer it to deeper oceanic layers, with potentially disastrous implications for global climate control [25]. It is essential to comprehend the intricate and multifaceted effects of plastic pollution in order to design and execute effective measures to mitigate their adverse effects on marine ecosystems and human health.

2.2.4 Pathways in Soil Ecosystems

Mismanaged plastic disposal releases toxic chemicals into surrounding soils, which enter groundwater and surrounding water bodies, causing serious harm to the surrounding ecosystems (see Figure 2.2). MPs in soils have been found to be toxic to soil invertebrates. Earthworms, for instance, have been found to have their burrowing activity impaired in microplastic-contaminated soils, impacting their body condition and soil structure. This is prominent in areas with high agricultural use, where plastic contamination has been linked to restriction of plant performance and soil parameter disruption. MPs have the ability to change soil properties such as texture, water content, and nutrient status, with a final effect on plant growth and soil health compromised [22].

The increased deposition of plastics in field ecosystems has raised global concern on soil fertility, productivity, and biodiversity worldwide. Furthermore, the potential entry of MPs into humans through the food chain *via* soil–plant–animal routes has dire health consequences. MPs and NPs inhibit critical processes in the soil, especially carbon sequestration, which is crucial for global carbon cycling. MP separation from organic

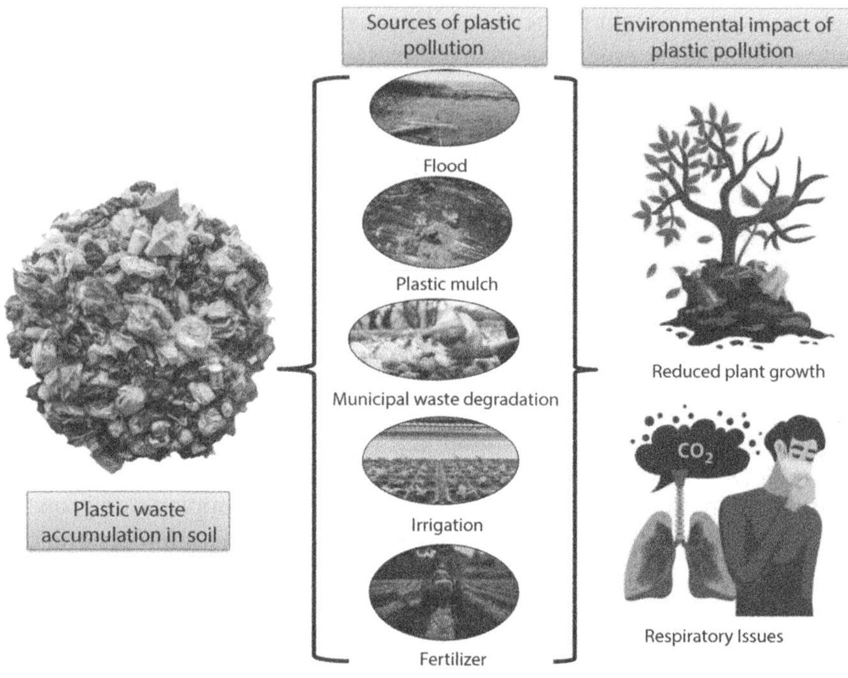

Figure 2.2 Sources and environmental impact of plastic waste accumulation in soil.

carbon in the soil is analytically challenging because of their similar chemical composition, hence complicating the assessment of environmental consequences. MPs also interfere with the nitrogen and phosphorus cycle, resulting in unbalanced nutrients, which decrease soil fertility and impede plant development [25].

Plastic particles also alter soil physical properties, including moisture, porosity, and oxygen diffusion, and thus the aerobic and anaerobic microbial communities. These alterations introduce polymer-specific effects to soil pH: HDPE decreases pH while PE, LDPE, and polylactic acid (PLA) increase it [26]. The chemical alterations to the soil negatively affect nutrient cycling, microbial diversity, and soil productivity with long-term effects on terrestrial ecosystem function and global food security.

In addition to their terrestrial impacts, plastics in soil serve as vectors for pollutants to reach aquatic systems. Some plastic particles remain on the soil surface, while others are mobilized by surface runoff or percolate deeper into the subsoil, potentially contaminating groundwater [27]. Land-based plastic pollution can also migrate to marine and freshwater environments through atmospheric transport, direct deposition along coastlines, and discharge from WWTPs. Among these pathways, rivers, creeks, and canals act as primary conduits for transporting plastic waste from inland sources to coastal regions [28]. Understanding the source of plastic pollution, transport processes, and environmental impacts in soil ecosystems is necessary for the design of effective mitigation measures. Specialized mitigation measures are necessary not only for soil fertility and agricultural productivity conservation but also for the conservation of larger environmental systems and for human health.

2.3 Environmental and Ecological Impacts

Plastics of diverse sources widely pollute the environment, particularly in agriculture where tire degradation and vinyl mulch add plastic debris to the soil ecosystems. After being released, these plastics may resurface, migrate into deeper depths, and break down under UV radiation and temperature fluctuations, producing MPs and NPs. Soil invertebrates such as insects, collembolans, and earthworms facilitate further vertical transport of MPs in the soil profile.

The widespread utilization and production of plastics have led to numerous environmental concerns, with far-reaching effects on aquatic life. Due to their size, MPs are readily ingested by many organisms in the ocean. Once ingested, MPs are likely to accumulate in tissues and biomagnify in food webs to ultimately reach upper trophic levels. Toxicity of MPs can be explained by both their physical properties and their capacity to adsorb and transfer toxic pollutants, including persistent organic pollutants and heavy metals [30].

Plastic pollution endangers coastal and marine ecosystems worldwide. Land- and sea-based plastic accumulation affects the function and organization of marine systems. Debris ranges from big to small pieces and is now ubiquitous in marine ecosystems, including seawater and organisms. Such pollution damages food webs and biodiversity dynamics through suffocation, entanglement, ingestion, starvation, and toxic exposure. Beyond that, plastics serve as vectors of invasive species by offering surfaces for colonization and transport, further degrading native biodiversity. Their persistent presence results in long-term ecological disruptions, reducing species diversity and ecosystem resilience. As marine ecosystems decline, their capacities such as biodiversity support, water filtration, and sustainable fisheries decline significantly [29].

Perhaps the most significant environmental impact of plastic waste mismanagement is the widespread pollution of natural habitats in terrestrial, freshwater, and marine ecosystems. Environmental plastic waste alters ecosystem structure and function and leads to the loss of biodiversity. Wildlife is most vulnerable, and the consequences are devastating, from ingestion, entanglement, and smothering to exposure to leached toxicants. As the plastics degrade to MPs and NPs, they infiltrate food webs and lead to ecosystem stress, compromising ecological health [31]. Plastic pollution is thus an environmental concern of major proportions, with widespread and enduring impacts on ecological compartments. Its persistence and ability to infiltrate soil, water, and biota add enormous risk to biodiversity, ecosystem integrity, and human health. Mitigation of these impacts requires a multicomponent approach, including effective waste management systems, the creation of alternative materials, and enhanced awareness worldwide. Such interventions are necessary in order to limit the ecological footprint of plastic pollution and ensure the long-term sustainability of ecosystems worldwide.

2.3.1 Biodiversity Threats

Though an estimated 80% of marine litter originates from land, a significant proportion of plastic litter enters river systems primarily in densely populated coastal regions, where it poses a vital risk to marine life, especially fish and birds. Inadequate waste collection and inefficient recycling habits have also contributed to the quantity of plastics in the ecosystems. A density estimate of plastic pellets in oceanic systems globally is between 1000 and 4000 particles per square kilometer, primarily along the seabed and coastal regions [32].

Plastic pollution kills marine life through ingestion, entanglement, and by facilitating the spread of pathogenic microorganisms. Apart from its ecological impact, plastic pollution negatively affects commercial activities such as fishing and tourism. These activities are threatened by damage to equipment, reduced fish quality, fish depletion, and the cost of operation, all of which lower the economic viability of activities depending on the sea. Further, plastic litter introduces nanosized plastic fragments and toxic pathogens into the sea, which can pose potential health threats to human beings through consumption of seafood and exposure to the environment [33].

Seabirds are susceptible to plastic ingestion because they have a low capacity to regurgitate indigestible objects. This leads to gastrointestinal blockage, abnormal feeding behavior, and altered behavior. Birds that feed on the surface of water, such as gulls, shearwaters, and fulmars, are likely to ingest floating plastic or become entangled in plastic litter [34].

Marine and coastal plastic pollution is a critical risk to human health, economic prosperity, and diversity. There should be international cooperation among nations in order to improve waste treatment plants, improve recycling, and limit the production and use of plastics.

2.3.2 Toxicological Risks

MNPs pose a serious threat to ecosystems due to their persistence, toxicity, and ability to interact with biological systems. Their minute size facilitates ingestion by microorganisms, leading to bioaccumulation in non-selective feeders that mistake these particles for food. MNPs can also penetrate biological membranes and accumulate within cells and tissues, making them difficult to eliminate. Such exposure disrupts oxidative balance and triggers abnormal metabolic and immune responses, ultimately compromising organismal health [26].

Although plastics have significantly contributed to technological and societal progress, their accumulation in macro-, micro-, and nanoforms is increasingly linked to a range of health concerns. Plastic pollution has been associated with thyroid dysfunction, obesity, diabetes, and reproductive complications. Nanoplastics, in particular, are known to disrupt the gut microbiota, influencing endocrine, immune, and neurological functions *via* the gut–brain axis. Of particular concern is their genotoxic potential: due to their nanoscale size, these particles can enter cell nuclei, where they may induce DNA damage, mutations, and possibly contribute to carcinogenesis effects that vary depending on particle shape, chemical structure, and surface properties.

Inhalation of plastic fibers is another emerging risk, especially in occupational settings such as the textile industry, where long-term exposure can lead to pulmonary inflammation, oxidative stress, and chronic respiratory diseases. Additionally, commonly used plastics such as polyethylene terephthalate (PET) can leach endocrine-disrupting chemicals like phthalates into food and beverages, even under standard usage conditions, thereby contributing to hormonal imbalances [35].

The widespread presence of MNPs in environmental matrices underscores a growing ecological and public health crisis. Urgent action is required to reduce their release, strengthen regulatory frameworks, and promote the development and adoption of safer, more sustainable alternatives to protect both ecosystems and human health.

2.4 Clean-Up Approach

Plastic pollution poses a growing threat to the environment, human health, and food safety due to the persistent nature of plastics, which resist degradation and remain in ecosystems for extended periods. Ranging from large macro-plastics to microscopic NPs, these particles can act as vectors for harmful chemicals and pathogenic microorganisms. To address this challenge, researchers and environmentalists are developing innovative clean-up strategies that combine technological advancements with biological solutions. Smart waste management tools such as AI-powered sorting systems, sensor-enabled bins, and biodegradable packaging enhance waste segregation and reduce plastic accumulation. Concurrently, microbial biodegradation offers a promising natural approach. Specific bacterial strains and enzymes have demonstrated the ability to break down commonly used plastics such as LDPE and PS. Studies from diverse geographical regions have identified microorganisms like *Rhodococcus*, *Exiguobacterium*, and *Bacillus cereus* as effective degraders under suitable environmental conditions. These biological and technological interventions offer scalable, sustainable solutions and support global initiatives such as the proposed international plastics treaty aimed at reducing plastic pollution throughout its entire life cycle [36].

Table 2.2 summarizes key global studies investigating MP contamination across various environmental matrices. It highlights prevalent polymer types, including PE, PU, and PS, originating from commercial, synthetic, and waste-derived sources. MP particle sizes range from <75 μm to several millimeters, with concentrations tailored to different exposure scenarios in soils, sediments, and aquatic systems. Experimental media include both synthetic and natural soils, plaster–charcoal composites, and leachates, with exposure durations varying from several days to multiple months. Biological endpoints assessed in soil invertebrates, such as earthworms and collembolans, include ingestion, tissue damage, reproductive effects, and MP transport behavior. Various analytical techniques were employed for accurate identification and quantification of MPs. The findings consistently reveal bioaccumulation, limited biodegradation, and particle mobility, all of which are influenced by polymer type and particle size. Evidence from agricultural fields and landfill environments underscores the widespread distribution of MPs, reinforcing the urgent need for standardized testing protocols and comprehensive ecological risk assessments.

Table 2.2 Overview of microplastic characterization, quantification, and experimental findings.

Polymer type/ sample	Material source	Particle size	Concentration	Exposure medium	Duration of exposure	Experimental design and procedure	Measured outcomes	Findings	Reference
Low-density polyethylene (LDPE)	Lab-prepared	200–300 μm	1–15% (v/v)	Sand-based soil	4, 14, 60 days	LDPE mixed with soil, effects on earthworm behavior and physiology examined	Ingestion, mortality, growth, reproduction, burrowing	Reduced growth rate, concentration of MPs in earthworm casts	[37]
High-density polyethylene (HDPE)	Recycled plastic bags	0.92 ± 1.09 mm^2	HDPE: 236–4505 mg/kg, zinc: 0.6–12 mg/kg	Agricultural and forest soils	28 days	Soil amended with HDPE and zinc, zinc uptake, and HDPE retention monitored in earthworms	Adsorption, biomass changes, ingestion	No zinc accumulation, no mortality, no weight changes, no HDPE retention in gut	[38]
Low-density polyethylene (LDPE)	Lab-prepared	200–300 μm	7–60% (w/w)	Sand-based soil	14 days	LDPE introduced in mesocosms, earthworm burrowing, and MP transport evaluated	Biomass, burrowing patterns, MP movement	Maximum burrowing and MP transport efficiency with 7% LDPE	[39]
Urea-formaldehyde microplastic	Commercially obtained	<100, 100–200 μm	5 mg (100–200 μm), 2.5 mg (<100 μm)	Plaster of Paris/ charcoal mixture	7 days	Collembolan species exposed, horizontal MP transport observed	MP transport	Larger particles transported further by F. candida	[40]

(Continued)

Table 2.2 Overview of microplastic characterization, quantification, and experimental findings. (*Continued*)

Polymer type/ sample	Material source	Particle size	Concentration	Exposure medium	Duration of exposure	Experimental design and procedure	Measured outcomes	Findings	Reference
Polyethylene (PE)	Commercially sourced	250–1000 μm	62.5–1000 mg/kg (dry weight)	OECD artificial soil	28, 56 days	Earthworm exposure, effects on reproduction, survival, growth, histology, and immunity assessed	Survival, reproduction, weight, tissue damage, immune response	No effects on survival, reproduction, or weight; evidence of tissue damage and immune response	[41]
Polyethylene (PE)	Plastic film and facial cleanser particles	183 ± 93 μm (cleanser), 137 ± 51 μm (film)	4 mg/g (dry weight)	Filter paper	14 days	Isopod exposure, effects on feeding, defecation, and energy reserves measured	Feeding, defecation, energy storage	No significant effects on endpoints	[42]
Soil	Saline solution processing	5 mm and 0.3 mm sieves	Saturated NaCl	-	Multiple extractions, ultrasonic	Floatation with saturated saline, ultrasonic treatment	Microplastic types extracted	Successful extraction of PP, PE, PA, PC, ABS, PMMA, PS	[43]
Soil	1,2,4-Trichloro-benzene (TCB)	-	TCB	120°C TCB	-	Pyrolysis-GC/MS	Polymer quantification	Fast quantification of PE, PP, PS	[44]

(*Continued*)

Table 2.2 Overview of microplastic characterization, quantification, and experimental findings. (*Continued*)

Polymer type/ sample	Material source	Particle size	Concentration	Exposure medium	Duration of exposure	Experimental design and procedure	Measured outcomes	Findings	Reference
20 different polymer types (including polypropylene, polyethylene, polystyrene)	Leachate from controlled landfills and open dumps.	Fibers, films, fragments	Fresh leachate: 8.80 MPs/L (landfill), 9.93 MPs/L (open dumping) Dewatered sludge: 37 MPs/g Return sludge: ~500 MPs/L	Leachate from solid waste disposal sites and treatment system effluents.	-	MP quantification from leachate and treatment systems, identifying polymer types, shapes, and concentrations	MP concentration and types in leachate/ effluent; treatment efficiency in MP removal	20 polymer types found. Fibers, films, and fragments dominate. PP, PE in landfill; PE, PS in open dumping. Biological treatment removed ~57.1% MPs (landfill), aerated lagoon/ membrane ~75.2% (open dumping). Membrane effluent had more fibers. MPs in dewatered and return sludge.	[45]

(*Continued*)

Table 2.2 Overview of microplastic characterization, quantification, and experimental findings. (*Continued*)

Polymer type/ sample	Material source	Particle size	Concentration	Exposure medium	Duration of exposure	Experimental design and procedure	Measured outcomes	Findings	Reference
LLDPE, HDPE, PP, PS, PET, uPVC, PA66, SBR	Commodity polymers	1 mm		Sewage sludge	Mesophilic (35°C) and thermophilic (55°C) anaerobic digestion	Polymers undergo mesophilic and thermophilic anaerobic digestion, analyzed visually, chemically, mechanically, and thermally	Visual, chemical, mechanical (embrittlement, plasticization), and thermal changes	Polymer degradation observed (smoothening, embrittlement, hydrolysis, plasticization, cracking). Anaerobic digestion was incomplete.	[46]

2.4.1 Technological Approaches

Plastics in the environment are typically classified by size, though there is no universally accepted standard for defining these categories [36]. Common groupings include macro-plastics, mesoplastics, MPs, and NPs. Macro-plastics, those larger than 2.5 cm such as plastic bags, bottles, fishing nets, and toys, are among the most visible forms of marine debris, frequently observed on beaches, floating in water, or resting on sea beds. Over time, these large items fragment into smaller particles: mesoplastics (2.5 cm to 5 mm), MPs (ranging from a few microns to 5 mm), and NPs (1–100 nm). Despite the widespread use of this framework, scientific debate persists over the precise size thresholds that define each category [47].

Plastics are extensively used across sectors such as packaging, agriculture, automotive, and healthcare. However, their high durability and resistance to degradation have led to growing environmental concerns. Many plastics contain toxic additives such as phthalates, BPA, and heavy metals like lead, cadmium, and methyl mercury, which amplify risks to ecosystems and human health [48–50]. With nearly 400 million metric tons of plastic waste produced annually, about half from single-use plastics, the need for sustainable waste management solutions is urgent [51].

Current remediation technologies for plastic waste include mechanical recycling, which involves sorting, cleaning, shredding, melting, and remolding plastics into new products. Landfilling remains widespread, though it poses significant environmental risks due to space constraints and leachate contamination. Incineration is also common, offering waste volume reduction and energy recovery but releasing hazardous emissions. Additional approaches include manual segregation, composting for biodegradable plastics, and co-processing in cement kilns, where plastics are used as alternative fuel. These methods form the backbone of waste management strategies, particularly in regions with limited access to advanced technologies.

The COVID-19 pandemic further strained plastic waste systems, especially due to increased biomedical and food-related plastic use. The surge in single-use plastics, driven by hygiene concerns, led to widespread improper disposal, heightening health risks particularly for sanitation workers [50]. In response, waste management firms such as Ecube, Enevo, and Bigbelly have introduced smart technologies to improve collection efficiency and reduce bin overflow. However, these efforts often emphasize technology over consumer education. For lasting impact, sustainable waste management must integrate innovation with public awareness and engagement [48]. As concerns over MPs pollution mount, there is growing momentum for comprehensive policies and regulatory frameworks. A proposed global plastics treaty aims to foster international collaboration to transition toward a circular economy and eliminate plastic pollution by 2040. While not a standalone solution, such a treaty could significantly strengthen global waste governance by promoting sustainable practices in plastic production, consumption, and disposal throughout the material's life cycle [51].

Plastics may also be categorized by durability. Durable plastics with lifespans exceeding three years are commonly used in construction, vehicles, and appliances. Non-durable plastics, which degrade more quickly, are typically found in packaging, disposable items, and medical products [52]. Since the early 1990s, recycling has gained prominence as an alternative to landfilling and incineration. The global recycling rate has grown at an average of 0.7% per year, reaching around 20% today [53]. Tackling the plastic pollution crisis requires a multifaceted approach that integrates technological innovation, public education, sustainable material alternatives, and robust policy frameworks. Only through coordinated global efforts can we transition to a circular economy that protects environmental and human health.

2.4.2 Case Studies

Plastic pollution poses a significant environmental threat due to its persistence and detrimental effects on ecosystems and human health. Synthetic polymers such as PE, PVC, PS, and PET degrade very slowly, breaking down into micro- and NPs that can infiltrate the food chain [54]. Microorganisms present a promising solution by producing enzymes such as laccases, proteases, cutinases, PETase, and MHETase, whose activity can be optimized through bioinformatics and protein engineering.

A study conducted in Jaipur, Rajasthan, demonstrated the potential of *Exiguobacterium* sp. (HSK30) in degrading LDPE, achieving a 16.89% reduction in film weight over 120 days. This biodegradation was confirmed through biofilm formation and analytical techniques including Fourier-transform infrared (FTIR) spectroscopy, SEM, and energy-dispersive X-ray (EDX) spectroscopy. Growing environmental concerns continue to drive research into biodegradable polymers and microbial approaches that support sustainable recycling and waste reduction [55]. Furthermore, four microbial consortia comprising bacterial and fungal strains were developed based on enzymatic profiles and evaluated for their ability to degrade both virgin and recycled plastics [56]. Additional details are provided in Table 2.3.

Table 2.3 Overview of microbial and enzymatic polymer degradation: mechanisms, operational conditions, and key findings.

Group of enzyme/ bacteria	Polymer type	Microbial degradation process	Working principles	Temperature	Degradation days	Findings	Reference
Exiguobacterium sp.	Vegetable packaging	Hydrolysis	Reduces LDPE film weight by 16.89%; confirmed *via* biofilm, FTIR, SEM, EDX	37°C	120 days	Effective in LDPE biodegradation with clear biofilm formation	[55]
Bacillus cereus	Polystyrene	Hydrolysis	Degraded LDPE and polystyrene, with polystyrene showing a higher rate	37°C	42 days	Polystyrene degrades faster than LDPE.	[56]
Micrococcus flavus	Polyethylene	Hydrolysis	Biofilm-assisted HDPE degradation confirmed *via* SEM, AFM, FTIR, TGA	30°C	30 days	HDPE weight reduced by 1.8% in 30 days	[57]
Bacillus cereus (E1) and *Rhodococcus equi (E3)*	Polyethylene	Oxidation	CH_2 chain breakdown reduces hydrophobicity and mass	37°C	110 days	Structural deterioration enhances degradation.	[58]

(Continued)

Table 2.3 Overview of microbial and enzymatic polymer degradation: mechanisms, operational conditions, and key findings. *(Continued)*

Group of enzyme/ bacteria	Polymer type	Microbial degradation process	Working principles	Temperature	Degradation days	Findings	Reference
Nitratireductor sp. Z-1 and Gordonia sp. Z-2	Polyethylene	Oxidation	LDPE degradation *via* microbial colonization, ROS, enzymatic breakdown	30°C	30 days	Oxidation promotes LDPE degradation.	[59]
C. fructicola	Low-density polyethylene	-	Uses LDPE as a carbon source, producing CO_2 and laccase enzymes	27.0 ± 2.0°C	90 days	Fungal biodegradation of LDPE confirmed.	[60]
P. knackmussii N1-2. P. aeruginosa RD1-3	Polyethylene	Hydrolysis	Enzymatic breakdown increases oxygenic groups, reducing mass	28°C	56 days	Bacteria show metabolic potential for PE degradation.	[61]
Rhodococcous	Polyethylene	oxidation	Uses multicopper oxidase, lipase, esterases for depolymerization	30°C	609 days	PE degradation linked to metabolism pathways	[62]

(Continued)

Table 2.3 Overview of microbial and enzymatic polymer degradation: mechanisms, operational conditions, and key findings. (*Continued*)

Group of enzyme/ bacteria	Polymer type	Microbial degradation process	Working principles	Temperature	Degradation days	Findings	Reference
Pseudomonas sp.	Polyethylene	oxidation	PEG degradation *via* oxidation, confirmed by MALDI-TOF MS	28°C	-	Bacterial activity alters PEG molecular weight.	[63]
Streptomyces sp. PU10	Polyurethane	Hydrolysis	Enzymes break PU into intermediates for metabolic pathways	30°C	7 days	PU degradation supports polyketide biosynthesis.	[64]

2.5 Mitigation and Prevention Approaches

Plastic pollution poses a significant environmental challenge affecting both marine and terrestrial ecosystems, demanding integrated mitigation and prevention strategies. MPs originating from synthetic textiles, packaging waste, and industrial effluents disrupt sedimentary microbial communities and biogeochemical cycles, particularly nitrogen cycling and seafloor metabolism. Research highlights the promise of biodegradable plastics and microbial degradation as potential solutions, while harmful plastic additives like DEHP have been shown to alter sediment oxygen levels, underscoring the need for targeted interventions [65–67].

Advanced analytical techniques such as spectroscopy and pyrolysis-GC–MS have deepened our understanding of microplastic distribution and its interactions with soil microbes. Degradation rates vary depending on environmental factors including temperature and salinity [68]. Plastic pollution further contributes to groundwater contamination, soil degradation, and loss of marine biodiversity, problems exacerbated by ineffective waste management and limited regulatory enforcement. Mitigation strategies include thermochemical methods like pyrolysis and gasification, which convert plastic waste into energy and valuable byproducts. Prevention efforts focus on replacing conventional plastics with sustainable alternatives such as bioplastics, glass, bamboo, ceramics, and stainless steel, alongside expanding recycling infrastructure and raising public awareness. Strengthening regulations and adopting innovative waste management practices are critical to curbing plastic input into the environment [69].

Microplastics (MPs) enter food chains through organisms like mollusks, causing bioaccumulation and ecological damage. Spread by natural forces and wildlife, they degrade soil and pose long-term health concerns. Tackling this issue demands eco-friendly plastic alternatives, unified research approaches, and stronger recycling systems, currently hindered by limited infrastructure, skill gaps, and weak regulation [70]. Of the 370 million tons of plastic produced globally in 2019, only 9% was recycled. Key mitigation measures include policy reforms, circular economy models, plastic bans, improved recycling technologies, and the development of eco-friendly materials. Public engagement and international collaboration are essential to effectively combat plastic pollution. Addressing this crisis requires a comprehensive approach that integrates policy, technology, community involvement, and sustainable alternatives to reduce plastic generation, enhance waste recovery, and protect ecosystems at every level.

2.5.1 Sustainable Materials and Product Design

The transition to sustainable materials and eco-conscious product design represents a crucial step in addressing the escalating plastic pollution crisis and enhancing environmental resilience. Plastic pollution poses a significant threat to both marine and terrestrial ecosystems, driving the urgent need for alternative materials and innovative design strategies. Research has demonstrated that bio-based and seaweed-derived biomaterials are effective in reducing environmental contamination, particularly in the remediation of polluted marine sediments [71].

The rapid increase in plastic consumption especially in packaging has resulted in widespread pollution of land, water, and the food chain. This highlights the pressing need for the food industry and other sectors to adopt reduce, reuse, and recycle strategies while embracing circular economy principles to minimize waste and its ecological impact [72]. Bioplastics have emerged as a promising alternative to conventional plastics; however, their broader adoption remains hindered by misconceptions, limited public awareness, and challenges related to biodegradability [73].

Despite growing efforts to promote recycling within circular economy frameworks, challenges such as poor material retention and mass loss during recycling processes continue to limit effectiveness. A study in Argentina examining six types of plastics across five scenarios reported low circularity rates; HDPE recycling reached only 13.6%, and PS was as low as 3%, highlighting the urgent need for improved recycling technologies and infrastructure [74].

Implementing sustainable materials and innovative packaging designs is essential for mitigating plastic pollution in both terrestrial and aquatic environments. While advances in biodegradable alternatives and circular models are promising, achieving long-term sustainability requires widespread adoption of eco-friendly practices and improvements throughout the entire product lifecycle. Ultimately, a shift toward sustainable material design must be accompanied by systemic changes in production, consumption, and waste management to secure a cleaner, healthier environment for future generations.

2.5.2 Community Driven Initiatives

Plastic pollution remains a pressing global challenge, characterized by significant regional disparities in waste management policies. Within the East African Community (EAC), Rwanda and Kenya have made progressive strides by implementing plastic bans; however, enforcement gaps, financial limitations, and the lack of viable alternatives continue to impede progress in other member states. Furthermore, weak regional coordination and low public awareness exacerbate the problem. Rwanda's success underscores the value of integrated, participatory approaches, emphasizing the need for harmonized legislation, enhanced infrastructure, and stronger community engagement to effectively mitigate plastic pollution and protect ecosystems [75].

Material flow analysis (MFA) has become a valuable tool for evaluating plastic waste management systems. Nevertheless, studies indicate that many developing countries suffer from limited data on waste generation and recovery. Policy reviews reveal that existing regulations often inadequately address the full lifecycle of plastics, focusing more on waste leakage than upstream production controls. In countries like Indonesia, insufficient local government support and minimal community involvement further hinder effective waste management. MFA can help inform strategies to reduce PET waste leakage, improve recycling efficiency, and lessen environmental impacts [76]. International collaboration is equally pivotal in combating plastic pollution. For instance, in January 2024, the Global Plastic Action Partnership (GPAP) partnered with the Colombian government to launch NPAP Colombia, implemented by WWF Colombia. This initiative unites government agencies, civil society, academia, and the private sector to develop an Action Roadmap for plastic reduction, scheduled for release in 2025. Similarly, programs like Plastic Free Schools work with educational institutions to eliminate single-use plastics and promote environmental awareness [77].

Community-driven actions have demonstrated transformative potential in addressing plastic waste. In El Manglito, Baja California, Mexico, the Guardianas del Conchalito, a local women's group restored a degraded mangrove ecosystem and established a sustainable oyster-farming enterprise. Their grassroots efforts exemplify the power of environmental restoration coupled with economic empowerment. In Senegal, Babacar Thiaw founded the Copacabana Surf Village, engaging youth in beach cleanups and sustainability initiatives. Meanwhile, in Bali, Indonesia, the Bencheghib siblings launched Sungai Watch, installing plastic barriers in rivers and removing over five million pounds of waste, illustrating how localized efforts can drive large-scale environmental impact [78].

In August 2024, large-scale community cleanups in Indonesia's Balikpapan and Deli Serdang regions, coordinated by the Regional Knowledge Centre for Marine Plastic Debris (ERIA) and JAPAN NUS CO., Ltd. (JANUS), mobilized over 100 volunteers and collected nearly a ton of waste.

These efforts highlighted the vital role of civic participation in protecting mangroves and marine environments [79].

On the policy front, the European Commission's 2015 Action Plan for a Circular Economy laid the foundation for enhanced plastic management through innovative policies, tools, and governance frameworks. The European Strategy for Plastics set ambitious targets under Directive 94/62/EC, aiming for all plastic packaging in the EU to be reusable or recyclable by 2030 [80]. Community-driven initiatives, when supported by robust policy frameworks and international partnerships, play a critical role in mitigating plastic pollution. Empowering local stakeholders, fostering grassroots innovation, and ensuring inclusive participation can generate sustainable, scalable solutions that safeguard ecosystems and human health alike.

2.6 Future Perspectives and Challenges

Addressing plastic contamination in marine and terrestrial ecosystems demands continuous innovation and a comprehensive, multifaceted strategy. Future efforts should prioritize the development of biodegradable and sustainable materials to reduce plastic dependency at the source. Advancements in biotechnology particularly microbial and enzymatic degradation require further research to enhance efficiency, scalability, and real-world applicability. Equally important is the need to expand waste management infrastructure, especially in developing regions, to minimize plastic leakage into the environment.

Nonetheless, several challenges persist. The inherently slow degradation rate of conventional plastics, the pervasive presence of MPs, and the logistical complexity of large-scale cleanup efforts pose significant obstacles. Regional disparities in policy implementation and enforcement further complicate mitigation efforts, underscoring the necessity of global cooperation. Moreover, public awareness and behavioral shifts are critical to reducing plastic waste generation at the individual and community levels. Strengthening interdisciplinary collaboration among scientists, policymakers, industry leaders, and civil society will be essential to overcoming these barriers.

Scientific advancements such as the development of biodegradable plastics, enhanced microbial and enzymatic degradation methods, and improved recycling technologies offer promising pathways for reducing plastic waste and accelerating its breakdown. However, these innovations must be supported by robust policy frameworks. This includes stricter regulations on plastic production and disposal, incentives for adopting sustainable alternatives, and effective enforcement of environmental laws. Equally important is the role of public engagement in raising awareness, fostering responsible consumption, and encouraging grassroots cleanup efforts. Educational initiatives and behavioral change programs can empower individuals to adopt more sustainable practices, while industries must take accountability by transitioning toward eco-friendly production methods and packaging materials. Ultimately, the integration of technological innovation, policy reform, and active public participation offers the most effective route to long-term, sustainable solutions. By uniting these efforts, we can significantly reduce plastic contamination and safeguard both marine and soil ecosystems for future generations.

2.7 Conclusion

Plastic pollution in both marine and terrestrial ecosystems is a critical issue that impacts the environment, economy, and human health. In this chapter, we discussed the source of plastic waste, plastic waste dispersion, and the damage inflicted by plastic waste. Emerging solutions such as microbial and enzymatic degradation, the development of sustainable alternatives, and improved recycling technologies offer promising avenues for mitigation. Technological innovations alone cannot resolve the crisis. Firm policies, community engagement, and public awareness are also needed in solving this problem. A combination of scientific innovation and responsible actions can lead us to strive for a cleaner and healthier future.

Acknowledgement

The authors gratefully acknowledge the support provided by the Director of CSIR–North East Institute of Science and Technology (CSIR–NEIST), Jorhat. We acknowledge the CSIR–NEIST in-house project OLP2503A and the GAP project GAP0836/GAP0837. Pooja Talukdar also wishes to thank the Department of Science and Technology (DST), Government of India, for financial assistance through the WISE-PhD program (File No. DST/WISE-PhD/LS/2024/3, TPN 104633).

AI disclosure statement: The author confirms that no artificial intelligence (AI)-generated content or AI-assisted writing tools were used in the preparation of this chapter (including summaries and the abstract). All text, analysis, and conclusions are the result of the author's original work.

References

1. Hsu, W.T., Domenech, T., McDowall, W., How circular are plastics in the EU?: MFA of plastics in the EU and pathways to circularity. *Clean. Environ. Syst.*, 2, 100004, 2021.
2. Nikiema, J. and Asiedu, Z., A review of the cost and effectiveness of solutions to address plastic pollution. *Environ. Sci. Pollut. Res.*, 29, 24547, 2022.
3. Bergmann, M., Collard, F., Fabres, J., Gabrielsen, G.W., Provencher, J.F., Rochman, C.M., Van Sebille, E., Tekman, M.B., Plastic pollution in the Arctic. *Nat. Rev. Earth Environ.*, 3, 323, 2022.
4. Li, P., Wang, X., Su, M., Zou, X., Duan, L., Zhang, H., Characteristics of plastic pollution in the environment: a review. *Bull. Environ. Contam. Toxicol.*, 107, 577, 2021.
5. Dube, E. and Okuthe, G.E., Plastics and micro/nano-plastics (MNPs) in the environment: occurrence, impact, and toxicity. *Int. J. Environ. Res. Public Health*, 20, 6667, 2023.
6. Alhazmi, H., Almansour, F.H., Aldhafeeri, Z., Plastic waste management: A review of existing life cycle assessment studies. *Sustainability*, 13, 5340, 2021.
7. Rai, P.K., Lee, J., Brown, R.J., Kim, K.H., Environmental fate, ecotoxicity biomarkers, and potential health effects of micro- and nano-scale plastic contamination. *J. Hazard. Mater.*, 403, 123910, 2021.
8. Bodor, A., Feigl, G., Kolossa, B., Mészáros, E., Laczi, K., Kovács, E., Rákhely, G., Soils in distress: The impacts and ecological risks of (micro)plastic pollution in the terrestrial environment. *Ecotoxicol. Environ. Saf.*, 269, 115807, 2024.
9. Gondal, A.H., Bhat, R.A., Gómez, R.L., Areche, F.O., Huaman, J.T., Advances in plastic pollution prevention and their fragile effects on soil, water, and air continuums. *Int. J. Environ. Sci. Technol.*, 20, 6897, 2023.
10. Proshad, R., Kormoker, T., Islam, M.S., Haque, M.A., Rahman, M.M., Mithu, M.M.R., Toxic effects of plastic on human health and environment: A consequences of health risk assessment in Bangladesh. *Int. J. Health*, 6, 1, 2018.
11. Kumar, R., Manna, C., Padha, S., Verma, A., Sharma, P., Dhar, A., *et al.*, Micro (nano) plastics pollution and human health: How plastics can induce carcinogenesis to humans? *Chemosphere*, 298, 134267, 2022.
12. Vega, G.C., Gross, A., Birkved, M., The impacts of plastic products on air pollution—A simulation study for advanced life cycle inventories of plastics covering secondary microplastic production. *Sustain. Prod. Consum.*, 28, 848–865, 2021.
13. Bhat, M.A., Gedik, K., Gaga, E.O., Atmospheric micro (nano) plastics: future growing concerns for human health. *Air Qual. Atmos. Health*, 16, 2, 233–262, 2023.
14. Haldar, S., Muralidaran, Y., Míguez, D., Mulla, S.I., Mishra, P., Eco-toxicity of nano-plastics and its implication on human metabolism: Current and future perspective. *Sci. Total Environ.*, 861, 160571, 2023.

15. Lee, Y., Cho, J., Sohn, J., Kim, C., Health effects of microplastic exposures: current issues and perspectives in South Korea. *Yonsei Med. J.*, 64, 5, 301, 2023.
16. Gündogdu, S., Rathod, N., Hassoun, A., Jamroz, E., Kulawik, P., Gokbulut, C., *et al.*, The impact of nano/micro-plastics toxicity on seafood quality and human health: facts and gaps. *Crit. Rev. Food Sci. Nutr.*, 63, 23, 6445–6463, 2023.
17. Borriello, L., Scivicco, M., Cacciola, N.A., Esposito, F., Severino, L., Cirillo, T., Microplastics, a global issue: human exposure through environmental and dietary sources. *Foods*, 12, 18, 3396, 2023.
18. Su, L., Xiong, X., Zhang, Y., Wu, C., Xu, X., Sun, C., Shi, H., Global transportation of plastics and microplastics: A critical review of pathways and influences. *Sci. Total Environ.*, 831, 154884, 2022.
19. Thushari, G.G.N. and Senevirathna, J.D.M., Plastic pollution in the marine environment. *Heliyon*, 6, 8, e04709, 2020.
20. Pathak, P., Sharma, S., Ramakrishna, S., Circular transformation in plastic management lessens the carbon footprint of the plastic industry. *Mater. Today Sustain.*, 22, 100365, 2023.
21. Piyathilake, U., Lin, C., Bundschuh, J., Herath, I., A review on constructive classification framework of research trends in analytical instrumentation for secondary micro (nano) plastics: What is new and what needs next? *Environ. Pollut.*, 335, 122320, 2023.
22. Zaini, N., Kasmuri, N., Mojiri, A., Kindaichi, T., Nayono, S.E., Plastic pollution and degradation pathways: a review on the treatment technologies. *Heliyon*, 10, e21198, 2024.
23. Oliveira, J., Belchior, A., Da Silva, V.D., Rotter, A., Petrovski, Z., Almeida, P.L., , Marine environmental plastic pollution: mitigation by microorganism degradation and recycling valorization. *Front. Mar. Sci.*, 7, 567126, 2020.
24. Waldschläger, K., Lechthaler, S., Stauch, G., Schüttrumpf, H., The way of microplastic through the environment–Application of the source-pathway-receptor model. *Sci. Total Environ.*, 713, 136584, 2020.
25. Kumar, R., Verma, A., Shome, A., Sinha, R., Sinha, S., Jha, P.K., *et al.*, Impacts of plastic pollution on ecosystem services, sustainable development goals, and need to focus on circular economy and policy interventions. *Sustainability*, 13, 17, 9963, 2021.
26. Liu, Y., Liu, J., Zhang, X., Sun, Y., Biodegradable polymers for sustainable packaging applications. *Polymers*, 14, 3, 546, 2022.
27. Skariyachan, S., Patil, A.A., Shankar, A., Manjunath, M., Bachappanavar, N., Kiran, S., Insights on biodegradation of plastics: an overview. *Sci. Total Environ.*, 796, 148837, 2021.
28. Su, L., Tan, M., Zhao, X., Liu, Y., Shi, H., Microbial degradation and valorization of plastic wastes. *Front. Microbiol.*, 14, 1113257, 2023.

29. Khan, A., Anees, M., Kausar, F., Siddique, M.H., Khan, A.M., Biodegradation of plastic waste: a step toward environmental safety. *Biodegrad. Pollut. Environ. Clean.*, 2020.
30. Webb, H.K., Arnott, J., Crawford, R.J., Ivanova, E.P., Plastic degradation and its environmental implications with special reference to poly (ethylene terephthalate). *Polymers*, 5, 1, 1–18, 2013.
31. Pathak, V.M., Navneet, Review on the current status of polymer degradation: a microbial approach. *Bioresour. Bioprocess.*, 4, 15, 2017.
32. Yousif, E. and Haddad, R., Photodegradation and photostabilization of polymers, especially polystyrene: review. *SpringerPlus*, 2, 398, 2013.
33. Agarwal, N., Solanki, V., Jonnalagadda, S., Ameta, K., Singh, N., Singh, A., Bind, V., Emerging Pollutants in Aquatic Environment: Critical Risk Assessment and Treatment Options, 2024, 10.2174/97898152389691240l0004.
34. Shah, A.A., Pathan, A.A., Qazi, I.A., Khan, Z.A., Inayatullah, Syed, Synthesis, applications, and microbial degradation of polylactic acid. *Pak. J. Bot.*, 50, 5, 2115–2122, 2018.
35. Mohanan, N., Montazer, Z., Sharma, P.K., Levin, D.B., Microbial and enzymatic degradation of synthetic plastics. *Front. Microbiol.*, 11, 580709, 2020.
36. Bagewadi, Z.K., Mulla, S.I., Ninnekar, H.Z., Polyethylene degradation by fungal isolate, *Aspergillus terreus* strain MANGF1, isolated from plastic contaminated soil. *Int. Biodeterior. Biodegrad.*, 122, 69–79, 2017.
37. Mehmood, M.A., Adnan, A., Khalid, A., Arslan, M., Hameed, A., Ahmed, S., Isolation, characterization and molecular identification of polyvinyl chloride (PVC) plastic degrading fungal isolates. *J. Basic Microbiol.*, 56, 9, 1055–1063, 2016.
38. Bandopadhyay, S., Dey, S., Mondal, N.K., Microbial degradation of low density polyethylene plastic by *Pseudomonas* sp. *J. Environ. Chem. Eng.*, 10, 1, 106847, 2022.
39. Balasubramanian, V., Natarajan, K., Hemambika, B., Ramesh, N., Sumathi, C.S., Kottaimuthu, R., Rajesh Kannan, V., High-density polyethylene (HDPE)-degrading potential bacteria from marine ecosystem of Gulf of Mannar, India. *Lett. Appl. Microbiol.*, 51, 2, 205–211, 2010.
40. Kumar Sen, S. and Raut, S., Microbial degradation of low density polyethylene (LDPE): a review. *J. Environ. Chem. Eng.*, 3, 1, 462–473, 2015.
41. Gajendiran, A., Krishnamoorthy, S., Abraham, J., Microbial degradation of low-density polyethylene (LDPE) by *Bacillus cereus*. *Appl. Biochem. Biotechnol.*, 183, 115–127, 2017.
42. Yanto, D.H.Y. and Tachibana, S., Potential utilization of lipolytic molds in high density polyethylene (HDPE) biodegradation. *J. Environ. Chem. Eng.*, 12, 106737, 2024.
43. Kumar, R., Shome, A., Verma, A., Sinha, S., Microbial strategies and tools to biodegrade polyethylene: current advances and future perspectives. *Sci. Total Environ.*, 865, 161142, 2023.

44. Kundu, D., Roy, S., Mandal, A., Microbial degradation of polyethylene and polypropylene: recent advances and challenges. *Environ. Sci. Pollut. Res.*, 30, 1010–1027, 2023.
45. Sabir, F., Asad, M., Ramzan, N., Shakir, H.A., Nawaz, R., Aslam, A., *et al.*, Biodegradation of polyethylene: a brief review. *J. Polym. Environ.*, 28, 445–462, 2020.
46. Janczak, K., Hrynkiewicz, K., Znajewska, Z., Malinowski, T., Bacterial degradation of plastics. *J. Environ. Manage.*, 354, 119128, 2024.
47. Khan, M.U., Nomani, M.Z.M., Basit, A., Khan, S., Biodegradation of plastics by bacteria isolated from municipal landfill site. *Polym. Bull.*, 77, 2655–2668, 2020.
48. Chatterjee, S. and Sharma, S., Microbial degradation of polyethylene (HDPE): a brief review. *J. Polym. Environ.*, 27, 2479–2490, 2019.
49. Pramila, R. and Ramesh, K.V., Screening of polyethylene degrading microbes from garbage soil. *Libyan Agric. Res. Cent. J. Int.*, 2, 200–204, 2011.
50. Agarwal, N., Solanki, V., Pare, B., Singh, N., Jonnalagadda, S., Current Trends in Nanocatalysis for Green Chemistry and its Applications- A mini-Review. *Curr. Opin. Green Sustainable Chem.*, 41, 100788, 2023, 10.1016/j.cogsc.2023.100788.
51. Ali, S.S., Elsamahy, T., Al-Tohamy, R., Zhu, D., Mahmoud, Y.A.G., Koutra, E., Kornaros, M., Plastic waste degradation by newly isolated *Paenibacillus glucanolyticus* strain SLM1. *Polymers*, 13, 3, 423, 2021.
52. Gayathri, P., Uma, L., Saranya, D., Shabana, R., Screening and identification of LDPE degrading fungi from marine debris. *Int. J. Pharm. Biol. Arch.*, 3, 966–971, 2012.
53. Ghosh, S.K., Pal, S., Ray, S., Study of microbes having potentiality for biodegradation of plastics. *Environ. Sci. Pollut. Res.*, 20, 4339–4355, 2013.
54. Miditana, S.R., Dokka, W.T., Babu, A.R., Ampolu, S., Agarwal, N., Simma, N., Cyclodextrins for UV protection in food, beverages, and agricultural products: A comprehensive review. *N. Afr. J. Food Nutr. Res.*, 9, 20, 219–235, 2025, https://doi.org/10.51745/najfnr.9.20.219-235.
55. Ojha, N., Pradhan, N., Singh, R., Barla, A., Shrivastava, A., Divecha, J., *et al.*, Evaluation of HDPE and LDPE degradation by fungus, implemented by statistical optimization. *Sci. Rep.*, 7, 39515, 2017.
56. Bhardwaj, H., Gupta, R., Tiwari, A., Microbial population associated with plastic degradation. *Open Access J. Biomed. Sci.*, 3, 4, 1611–1615, 2021.
57. Chandra, R. and Rustgi, R., Biodegradable polymers. *Prog. Polym. Sci.*, 23, 7, 1273–1335, 1998.
58. Sankhla, M.S., Shahnawaz, M., Hussain, I., Biodegradation of low density polyethylene (LDPE) by fungi isolated from polythene polluted site. *J. Bioremed. Biodegrad.*, 7, 343, 2016.
59. Sudhakar, M., Doble, M., Murthy, P.S., Venkatesan, R., Marine microbe-mediated biodegradation of low- and high-density polyethylenes. *Int. Biodeterior. Biodegrad.*, 61, 3, 203–213, 2008.

60. Ammendolia, J., Saturno, J., Brooks, A.L., Jacobs, S., Jambeck, J.R., An emerging source of plastic pollution: environmental presence of plastic personal protective equipment (PPE) debris related to COVID-19 in a metropolitan city. *Environ. Pollut.*, 269, 116160, 2021.
61. Hou, L., Xi, J., Liu, J., Wang, P., Xu, T., Liu, T., Qu, W., Lin, Y.B., Biodegradability of polyethylene mulching film by two *Pseudomonas* bacteria and their potential degradation mechanism. *Chemosphere*, 286, 131758, 2022.
62. Tao, X., Ouyang, H., Zhou, A., Wang, D., Matlock, H., Morgan, J.S., Ren, A.T., Mu, D., Pan, C., Zhu, X., Han, A., Polyethylene degradation by a *Rhodococcus* strain isolated from naturally weathered plastic waste enrichment. *Environ. Sci. Technol.*, 57, 37, 13901, 2023.
63. Geisler, M., Khan, J., Heine, T., Ansorge-Schumacher, M.B., Thiele, J., Kaufmann, A., Analysis of high-molecular weight polyethylene glycol degradation by *Pseudomonas* sp. *Polym. Degrad. Stab.*, 232, 111144, 2025.
64. Pantelic, B., Siaperas, R., Budin, C., de Boer, T., Topakas, E., Nikodinovic-Runic, J., Proteomic examination of polyester-polyurethane degradation by *Streptomyces* sp. PU10: Diverting polyurethane intermediates to secondary metabolite production. *Microb. Biotechnol.*, 17, 3, e14445, 2024.
65. Roy, P., Mohanty, A.K., Misra, M., Microplastics in ecosystems: their implications and mitigation pathways. *Environ. Sci. Adv.*, 1, 1, 9, 2020.
66. Seeley, M.E., Song, B., Passie, R., Hale, R.C., Microplastics affect sedimentary microbial communities and nitrogen cycling. *Nat. Commun.*, 11, 1, 2372, 2020.
67. Ladewig, S.M., Bartl, I., Rindelaub, J.D., Thrush, S.F., Rapid effects of plastic pollution on coastal sediment metabolism in nature. *Sci. Rep.*, 14, 1, 17963, 2024.
68. Bhattacharjee, U., Baruah, K.N., Shah, M.P., Exploring sustainable strategies for mitigating microplastic contamination in soil, water, and the food chain: A comprehensive analysis. *Environ. Chem. Ecotoxicol.*, 2025.
69. Fayshal, M.A., Current practices of plastic waste management, environmental impacts, and potential alternatives for reducing pollution and improving management. *Heliyon*, 10, 23, 2024.
70. Kibria, M.G., Masuk, N.I., Safayet, R., *et al.*, Plastic Waste: Challenges and Opportunities to Mitigate Pollution and Effective Management. *Int. J. Environ. Res.*, 17, 20, 2023, https://doi.org/10.1007/s41742-023-00507-z.
71. Di Clemente, M.E., Barjoveanu, G., Todaro, F., Notarnicola, M., Teodosiu, C., Bio-Based Materials as a Sustainable Solution for the Remediation of Contaminated Marine Sediments: An LCA Case Study. *Polymers*, 16, 15, 2101, 2024.
72. Ncube, L.K., Ude, A.U., Ogunmuyiwa, E.N., Zulkifli, R., Beas, I.N., An overview of plastic waste generation and management in food packaging industries. *Recycling*, 6, 1, 12, 2021.

73. Ghasemlou, M., Barrow, C.J., Adhikari, B., The future of bioplastics in food packaging: An industrial perspective. *Food Packag. Shelf Life*, 43, 101279, 2024.
74. Zapata, M.A., Bastias, F.A., Civit, B., Arena, P., Assessing the circularity potential of plastics with a substitutability approach: A case study in Argentina. *Rev. UDCA Act. Divulg. Cient.*, 28, 1, 2025.
75. Nduwimana, J., Ndikumana, T., Luis, P., Manirakiza, N., Plastic waste management policies in the East African Community (EAC) region: challenges and prospects. *J. Mater. Environ. Sci.*, 16, 1, 112, 2025.
76. Widiyanti, A., Rancak, G.T., Ramadan, B.S., Hanaseta, E., Comprehensive literature review of material flow analysis (MFA) of plastics waste: recent trends, policy, management, and methodology. *J. Mater. Cycles Waste Manag.*, 27, 1, 1, 2025.
77. ForestNation, Community-led initiatives making strides in plastic waste reduction, https://forestnation.com/blog/community-led-initiatives-making-strides-in-plastic-wastereduction/.
78. Conde Nast Traveler, Surfers pushing new boundaries in ocean conservation, https://www.cntraveler.com/story/surfers-pushing-new-boundaries-in-ocean-conservation.
79. ERIA, Marine plastic debris, https://www.eria.org/news-and-views/category/all/marine-plastic-debris.
80. EU, A European Strategy for Plastics in a Circular Economy, https://www.europarc.org/wp-content/uploads/2018/01/Eu-plastics-strategy-brochure.pdf.

3

Impact of Plastic Remediation on Biodiversity and Ecosystem Health

Kartik Adake[1,2]*, Shreyas Joshi[3], Kedar Ahire[1] and Samir Terdalkar[4]

[1]*Department of Zoology, Savitribai Phule Pune University, Pune, Maharashtra, India*
[2]*Department of Zoology, St. Xavier's College (Empowered Autonomous Institute), Mumbai, Maharashtra, India*
[3]*Department of Microbiology, Fergusson College (Autonomous), Pune, Maharashtra, India*
[4]*Department of Zoology, Willingdon College (Autonomous), Sangli, Maharashtra, India*

Abstract
Plastic pollution has emerged as a defining environmental crisis of the Anthropocene, with profound implications for both biodiversity and ecosystem health. This chapter presents a comprehensive analysis of plastic pollutants, ranging from macroplastics to micro- and nanoplastics, with a focus on their origins, classifications, and detrimental effects on terrestrial and aquatic ecosystems. Emphasis is placed on the silent yet severe physiological impacts caused by microplastics (MPs), including cytotoxicity, oxidative stress, bioaccumulation, and biomagnification across trophic levels. The review explores diverse remediation strategies categorized as physicochemical, physicochemical, and biochemical, including coagulation, membrane filtration, photo-Fenton reactions, and enzymatic degradation by naturally occurring and engineered organisms. The emergence of plastic-degrading species like *Piscinibacter sakaiensis* and the development of bioplastics from natural sources highlight the adaptive and innovative responses to plastic pollution. Moreover, the chapter addresses the need for biodiversity recovery monitoring and advocates for unified sampling protocols to enhance scientific rigor in MP studies. Ultimately, it underscores the shared

**Corresponding author*: kartik.adake@xaviers.edu

Neha Agarwal, Brijesh Pare, Vijendra Singh Solanki and Sankara Rao Miditana (eds.) Plastic Remediation: Transformative Pathways for a Rejuvenated Sustainable Environment, (55–78)

responsibility of policymakers, scientists, and citizens in implementing sustainable practices, fostering innovation in remediation technologies, and restoring ecological balance.

Keywords: Plastic pollution, MPs, biodiversity loss, bioremediation, physicochemical degradation, bioaccumulation, ecosystem health

3.1 Introduction

In today's world, plastics have become an inseparable part of life. From MP beads intentionally added to products like cosmetics [1] to synthetic fibers used in clothing [2], our lives have become, literally and metaphorically, wrapped in plastic. The invention of petroleum-derived polymers revolutionized multiple industries due to plastics being easily produced, impermeable to fluids, and perhaps most importantly, cheap. Additionally, various types of plastics with a wide array of properties diversified its applications and uses, ranging from disposable carrybags to industrial grade components. However, this ubiquitous implementation has also led to widespread plastic pollution. The extent of accumulation of any pollutant can be attributed to two factors: rate of introduction of the pollutant into the environment and the rate of degradation of the pollutant in said environment. The amount of plastic waste generated is unprecedented, primarily due to the discarding of "single-use" plastics [3]. Furthermore, plastics are high-molecular-weight long chain hydrocarbon hydrophobic polymers, making them highly unreactive [4]. Any functional groups present in the chain are likely masked by the complex structure-limiting sites for chemical reactions, making them practically inert to chemical, as well as biological degradation.

It would be inaccurate to say plastics are completely inert. In fact, the plastic issue would not be as problematic if they were completely inert. Macroplastics degrade partially and smaller MPs and nanoplastics leach off into the environment over time. Although macroplastics are the most infamous culprits for loss of biodiversity due to direct impacts, as discussed further in the chapter, MPs and nanoplastics act as silent killers, arguably being more detrimental to whole-animal physiology, particularly due to biomagnification and bioaccumulation, also discussed further in the chapter. MPs and nanoplastics are microscopic plastic pollutants. These can be generated by the incomplete breakdown of macroplastics as mentioned above; these are termed as secondary MPs. Alternatively, primary MPs are microscopic plastic items synthesized at that scale specifically for some particular use, like the aforementioned MP beads used in cosmetics, or elsewhere as cheap bulking agents.

The delineation of these microscopic pollutants into MPs and nanoplastics based on size was probably first done for convenience, but today, it is vital to understand the permeability of these pollutants to bodily barriers like the gut cell lining, the blood–brain barrier, and the placenta to name a few. In this chapter, we aim to explore the impacts of plastic pollutants on ecology and biodiversity while understanding its effects at a molecular and cellular level. We also aim to investigate various mechanisms of plastic degradation such as physicochemical, as well as biochemical, as life adapts to the omnipresence of petroleum plastics. Moreover, we outline its applications in the methods devised for bioremediation of plastic pollution.

3.2 Sources and Types of Plastic Pollution

Plastics as pollutants can get introduced into the environment *via* various offlets. Mismanaged waste, rather intuitively, emerges as the largest source [5, 6]. Other significant contributors include agriculture [7] and construction [8]. Identifying and quantifying the sources of plastic pollution is important to formulate action plans to choke out the introduction of new plastic wastes in developing pollution remediation and control strategies (Figure 3.1). Simply clearing pre-existing waste holds no long-term significance if the introduction of new waste is not controlled.

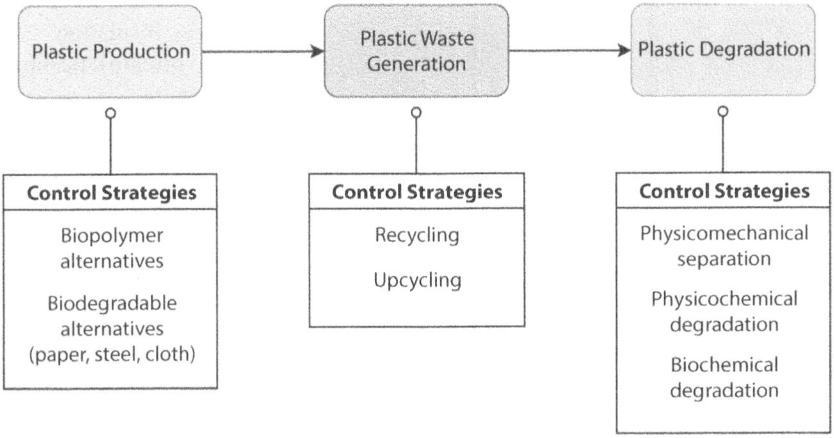

Figure 3.1 Overview of control strategies for plastic waste management.

Table 3.1 Types of plastic polymers based on functional groups.

Plastic type	Functional group(s)	Monomer(s)	Typical uses
Polyethylene (PE)	Alkane	Ethylene	Plastic bags, milk jugs, detergent bottles, piping, films, toys, electrical insulation
Polypropylene (PP)	Alkane	Propylene	Food containers, bottle caps, car parts, carpets, ropes, medical components, disposable diapers
Polyethylene terephthalate (PET)	Ester, arene	Ethylene, MHET (major), BHET (minor)	Beverage bottles, food jars, synthetic fibers, food packaging
Polyvinyl chloride (PVC)	Chlorinated alkanes	Vinyl chloride	Pipes and fittings, window frames, flooring, electrical cable insulation, blood bags, credit cards
Polystyrene (PS)	Arene	Styrene	Disposable cups and plates, food containers, CD/DVD cases, insulation (Styrofoam), packing peanuts
Polycarbonate (PC)	Carbonate ester, carbonate, arene	Bisphenol A, phosgene	CDs/DVDs, eyeglass lenses, safety goggles, riot shields, car headlights, electronic housings, baby bottles

(*Continued*)

Table 3.1 Types of plastic polymers based on functional groups. (*Continued*)

Plastic type	Functional group(s)	Monomer(s)	Typical uses
Polyamide (PA)	Amide	Diamine and diacid or caprolactam	Textiles, engineering plastics, ropes, fishing lines, bristles for brushes
Polymethyl methacrylate (PMMA)	Ester	Methyl methacrylate	Transparent sheets, signs, display cases, automotive lights, dental fillings, bone cement

Research continues on identifying and screening methods for biochemical degradation, naturally evolved and genetically modified organisms have emerged as efficient plastic-degrading biological systems. However, it is difficult to find a "one-fit-all" plastic degrader. This is due to the vast differences between the structural and functional chemistries of different plastic types (Table 3.1).

Plastic waste can also be classified based on its size and shape. Different sources report different ranges for classifying plastic waste based on its size. We chose to showcase the classification by [9] as it correlates the size of the plastic waste generated to the sizes of different aquatic organisms (Table 3.2). Considering these parallels is essential to understand the entry of these plastics into the food web. It is especially significant to understand the sizes of MPs to the corresponding phytoplankton and zooplanktons as their alpha diversities are a popular indicator of the health of an aquatic ecosystem [10, 11].

Of these, meso- and micro-sized plastics, or MPs for the sake of consistency, can be further classified based on different morphologies, which can help us understand the distribution of plastic wastes in the environment. Based on morphology, MPs can be classified into the following groups:

1. Fragments: irregularly shaped [12]
2. Pellets: irregular spherical, or cylindrical or discoid [12, 13]
3. Films: irregularly shaped, thin sheet-like [13, 14]
4. Filaments: hair-like with sharp ends [12]
5. Fibers: hair-like with frayed ends [12, 15]
6. Beads: almost perfectly spherical [12, 16]
7. Foams: irregularly shaped porous [14]

Table 3.2 An MP size classification scheme aligned with universal plankton survey methods [9].

Current size categories	Size range	Proposed size categories	Size range	Organism of equivalent size
Nanoplastic	0.001–1 µm	Femto-size plastics	0.02–0.2 µm	Virus
		Pico-size plastics	0.2–2 µm	Bacteria
		Nano-size plastics	2–20 µm	Flagellates
MP	1–1000 µm	Micro-size plastics	20–200 µm	Diatoms, dinoflagellates, ciliates, daphnids
		Meso-size plastics	200–2000 µm	Amphipods, appendi-cularians, chetognatos, copepods, thaliaceans
Mesoplastic	1–10 mm	Macro-size plastic	0.2–20 cm	Euphausiids, heteropods, jellyfish, larval fish, mysids, pteropods, solitary salps
Macroplastic	> 1 cm	Mega-size plastic	20–200 cm	Jellyfish, colonial salps

3.3 Impact of Plastic Pollution on Biodiversity

The visual mimesis between a floating plastic carry-bag in the ocean and a jellyfish posing a lethal threat to sea turtles is probably the most infamous impact of plastic pollution on biodiversity. Sea turtles are far from the only marine animals affected by this; seabirds, cetaceans, fish, manatees, etc. all suffer from the same problem, *id est*, the blockage of the alimentary canal due to the inability to regurgitate accidentally swallowed macroplastics [17–20]. Discarded and lost fishing equipment, especially fishing nets, are also potentially fatal for marine megafauna. Such equipment, aptly termed "Ghost Gear," increase their mortality rates by trapping and immobilizing them [18, 21]. These direct damages reduce alpha diversity not only by directly killing the animals, but also by exerting migratory pressure due to said problems.

Microfauna are not exempt from the detriments of plastic pollution. [22, 23], etc. reported an increased mortality of filter-feeding zooplanktons on ingestion of MPs. Some zooplanktons are able to partially degrade ingested MPs and excrete them in plankton pellets. MP incorporation into these fecal pellets reportedly elevated its settling velocity, increasing MP distribution and abundance in different strata of the waterbody. Some zooplanktons can extensively degrade MPs, which, at first, appears to be a solution, but is a significant contributor of nanoplastic pollutants to the environment [24]. Zooplanktons grazing on MPs also have a rather pernicious influence on global photosynthesis efficiency. Heightened MP ingestion by zooplanktons indirectly reduces global net oxygen production by decreasing grazing pressure on phytoplankton. In nutrient limited environments, where phytoplankton blooms are not feasible, organic matter overturn is diminished reducing overall oxygen production. In nutrient replete regions, phytoplankton population can experience overgrowth leading to an increased introduction of organic matter in the environment, which naturally undergoes oxygen-dependent mineralization or aerobic assimilation [25].

Studies evaluate direct cytotoxicity of MPs on mammalian cells, albeit with the emphasis on human safety at the forefront; the findings can very easily be extrapolated to probe the impacts on whole-animal physiology and cellular biology of other animals [26]. MP-induced cytotoxicity can arise from the oxidative stress caused by the presence of plastic polymers [27]. Gut epithelial cells are the body's first line of defense against ingested hazards, and thus naturally bear the initial direct toll of the oxidative stress induced by MPs. Oxidative stress is known to be detrimental to pre-formed tight junction proteins [28], creating a positive feedback loop of influx of smaller MPs *via* paracellular transport [29].

This, combined with the inability of immune phagocytes to degrade petroleum polymers [30], leads to progressive accumulation of plastics and plastic derivatives. These remain persistent in the body and experience trophic transfer leading to biomagnification.

Macroplastics and MPs, once exposed to aqueous or moist environments, allow spontaneous accumulation of organic matter on its surface, paving way for bacterial biofilm formation. This further enhances adherence of small animals like crustaceans [31, 32], mollusks [31], tunicates [33], and nudibranchs [34]. This is called biofouling [35] and can facilitate dispersion of invasive species [33, 36]. The aforementioned biofilms formed on plastics, especially on MPs, can accelerate bacterial colonization of the gut serving as fomites for potentially pathogenic microbes [37]. This poses a threat not only to human health and safety, but also that of other animals in the ecosystem. The direct impact of plastic-mediated biodiversity loss is not exclusive to animals. For instance, a recent study reported indicators of anoxic stress in mangroves due to macroplastics enveloping the pneumatophores, further threatening a globally endangered ecosystem [38].

3.4 Plastic Remediation Strategies

The overuse of single-use products combined with the inability to efficiently degrade petroleum plastics has led to ubiquitous macroplastic, MP, and nanoplastic pollution on the planet. MPs, especially, have shown daunting degrees of biomagnification and bioaccumulation, being detected in human blood [39], placenta [40], semen [41], and even the brain [42]. MP contamination in water bodies [43, 44] and the resulting biomagnification from consuming contaminated foods, especially seafood [45, 46], fruits and vegetables [47], and even human breast milk [48] and cow milk [49].

These discoveries and the ensuing health and safety concerns have facilitated a shift towards researching alternatives and remediation methods. The 5 "Rs" (refuse, reduce, reuse, repurpose, and recycle) of waste management encompass well-known strategies of controlling the ever-increasing waste generation. Although equally vital, these do not outline a method for eliminating the pre-existing plastic waste in the environment. Plastic waste collection drives are a popular method of cleanup of environmental plastic waste; however, it is limited to the removal of macroplastic contamination from environments. The fate of this collected plastic is variable [50]. Majority of the plastic is collected in landfills, ultimately causing MP and nanoplastic pollution by leaching, or is incinerated, contributing to atmospheric pollution.

Another strategy of repurposing plastic waste is by upcycling. Upcycling refers to the methods of creating products of a higher value than that of the waste product. Certain NGOs in India focus on repurposing plastic waste to create products like bags. Although much welcome, it is also important to note that upcycling is barely a way of delaying the introduction of plastic pollutants and not a solution to the plastic problem. Furthermore, albeit mere speculations lacking backing from scientific reports, the processing steps required in upcycling may also serve as an offlet of immediate MP and nanoplastic pollution. Another example of upcycling could be the recycling of plastic waste as construction material, particularly as a filler in concrete [6, 51].

Macroplastics can easily be screened physically and removed, which is a common step in all primary treatment processes for wastewater and raw water treatment [52]. However, the same is not true for MPs and nanoplastics. As outlined previously in Figure 3.1, the methods for removal of MPs can be broadly divided into three categories: (1) physicomechanical, (2) physicochemical, and (3) biochemical.

3.4.1 Physicomechanical Methods

Physicomechanical methods for plastic remediation are means of isolating and concentrating plastic contaminants. Agglomeration of MPs, although not necessary, contributes by increasing the efficiency of MP degradation by physicochemical and/or biochemical methods.

3.4.1.1 Coagulation

Coagulation is the most commonly used method of clarifying water. Coagulants are chemicals which facilitate agglomeration and flocculation of particles suspended in water, increasing their settling velocity and filtration efficiency [53]. Polyaluminum chloride (PAC) [54] and $FeCl_3{\cdot}6H_2O$ combined with anionic polyacrylamide (PAM) [55] showcased highly efficient flocculation and coagulation, notably increasing at a higher pH (pH = 9).

3.4.1.2 Adsorption

Activated charcoal is a commonly used clarifying agent for aqueous systems. Activated charcoal is characterized by a topologically complex surface and numerous pores contributing to a large surface area for efficient adsorption of suspended as well as dissolved molecules, primarily by weak intermolecular interactions like Van der Waal's forces [56]. Recent studies indicate potential for biomass-derived charcoal (biochar) in sequestering MPs from aqueous systems [57, 58]. Dutta *et al.* described a novel POM (polyoxometalate) nanocluster-infused triple IPN (interpenetrating polymer network) hydrogel, which showed phenomenal MP chelating and degrading capability by promoting photodegradation and redox-dependent chemidegradation on the hydrogel surface [59].

3.4.1.3 Filtration

Coarse mesh and fine mesh filters are used for screening of water for large particulate waste including macroplastics and some MPs [52]. Microfiltration (MF), ultrafiltration (UF), and nanofiltration (NF) are pressure-driven membrane-based filtration techniques, each specializing in removal of particles within specific size ranges: 0.1–10 μm, 1–100 nm, and 1–10 nm respectively to be exact. Reverse osmosis (RO) filters use dense membranes (<1 nm pore size) and can not only eliminate suspended particulate matter including MPs, but also some solutes and ions [60]. Membrane filters could be used in sequential combination to minimize clogging and maintain filtering efficiency of the filtration system.

3.4.1.4 Centrifugal Separation

Hydrocyclones employ centrifugal force to separate suspended solids from liquids. Unlike centrifuges, which utilize dynamic vessels like rotatory drums. Hydrocyclones feature a static vessel with the pressurized liquid fed tangentially to the hydrocyclone wall, which creates a vortex separating suspended solids according to their masses [61]. Yuan *et al.* reported a higher MP separation efficiency, regardless of the particle size, by aerating the hydrocyclone feed facilitating flotation separation in a miniature hydrocyclone setup [62].

Centrifugation is also a viable method. Most laboratory protocols for isolating MPs incorporate centrifugation [63–65]. However, these and similar reports only mention batch centrifugation, ideal for laboratory experiments, and reports outlining parameters for separation by continuous centrifugation, necessary for efficient large-scale remediation, were not found during our review.

NOTE: The abovementioned methods are not necessarily alternatives of each other, but can be used in combination to increase the efficacy of MP isolation [52–54, 57, 60].

3.4.2 Physicochemical Methods

3.4.2.1 Redox Chemidegradation

Reactive oxygen species (ROS) like superoxide radicals ($O_2^{\cdot -}$) or hydroxide ions (OH^-) generated by various methods can accelerate plastic degradation. The Fenton process is a classical reaction wherein ferrous ions (Fe^{2+}) reduce H_2O_2 to produce hydroxide ions (OH^-) and/or hydroxyl radicals ($OH^{\cdot}$)

and hydroperoxyl radicals ($O_2H^{\cdot}$) which further spontaneously oxidize to form superoxide radicals ($O_2^{\cdot-}$) [66]. This also produces ferric ions (Fe^{3+}) that precipitate as its hydroxides (Fe $(OH)_3$) or oxides (Fe_2O_3, Fe_3O_4, etc.) at neutral and alkaline pH reducing the ROS generation efficacy over time [67]. Photoreduction of Fe^{3+} is a popular strategy to replenish Fe^{2+} using photoresponsive materials like TiO_2 n-type semiconductors as electron donors [68]. UV irradiation of TiO_2 semiconductors generate electron-hole (e^--h^+), pairs which are used to dissipate ferric complexes and replenish the ferrous ion pool for sustaining the Fenton process and lyse H_2O or H_2O_2 to form additional hydroxide ions and free hydroxyl radicals [69]. This coupled photo-Fenton process was being investigated primarily for chemical reductions of dyes in wastewater treatment and are worth exploring for plastic chemidegradation in stand-alone and combined remediation strategies.

Alternatively, a recent study suggested the use of microbial-driven replenishment of ferrous ions in Fenton-coupled degradation of polystyrene MPs. It is, however, time-consuming with alternating aerobic Fenton processes and anaerobic biological reduction steps both using *Shewanella putrefaciens* [70].

3.4.2.2 Photodegradation

Photodegradation is a special form for redox degradation induced by the direct incidence of the high-energy UV light (295–400 nm) catalyzing the cleavage of the otherwise unreactive bonds within the plastic polymer. This causes formation of chemical radicals, which spontaneously react with water (H_2O) or molecular oxygen (O_2), therefore also termed as photo-oxidation [71–73].

Photodegradation is a piecemeal process of plastic degradation. While it can be used for eliminating plastic pollutants under controlled conditions, normally, it is a major contributor of slow leaching of MPs and nanoplastics. Photodegradation of plastics may lead to chain fragmentation and/or functional group substitution, making it more susceptible to enzymatic degradation [74].

3.4.3 Biochemical Methods

3.4.3.1 Enzymatic Degradation

Enzymes are biomolecules, usually proteins, which function as biocatalysts by reducing the activation energy required for chemical reactions to facilitate cellular processes. *Pseudomonas putida*, already a species of interest for clearing marine oil spills [75], primarily by utilizing GPo1 plasmid-borne membrane-bound AlkB hydroxylase [76], can have an extended functionality in polyethylene degradation [77]. Fungal esterases and cutinases, produced to decompose plant waste, can also degrade petroleum polymers, particularly polyethylene terephthalate (PET) [78]. Such plastic-degrading enzymes are called secondary enzymes as their intended target substrate is not the plastic itself.

Primary plastic-degrading enzymes are enzymes produced specifically to degrade plastics. *Piscinibacter sakaiensis* [79] syn. *Ideonella sakaiensis* [80] is an aerobic, Gram negative, monotrichous motile, oxidase and catalase-positive, rod-shaped bacteria isolated from a plastic recycling facility in Sakai, Japan [81].

P. sakaiensis 201-F6 operates a novel two-enzyme pathway for catabolism of polyethylene terephthalate (PET) plastics (Figure 3.2). Enzyme PETase (ISF6_4831), a serine esterase [82], functions as a depolymerizing enzyme, hydrolyzing PET to yield its constituent monomer 2-hydroxymethyl ethylene terephthalate (MHET) as the major product [83], which further gets hydrolyzed by MHETase to produce terephthalate and ethylene glycol [84], both of which are easily biodegradable [85, 86]. Bombelli *et al.* described the plastic degrading capacity of *Galleria mellonella* (greater wax moth) larvae employing a combination of own enzymes and gut microbiome enzyme [87–90]. This plastic eating capacity arises due to the structural similarity between plastics and its natural substrate: beeswax, both long-chain saturated hydrocarbons.

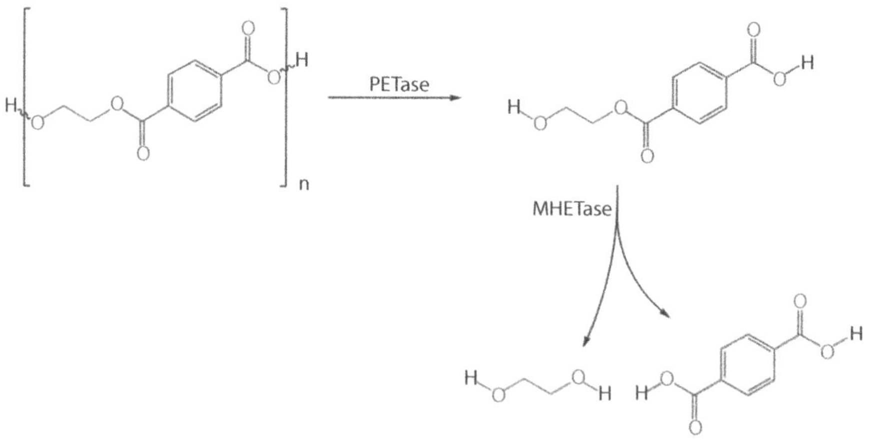

Figure 3.2 Pathway for catabolism of polyethylene terephthalate (PET) plastics.

Table 3.3 Biopolymer types and potential replacements.

Biopolymer type	Source organisms	Biological purpose	Plastic alternative	Ref.
Polyhydroxy-alkanoates (PHAs)	*Cupriavidus necator*, *Pseudomonas putida* etc.	Intracellular energy and carbon storage under nutrient limitation	PP, PE, PS, PET	[91, 92]
Polylactic acid (PLA)	Microbially fermented plant starch	Byproduct of fermentation	PP, PE, PS, PET	[93]
Starch-based bioplastics	Plants	Plant reserve energy source polysaccharide	PS, PE, PP	[94]
Cellulose-based bioplastics	Plants	Plant cell wall structural polysaccharide	PS, PP, PE, PVC	[94]
Chitin/chitosan bioplastics	Crustacean exoskeletons	Exoskeleton structural polysaccharide	PE, PET	[93]

3.4.3.2 Biopolymers

Biopolymer plastics, or simply bioplastics, are relatively biodegradable alternatives to petroleum plastics derived from natural sources (Table 3.3). These can potentially replace certain petroleum plastics for an eco-friendly option, but it falls short of shining as an economic attraction due to higher large-scale production costs compared to their synthetic counterparts.

3.5 Ecological and Evolutionary Responses to Plastic Remediation

The natural occurrence of potent plastic-degrading bacteria and fungi in ecosystems with high plastic pollution, like in the case of *Piscinibacter sakaiensis*, "Enzymatic degradation," shows some indication to the fact that life is already evolving to cope with plastic pollution without human intervention. But, to state the obvious, it would be sheer stupidity to use this as a justification to ignore the pellucid problem of plastic pollution. The rate of plastic waste accumulation far surpasses the net mass degraded by the wild-type degraders, as well as the genetically modified strains [95]. Therefore, plastic pollution remains a critical threat to global diversity. It is already regarded as one of the drivers of the proposed on-going Anthropocene mass extinction event [96, 97].

As mentioned previously, plastic pollution can force migration of species to avoid its detrimental effects. Simply clearing plastics can increase the habitability of the environment, like seen in the case of Mithi River in Powai, Mumbai, where removing macroplastic waste facilitated natural reintroduction of native fish to the river [98]. Although these anecdotal reports could be good indicators for progress in terms of recovery of biodiversity, these lack the significance of scientifically rigorous monitoring required to unequivocally declare the success of environmental remediation. A recent study reported recovery of live coral in the Palk Bay upon cleaning of human-generated debris, including fishing gear [99]. This study, however, was much localized and we were unable to find any scientific reports on the recovery of biodiversity in a broader ecosystem post-large-scale remediation efforts; we hope studies are under way and yield positive results.

3.6 Challenges and Future Perspectives

MP abundance, distribution, and impacts have become one of the most studied aspects in environmental and health sciences [100]. With a sudden boom in MP studies, a major problem faced by the scientific community as of now, according to us, is the lack of a standardized sampling and processing protocol. We believe these disparities in methodologies generate heterogeneity in results leading to incomparable findings. We believe it is vital to create a single standard methodology, or different indices to normalize the evaluation of abundance of MPs collected and processed using different methods. There is a scope for studying the impacts of plastic pollutants at a molecular level and the corresponding biological adaptations to discover better bioremediation strategies. Similarly, probing biological systems for industrial production of eco-friendly alternatives like biopolymers and bioplastics is steadily gaining popularity. While discoveries of alternatives and degradation methods are underway, innovative ways of recycling and upcycling also seem to be in demand to manage pre-existing waste.

3.7 Conclusion

The pervasive presence of plastics in abiotic, as well as biotic, elements of the ecosystem stands a key milestone in the environmental crisis of the anthropocene era. In this chapter, we briefly discussed the consequences of plastic pollution and the remediation methods. We also presented the case of natural occurrence of plastic-degrading bacteria like *Piscinibacter sakaiensis.* We also highlighted the need for long-term biodiversity monitoring projects to screen for recovery of biodiversity. While we acknowledge the fact that some ecosystems may be irreversibly damaged, remediation efforts will not be in vain if we could even restore a small but healthy portion of what's lost. Ultimately, we would like to invite scientists and citizens to shed light on the issue and urge policymakers to focus on sustainable development and cutting down on plastic production and responsible waste management. Needless to say, policy making is only one half of the effort; self-moderation for reducing plastic usage and proper disposal at an individual irrespective of laws are a must for ensuring a safe future.

AI disclosure statement: The author confirms that no artificial intelligence (AI)-generated content or AI-assisted writing tools were used in the preparation of this chapter (including summaries and the abstract). All text, analysis, and conclusions are the result of the author's original work.

References

1. Losetty, V., Reddy, G.N., Yadav, C.H., Sivakumar, K., Study on MP content in cosmetic products and their detrimental effect on human health. *Hazard. Eff. Micro- Nanoplast.*, 253–264, 2024, https://doi.org/10.1002/9781394238163.ch11.
2. De Falco, F., Di Pace, E., Cocca, M., Avella, M., The contribution of washing processes of synthetic clothes to MP pollution. *Sci. Rep.*, 9, 6633, 2019, https://doi.org/10.1038/s41598-019-43023-x.
3. Jambeck, J.R., Geyer, R., Wilcox, C., Siegler, T.R., Perryman, M., Andrady, A., Narayan, R., Law, K.L., Plastic waste inputs from land into the ocean. *Science*, 347, 768–771, 2015, https://doi.org/10.1126/science.1260352.
4. Lower, S., Polymers and plastics, Chemistry LibreTexts, 2025, https://chem.libretexts.org/Bookshelves/General_Chemistry/Chem1_(Lower)/07%3A_Solids_and_Liquids/7.09%3A_Polymers_and_Plastics.
5. Kibria, M.G., Masuk, N.I., Safayet, R., Nguyen, H.Q., Mourshed, M., Plastic waste: challenges and opportunities to mitigate pollution and effective

management. *Int. J. Environ. Res.*, 17, 20, 2023, https://doi.org/10.1007/s41742-023-00507-z.

6. Rafey, A. and Siddiqui, F.Z., A review of plastic waste management in India – challenges and opportunities. *Int. J. Environ. Anal. Chem.*, 103, 3971–3987, 2023, https://doi.org/10.1080/03067319.2021.1917560.
7. Salama, K. and Geyer, M., Plastic mulch films in agriculture: their use, environmental problems, recycling and alternatives. *Environ. Sci. Proc.*, 10, 179, 2025, https://www.mdpi.com/2076-3298/10/10/179.
8. Hernandez, G., Low, J., Nand, A., Bu, A., Wallis, S.L., Kestle, L., Berry, T.-A., Quantifying and managing plastic waste generated from building construction in Auckland, New Zealand. *Waste Manag. Res.*, 41, 205–213, 2023, https://doi.org/10.1177/0734242X221105425.
9. Bermúdez, J.R. and Swarzenski, P.W., A MP size classification scheme aligned with universal plankton survey methods. *MethodsX*, 8, 101516, 2021, https://doi.org/10.1016/j.mex.2021.101516.
10. Jakhar, P., Role of Phytoplankton and Zooplankton as Health Indicators of Aquatic Ecosystem: A Review. *Int. J. Innovative Res. Stud.*, 2, 489–500, 2013.
11. Lim, A.S., Park, N.Y., Choi, H.S., Kang, S.U., Ok, J.H., Yoo, Y.D., Influence of environmental parameters on marine plankton diversity in the southern coastal waters of Korea: emphasis on thermal stratification. *Mar. Environ. Res.*, 203, 106829, 2025, https://doi.org/10.1016/j.marenvres.2024.106829.
12. Rosal, R., Morphological description of MP particles for environmental fate studies. *Mar. Pollut. Bull.*, 171, 112716, 2021, https://doi.org/10.1016/j.marpolbul.2021.112716.
13. Hebner, T.S. and Maurer-Jones, M.A., Characterizing MP size and morphology of photodegraded polymers placed in simulated moving water conditions. *Environ. Sci. Process. Impacts.*, 22, 2020, https://pubs.rsc.org/en/content/articlelanding/2020/em/c9em00475k.
14. Lozano, Y.M., Lehnert, T., Linck, L.T., Lehmann, A., Rillig, M.C., MP shape, polymer type, and concentration affect soil properties and plant biomass. *Front. Plant Sci.*, 12, 616645, 2021, https://doi.org/10.3389/fpls.2021.616645.
15. Magni, S., Binelli, A., Pittura, L., Avio, C.G., Della Torre, C., Parenti, C.C., Gorbi, S., Regoli, F., The fate of MPs in an Italian wastewater treatment plant. *Sci. Total Environ.*, 652, 602–610, 2019, https://doi.org/10.1016/j.scitotenv.2018.10.269.
16. Magni, S., Gagné, F., André, C., Della Torre, C., Auclair, J., Hanana, H., Parenti, C.C., Bonasoro, F., Binelli, A., Evaluation of uptake and chronic toxicity of virgin polystyrene microbeads in freshwater zebra mussel Dreissena polymorpha (Mollusca: Bivalvia). *Sci. Total Environ.*, 631–632, 778–788, 2018, https://doi.org/10.1016/j.scitotenv.2018.03.075.
17. Provencher, J.F., Bond, A.L., Avery-Gomm, S., Borrelle, S.B., Rebolledo, E.L.B., Hammer, S., Kühn, S., Lavers, J.L., Mallory, M.L., Trevail, A., van Franeker, J.A., Quantifying ingested debris in marine megafauna: a review

and recommendations for standardization. *Anal. Methods*, 9, 1454–1469, 2017, https://doi.org/10.1039/C6AY02419J.

18. Gregory, M.R., Environmental implications of plastic debris in marine settings—entanglement, ingestion, smothering, hangers-on, hitch-hiking and alien invasions. *Philos. Trans. R. Soc. B Biol. Sci.*, 364, 2013–2025, 2009, https://doi.org/10.1098/rstb.2008.0265.
19. Agarwal, N., Srivastava, A., Solanki, V., Singh, N., Yadav, R., Emerging trends in the use of microbial isolates for the treatment of dye containing wastewater, 2024, 10.1016/B978-0-443-13561-3.00010-7.
20. Roman, L., Schuyler, Q., Wilcox, C., Hardesty, B.D., Plastic pollution is killing marine megafauna, but how do we prioritize policies to reduce mortality? *Conserv. Lett.*, 14, e12781, 2021, https://doi.org/10.1111/conl.12781.
21. Stelfox, M., Hudgins, J., Sweet, M., A review of ghost gear entanglement amongst marine mammals, reptiles and elasmobranchs. *Mar. Pollut. Bull.*, 111, 1, 6–17, 2016, https://doi.org/10.1016/j.marpolbul.2016.06.034.
22. Cole, M., Lindeque, P.K., Fileman, E., Clark, J., Lewis, C., Halsband, C., Galloway, T.S., MPs alter the properties and sinking rates of zooplankton faecal pellets. *Environ. Sci. Technol.*, 50, 6, 3239–3246, 2016, https://doi.org/10.1021/acs.est.5b05905.
23. Jemec, A., Horvat, P., Kunej, U., Bele, M., Kržan, A., Uptake and effects of MP textile fibers on freshwater crustacean Daphnia magna. *Environ. Pollut.*, 219, 201–209, 2016, https://doi.org/10.1016/j.envpol.2016.10.037.
24. Zhao, J., Lan, R., Wang, Z., Su, W., Song, D., Xue, R., Liu, Z., Liu, X., Dai, Y., Yue, T., Xing, B., MP fragmentation by rotifers in aquatic ecosystems contributes to global nanoplastic pollution. *Nat. Nanotechnol.*, 19, 3, 406–414, 2024, https://doi.org/10.1038/s41565-023-01534-9.
25. Kvale, K., Prowe, A.E.F., Chien, C.-T., Landolfi, A., Oschlies, A., Zooplankton grazing of MP can accelerate global loss of ocean oxygen. *Nat. Commun.*, 12, 1, 2358, 2021, https://doi.org/10.1038/s41467-021-22554-w.
26. Liu, Z. and You, X., Recent progress of MP toxicity on human exposure based on *in vitro* and *in vivo* studies. *Sci. Total Environ.*, 903, 166766, 2023, https://doi.org/10.1016/j.scitotenv.2023.166766.
27. Kadac-Czapska, K., Ośko, J., Knez, E., Grembecka, M., MPs and oxidative stress—current problems and prospects. *Antioxidants*, 13, 5, 579, 2024, https://doi.org/10.3390/antiox13050579.
28. Rao, R., Oxidative stress-induced disruption of epithelial and endothelial tight junctions. *Front. Biosci. J. Virtual Libr.*, 13, 7210–7226, 2008, https://doi.org/10.2741/3223.
29. Sofield, C.E., Anderton, R.S., Gorecki, A.M., Mind over MPs: exploring MP-induced gut disruption and gut-brain-axis consequences. *Curr. Issues Mol. Biol*, 46, 5, 4186–4202, 2024, https://doi.org/10.3390/cimb46050256.
30. Abihssira-García, I.S., Park, Y., Kiron, V., Olsvik, P.A., Fluorescent MP uptake by immune cells of Atlantic salmon (Salmo salar L.). *Front. Environ. Sci.*, 8, 2020, https://doi.org/10.3389/fenvs.2020.560206.

31. Rech, S., Pichs, Y.J.B., García-Vazquez, E., Anthropogenic marine litter composition in coastal areas may be a predictor of potentially invasive rafting fauna. *PLoS One*, 13, 1, e0191859, 2018, https://doi.org/10.1371/journal.pone.0191859.
32. Tutman, P., Kapiris, K., Kirinčić, M., Pallaoro, A., Floating marine litter as a raft for drifting voyages for Planes minutus (Crustacea: Decapoda: Grapsidae) and Liocarcinus navigator (Crustacea: Decapoda: Polybiidae). *Mar. Pollut. Bull.*, 120, 1, 217–221, 2017, https://doi.org/10.1016/j.marpolbul.2017.04.063.
33. González-Ortegón, E., Demmer, J., Robins, P., Jenkins, S., Floating plastics as a potential dispersal vector for rafting marine non-native species. *Mar. Pollut. Bull.*, 207, 116919, 2024, https://doi.org/10.1016/j.marpolbul.2024.116919.
34. Goldstein, M.C., Carson, H.S., Eriksen, M., Relationship of diversity and habitat area in North Pacific plastic-associated rafting communities. *Mar. Biol.*, 161, 6, 1441–1453, 2014, https://doi.org/10.1007/s00227-014-2432-8.
35. Pauli, N.-C., Petermann, J.S., Lott, C., Weber, M., Macrofouling communities and the degradation of plastic bags in the sea: an *in situ* experiment. *R. Soc. Open Sci.*, 4, 10, 170549, 2017, https://doi.org/10.1098/rsos.170549.
36. Carlton, J.T., Chapman, J.W., Geller, J.B., Miller, J.A., Carlton, D.A., McCuller, M.I., Treneman, N.C., Steves, B.P., Ruiz, G.M., Tsunami-driven rafting: transoceanic species dispersal and implications for marine biogeography. *Science*, 357, 6358, 1402–1406, 2017, https://doi.org/10.1126/science.aao1498.
37. Cholewińska, P., Moniuszko, H., Wojnarowski, K., Pokorny, P., Szeligowska, N., Dobicki, W., Polechoński, R., Górniak, W., The occurrence of MPs and the formation of biofilms by pathogenic and opportunistic bacteria as threats in aquaculture. *Int. J. Environ. Res. Public Health*, 19, 13, 2022, https://doi.org/10.3390/ijerph19138137.
38. van Bijsterveldt, C.E.J., van Wesenbeeck, B.K., Ramadhani, S., Raven, O.V., van Gool, F.E., Pribadi, R., Bouma, T.J., Does plastic waste kill mangroves? A field experiment to assess the impact of macro plastics on mangrove growth, stress response and survival. *Sci. Total Environ.*, 756, 143826, 2021, https://doi.org/10.1016/j.scitotenv.2020.143826.
39. Leslie, H.A., van Velzen, M.J.M., Brandsma, S.H., Vethaak, A.D., Garcia-Vallejo, J.J., Lamoree, M.H., Discovery and quantification of plastic particle pollution in human blood. *Environ. Int.*, 163, 107199, 2022, https://doi.org/10.1016/j.envint.2022.107199.
40. Ragusa, A., Svelato, A., Santacroce, C., Catalano, P., Notarstefano, V., Carnevali, O., Papa, F., Rongioletti, M.C.A., Baiocco, F., Draghi, S., D'Amore, E., Rinaldo, D., Matta, M., Giorgini, E., Plasticenta: first evidence of MPs in human placenta. *Environ. Int.*, 146, 106274, 2021, https://doi.org/10.1016/j.envint.2020.106274.
41. Zhao, Q., Zhu, L., Weng, J., Jin, Z., Cao, Y., Jiang, H., Zhang, Z., Detection and characterization of MPs in the human testis and semen. *Sci. Total Environ.*, 877, 162713, 2023, https://doi.org/10.1016/j.scitotenv.2023.162713.

42. Agarwal, N., Solanki, V., Jonnalagadda, S., Ameta, K., Singh, N., Singh, A., Bind, V., Emerging Pollutants in Aquatic Environment: Critical Risk Assessment and Treatment Options, 2024, 10.2174/9789815238969124010004.
43. Anthony, J., Varalakshmi, S., Kumar Sekar, A., Thalavai Sivasankarasubbiah, K., Harikrishnan, T., Rangamaran, V.R., Gopal, D., Ramalingam, K., MPs pollution in Indian marine environment: sources, effects and solutions. *Front. Mar. Sci.*, 11, 2024, https://doi.org/10.3389/fmars.2024.1512802.
44. Agarwal, N., Solanki, V., Pare, B., Singh, N., Jonnalagadda, S., Current Trends in Nanocatalysis for Green Chemistry and its Applications- A mini-Review. *Curr. Opin. Green Sustainable Chem.*, 41, 100788, 2023, 10.1016/j.cogsc.2023.100788.
45. Alberghini, L., Truant, A., Santonicola, S., Colavita, G., Giaccone, V., MPs in fish and fishery products and risks for human health: a review. *Int. J. Environ. Res. Public Health*, 20, 1, 789, 2022, https://doi.org/10.3390/ijerph20010789.
46. Van Cauwenberghe, L. and Janssen, C.R., MPs in bivalves cultured for human consumption. *Environ. Pollut.*, 193, 65–70, 2014, https://doi.org/10.1016/j.envpol.2014.06.010.
47. Oliveri Conti, G., Ferrante, M., Banni, M., Favara, C., Nicolosi, I., Cristaldi, A., Fiore, M., Zuccarello, P., Micro- and nano-plastics in edible fruit and vegetables: the first diet risks assessment for the general population. *Environ. Res.*, 187, 109677, 2020, https://doi.org/10.1016/j.envres.2020.109677.
48. Ragusa, A., Notarstefano, V., Svelato, A., Belloni, A., Gioacchini, G., Blondeel, C., Zucchelli, E., De Luca, C., D'Avino, S., Gulotta, A., Carnevali, O., Giorgini, E., Raman microspectroscopy detection and characterisation of MPs in human breastmilk. *Polymers*, 14, 13, 2700, 2022, https://doi.org/10.3390/polym14132700.
49. Da Costa Filho, P.A., Andrey, D., Eriksen, B., Peixoto, R.P., Carreres, B.M., Ambühl, M.E., Descarrega, J.B., Dubascoux, S., Zbinden, P., Panchaud, A., Poitevin, E., Detection and characterization of small-sized MPs (≥5 μm) in milk products. *Sci. Rep.*, 11, 1, 24046, 2021, https://doi.org/10.1038/s41598-021-03458-7.
50. Parker, L., A whopping 91% of plastic isn't recycled. *Natl. Geogr.*, 2025, https://www.nationalgeographic.com/science/article/plastic-produced-recycling-waste-ocean-trash-debris-environment.
51. Bansal, S., Singh Kushwah, S., Garg, A., Sharma, K., Utilization of plastic waste in construction industry in India – a review. *Mater. Today Proc.*, 83, 2334–2342, 2023, https://doi.org/10.1016/j.matpr.2023.03.693.
52. Wastewater Technology Fact Sheet: Screening and grit removal, U.S. Environmental Protection Agency, EPA 832-F-03-011, 2003.
53. Leiknes, T., The effect of coupling coagulation and flocculation with membrane filtration in water treatment: a review. *J. Environ. Sci.*, 21, 1, 8–12, 2009, https://doi.org/10.1016/S1001-0742(09)60003-6.

54. Lee, I., Khujaniyoz, S., Oh, H., Kim, H., Hong, T., The removal of MPs from reverse osmosis wastewater by coagulation. *J. Environ. Chem. Eng.*, 12, 4, 113198, 2024, https://doi.org/10.1016/j.jece.2024.113198.
55. Ma, B., Xue, W., Ding, Y., Hu, C., Liu, H., Qu, J., Removal characteristics of MPs by Fe-based coagulants during drinking water treatment. *J. Environ. Sci.*, 78, 267–275, 2019, https://doi.org/10.1016/j.jes.2018.10.006.
56. Mohammad-Khah, A. and Ansari, R., Activated charcoal: preparation, characterization and applications – a review article. *J. Chem.*, 6, 1, 1–14, 2009.
57. Abuwatfa, W.H., Al-Muqbel, D., Al-Othman, A., Halalsheh, N., Tawalbeh, M., Insights into the removal of MPs from water using biochar in the era of COVID-19: a mini review. *Case Stud. Chem. Environ. Eng.*, 4, 100151, 2021, https://doi.org/10.1016/j.cscee.2021.100151.
58. Alom, N., Roy, T., Sarkar, T., Rasel, M., Hossain, M.S., Jamal, M., Removal of MPs from aqueous media using activated jute stick charcoal. *Heliyon*, 10, 18, e37380, 18, 2024, https://doi.org/10.1016/j.heliyon.2024.e37380.
59. Dutta, S., Misra, A., Bose, S., Polyoxometalate nanocluster-infused triple IPN hydrogels for excellent MP removal from contaminated water: detection, photodegradation, and upcycling. *Nanoscale*, 16, 10, 5188–5205, 2024, https://doi.org/10.1039/D3NR06115A.
60. Acarer, S., A review of MP removal from water and wastewater by membrane technologies. *Water Sci. Technol.*, 88, 1, 199–219, 2023, https://doi.org/10.2166/wst.2023.186.
61. Endres, E., Dueck, J., Neesse, T., Hydrocyclone classification of particles in the micron range. *Miner. Eng.*, 31, 42–45, 2012, https://doi.org/10.1016/j.mineng.2011.10.013.
62. Yuan, F., Li, X., Yu, W., Du, J., Wang, D., Yang, X., Zhou, C., Wang, J., Yuan, H., A high-efficiency mini-hydrocyclone for MP separation from water *via* air flotation. *J. Water Process Eng.*, 49, 103084, 2022, https://doi.org/10.1016/j.jwpe.2022.103084.
63. Grause, G., Kuniyasu, Y., Chien, M.-F., Inoue, C., Separation of MP from soil by centrifugation and its application to agricultural soil. *Chemosphere*, 288, 132654, 2022, https://doi.org/10.1016/j.chemosphere.2021.132654.
64. Majcen, A., Gohla, J., Steinhoff, A.S., Meißner, L., Tassoti, S., Spitzer, P., Fractionating MPs by density gradient centrifugation: a novel approach using LuerLock syringes in a low-cost density gradient maker. *Chem. Teach. Int.*, 6, 3, 259–267, 2024, https://doi.org/10.1515/cti-2023-0079.
65. Queiroz, L.G., De Oliveira Hallai, L., Rocha De Moraes, B., Ando, R.A., Pompêo, M., Rani-Borges, B., A straightforward protocol for extracting MPs from freshwater sediment with high organic content. *Knowl. Manag. Aquat. Ecosyst.*, 426, 6, 2025, https://doi.org/10.1051/kmae/2025001.
66. Kremer, M.L., The Fenton reaction: dependence of the rate on pH. *J. Phys. Chem. A*, 107, 11, 1734–1741, 2003, https://doi.org/10.1021/jp020654p.

67. Hamd, W., Daher, E.A., Tofa, T.S., Dutta, J., Recent advances in photocatalytic removal of MPs: mechanisms, kinetic degradation, and reactor design. *Front. Mar. Sci.*, 9, 885614, 2022, https://doi.org/10.3389/fmars.2022.885614.
68. Ameta, R., Chohadia, A.K., Jain, A., Punjabi, P.B., Fenton and photo-Fenton processes, in: *Advanced Oxidation Processes for Waste Water Treatment*, S.C. Ameta, and R. Ameta, (Eds.), pp. 49–87, Academic Press, 2018, https://doi.org/10.1016/B978-0-12-810499-6.00003-6.
69. Shen, M., Song, B., Zhou, C., Hu, T., Zeng, G., Zhang, Y., Advanced oxidation processes for the elimination of MPs from aqueous systems: assessment of efficiency, perspectives and limitations. *Sci. Total Environ.*, 842, 156723, 2022, https://doi.org/10.1016/j.scitotenv.2022.156723.
70. Yang, Y., Chen, J., Chen, Z., Yu, Z., Xue, J., Luan, T., Chen, S., Zhou, S., Mechanisms of polystyrene MP degradation by the microbially driven Fenton reaction. *Water Res.*, 223, 118979, 2022, https://doi.org/10.1016/j.watres.2022.118979.
71. Mohanan, N., Montazer, Z., Sharma, P.K., Levin, D.B., Microbial and enzymatic degradation of synthetic plastics. *Front. Microbiol.*, 11, 580709, 2020, https://doi.org/10.3389/fmicb.2020.580709.
72. Tuttle, E., Wiman, C., Muñoz, S., Law, K.L., Stubbins, A., Sunlight-driven photochemical removal of polypropylene MPs from surface waters follows linear kinetics and does not result in fragmentation. *Environ. Sci. Technol.*, 58, 5461–5471, 2024, https://doi.org/10.1021/acs.est.3c07161.
73. Andrady, A.L., Biodegradation of plastics: monitoring what happens, in: *Plastics Additives: An A-Z Reference*, Pritchard and G. (Eds.), pp. 32–40, Springer Netherlands, Dordrecht, 1998, https://doi.org/10.1007/978-94-011-5862-6_5.
74. Kinyua, E., William, G., Nyakairu, A., Tebandeke, E., Odume, N., Photocatalytic degradation of MPs: parameters affecting degradation. *Adv. Environ. Eng. Res.*, 4, 1–15, 2023, https://doi.org/10.21926/aeer.2303039.
75. Raghavan, P.U.M. and Vivekanandan, M., Bioremediation of oil-spilled sites through seeding of naturally adapted Pseudomonas putida. *Int. Biodeterior. Biodegrad.*, 44, 29–32, 1999, https://doi.org/10.1016/S0964-8305(99)00048-7.
76. Beilen, J.B.V., Li, Z., Duetz, W.A., Smits, T.H.M., Witholt, B., Diversity of alkane hydroxylase systems in the environment. *Oil Gas Sci. Technol.*, 58, 427–440, 2003, https://doi.org/10.2516/ogst:2003026.
77. Nedi, E.G., Ebu, S.M., Somboo, M., Identification and molecular characterization of polyethylene degrading bacteria from garbage dump sites in Adama, Ethiopia. *Environ. Technol. Innov.*, 33, 103441, 2024, https://doi.org/10.1016/j.eti.2023.103441.
78. Temporiti, M.E.E., Nicola, L., Nielsen, E., Tosi, S., Fungal enzymes involved in plastics biodegradation. *Microorganisms*, 10, 1180, 2022, https://doi.org/10.3390/microorganisms10061180.
79. Liu, Y., Du, J., Pei, T., Du, H., Feng, G.-D., Zhu, H., Genome-based taxonomic classification of the closest-to-Comamonadaceae group supports a

new family Sphaerotilaceae fam. nov. and taxonomic revisions. *Syst. Appl. Microbiol.*, 45, 126352, 2022, https://doi.org/10.1016/j.syapm.2022.126352.
80. Tanasupawat, S., Takehana, T., Yoshida, S., Hiraga, K., Oda, K., Ideonella sakaiensis sp. nov., isolated from a microbial consortium that degrades poly(ethylene terephthalate). *Int. J. Syst. Evol. Microbiol.*, 66, 2813–2818, 2016, https://doi.org/10.1099/ijsem.0.001058.
81. Yoshida, S., Hiraga, K., Takehana, T., Taniguchi, I., Yamaji, H., Maeda, Y., Toyohara, K., Miyamoto, K., Kimura, Y., Oda, K., A bacterium that degrades and assimilates poly(ethylene terephthalate). *Science*, 351, 1196–1199, 2016, https://doi.org/10.1126/science.aad6359.
82. Burgin, T., Pollard, B.C., Knott, B.C., Mayes, H.B., Crowley, M.F., McGeehan, J.E., Beckham, G.T., Woodcock, H.L., The reaction mechanism of the Ideonella sakaiensis PETase enzyme. *Commun. Chem.*, 7, 1–14, 2024, https://doi.org/10.1038/s42004-024-01154-x.
83. Duan, S., Zhang, N., Chao, T., Wu, Y., Wang, M., The structural and molecular mechanisms of type II PETases: a mini review. *Biotechnol. Lett*, 45, 1249–1263, 2023, https://doi.org/10.1007/s10529-023-03418-3.
84. Pinto, A.V., Ferreira, P., Neves, R.P.P., Fernandes, P.A., Ramos, M.J., Magalhães, A.L., Reaction mechanism of MHETase, a PET degrading enzyme. *ACS Catal.*, 11, 10416–10428, 2021, https://doi.org/10.1021/acscatal.1c02444.
85. Aksu, D., Vural, C., Karabey, B., Ozdemir, G., Biodegradation of terephthalic acid by isolated active sludge microorganisms and monitoring of bacteria in a continuous stirred tank reactor. *Braz. Arch. Biol. Technol.*, 64, 1–10, 2021.
86. Willetts, A., Bacterial metabolism of ethylene glycol. *Biochim. Biophys. Acta*, 677, 194–199, 1981, https://doi.org/10.1016/0304-4165(81)90085-4.
87. Bombelli, P., Howe, C.J., Bertocchini, F., Polyethylene bio-degradation by caterpillars of the wax moth Galleria mellonella. *Curr. Biol.*, 27, R292–R293, 2017, https://doi.org/10.1016/j.cub.2017.02.060.
88. Cassone, B.J., Grove, H.C., Elebute, O., Villanueva, S.M.P., LeMoine, C.M.R., Role of the intestinal microbiome in low-density polyethylene degradation by caterpillar larvae of the greater wax moth, Galleria mellonella. *Proc. R. Soc. B Biol. Sci.*, 287, 20200112, 2020, https://doi.org/10.1098/rspb.2020.0112.
89. Sanluis-Verdes, A., Colomer-Vidal, P., Rodriguez-Ventura, F., Bello-Villarino, M., Spinola-Amilibia, M., Ruiz-Lopez, E., Illanes-Vicioso, R., Castroviejo, P., Aiese Cigliano, R., Montoya, M., Falabella, P., Pesquera, C., Gonzalez-Legarreta, L., Arias-Palomo, E., Solà, M., Torroba, T., Arias, C.F., Bertocchini, F., Wax worm saliva and the enzymes therein are the key to polyethylene degradation by Galleria mellonella. *Nat. Commun.*, 13, 5568, 2022, https://doi.org/10.1038/s41467-022-33127-w.
90. Young, R., Ahmed, K.A., Court, L., Castro-Vargas, C., Marcora, A., Boctor, J., Paull, C., Wijffels, G., Rane, R., Edwards, O., Walsh, T., Pandey, G., Improved reference quality genome sequence of the plastic-degrading greater wax moth, Galleria mellonella. *G3 Genes Genomes Genet.*, 14, jkae070, 2024, https://doi.org/10.1093/g3journal/jkae070.

91. Morlino, M.S., Serna García, R., Savio, F., Zampieri, G., Morosinotto, T., Treu, L., Campanaro, S., Cupriavidus necator as a platform for polyhydroxyalkanoate production: an overview of strains, metabolism, and modeling approaches. *Biotechnol. Adv.*, 69, 108264, 2023, https://doi.org/10.1016/j.biotechadv.2023.108264.
92. Sikkema, W.D., Cal, A.J., Hathwaik, U.I., Orts, W.J., Lee, C.C., Polyhydroxyalkanoate production in Pseudomonas putida from alkanoic acids of varying lengths. *PLoS One*, 18, e0284377, 2023, https://doi.org/10.1371/journal.pone.0284377.
93. Al-Tayyar, N.A., Youssef, A.M., Al-Hindi, R., Antimicrobial food packaging based on sustainable bio-based materials for reducing foodborne pathogens: a review. *Food Chem.*, 310, 125915, 2020, https://doi.org/10.1016/j.foodchem.2019.125915.
94. Hossain, M.T., Shahid, M.A., Akter, S., Ferdous, J., Afroz, K., Refat, K.R.I., Faruk, O., Jamal, M.S.I., Uddin, M.N., Samad, M.A.B., Cellulose and starch-based bioplastics: a review of advances and challenges for sustainability. *Polym.-Plast. Technol. Mater.*, 63, 1329–1349, 2024, https://doi.org/10.1080/25740881.2024.2329980.
95. Martín-González, D., de la Fuente Tagarro, C., De Lucas, A., Bordel, S., Santos-Beneit, F., Genetic modifications in bacteria for the degradation of synthetic polymers: a review. *Int. J. Mol. Sci.*, 25, 5536, 2024, https://doi.org/10.3390/ijms25105536.
96. Miditana, S.R., Dokka, W.T., Babu, A.R., Ampolu, S., Agarwal, N., Simma, N., Cyclodextrins for UV protection in food, beverages, and agricultural products: A comprehensive review. *N. Afr. J. Food Nutr. Res.*, 9, 20, 219–235, 2025, https://doi.org/10.51745/najfnr.9.20.219-235.
97. Schickhoff, U., Bobrowski, M., Offen, I., Mal, S., The biodiversity crisis in the Anthropocene, in: *Geography and the Anthropocene*, pp. 79–111, Istanbul University Press, 2024, https://doi.org/10.26650/B/SS19.2023.020.05.
98. Fish being sighted now in Mithi, claims BMC, The Times of India, May 28, 2023, https://timesofindia.indiatimes.com/city/mumbai/fish-being-sighted-now-in-mithi-claims-bmc/articleshow/100561081.cms.
99. Pillai, R.R., Lakshmanan, S., Mayakrishnan, M., George, G., N, N. M., Impact of marine debris on coral reef ecosystem and effectiveness of removal of debris on ecosystem health – baseline data from Palk Bay. *Indian Ocean. Res. Sq.*, 2023, https://doi.org/10.21203/rs.3.rs-2720380/v1.
100. Farale, H., Sreevidhya, K.B., Bathinapatla, A., Kanchi, S., Impact of plastic contaminants on marine ecosystems and advancement in the detection of micro/nano plastics: a review. *J. Hazard. Mater. Adv.*, 18, 100736, 2025, https://doi.org/10.1016/j.hazadv.2025.100736.

Part II

PIONEERING INNOVATIONS: CUTTING-EDGE SCIENTIFIC AND TECHNOLOGICAL FRONTIERS IN PLASTIC REMEDIATION

4

Nanotechnological Advances in Plastic Remediation: Cutting Edge Solutions

Deepika Geedkar*, **Ashok Kumar, Rashmi Sharma and Pratibha Sharma**

School of Chemical Sciences, Devi Ahilya Vishwavidyalaya, Indore, Madhya Pradesh, India

Abstract

Plastic pollution is a serious environmental crisis that affects ecosystems, biological diversity, and human health. As the world produces more than 400 million tons of plastic annually, most of this trash is not adequately handled, and it keeps piling up in the environment. The problem is provoked by the durability and resistance of plastics to natural degradation, resulting in the build-up of plastics on land, in waterways, and in the ocean. The traditional approaches, including landfilling, incineration, and mechanical recycling, have challenges. Landfills lead to contamination of the environment, whereas incineration releases greenhouse gases and other toxic byproducts. Recycling, despite being necessary, is impeded by polymer multiplicity, contamination, and insufficient infrastructure. Such restrictions reveal the necessity of an effective, scalable, and environmentally friendly solution.

Bionanomaterials (BNMs) have become an innovative way of addressing these challenges. These materials, synthesized using biological materials or produced in keeping with green chemistry principles, combine the potential benefits of nanotechnology, such as high reactivity and surface area, with the sustainability and biocompatibility of biological materials. This chapter discusses various applications of BNMs in reducing plastic pollution. It deals with their synthesis by utilizing sustainable processes, including the use of plants, microorganisms, and biopolymers, along with their mechanisms of action. It also seeks to motivate creative research and partnership in addressing plastic pollution by developing the knowledge and use of BNMs. By introducing BNMs into policy and practice, we

**Corresponding author*: dgeedkar96@gmail.com

Neha Agarwal, Brijesh Pare, Vijendra Singh Solanki and Sankara Rao Miditana (eds.) Plastic Remediation: Transformative Pathways for a Rejuvenated Sustainable Environment, (81–122)

pave the way for sustainable, scalable solutions to one of the most pressing environmental challenges facing our planet today.

Keywords: Plastic pollution, BNMs, nanotechnology, sustainable solutions, plastic degradation, environmental remediation

4.1 Introduction

Plastic pollution is the hallmark of the environmental challenge of the 21st century. Since the invention of synthetic polymers at the beginning of the 20th century, the use of plastics has become an inseparable part of contemporary society [1]. Their multifunctionality, strength, and cheap production have facilitated their extensive use in a wide range of industries, including packaging, construction, healthcare, electronics, and agriculture [2]. Nevertheless, their stability against natural degradation has resulted in their accumulation in the environment [3]. The world produces over 400 million tonnes of plastic each year, and about 40 percent are single-use applications [4]. A substantial percentage of this plastic waste is poorly managed leading to the extensive contamination of terrestrial and aquatic systems [5]. An increase in global plastic production in recent decades is evidence of the magnitude of this challenge (Figure 4.1).

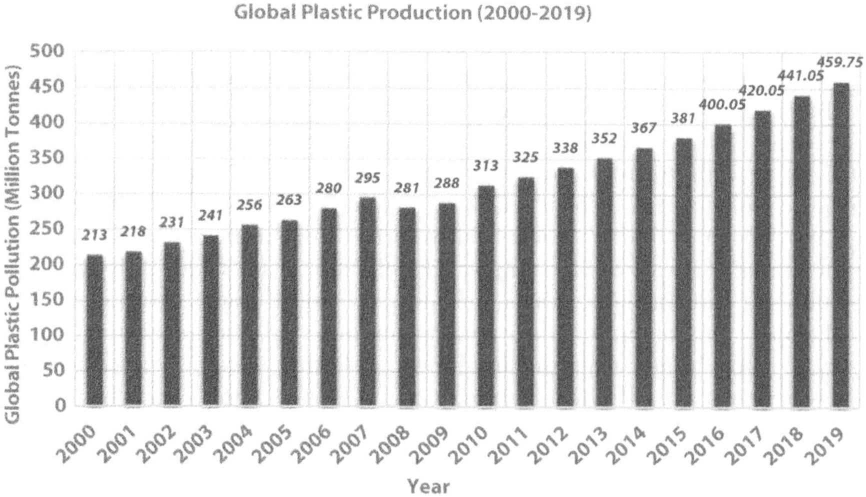

Figure 4.1 Global plastic production increased significantly between 2000 and 2019, reflecting the growing scale of the plastic pollution crisis. (Source: Our World in Data; data adapted from PlasticsEurope and OECD).

Plastic debris poses a major threat to biodiversity in the marine environment. These particles, widespread in marine environments and on land, can absorb toxic pollutants, worsening their negative effect on both ecosystems and human health [6]. Plastic pollution also threatens terrestrial ecosystems. In agrarian environments, plastic and microplastic litter damage soil structure and functionality, reducing soil fertility, water retention, and plant growth [7]. Chemical additives involved in plastic production, such as phthalates and bisphenols, leach into surrounding areas, contaminating groundwater and causing health hazards.

Although the scale of the issue is overwhelming, traditional plastic waste management approaches are insufficient. Recycling remains a significant strategy, but efficient recycling is logistically and economically challenging. A large number of plastic products are made of mixed polymers or are contaminated, making them unfit for traditional recycling methods [8]. Furthermore, the volume of plastic waste exceeds the capacity of recycling systems, where only 9 percent of all plastic waste generated so far has been recycled [9]. To address these issues, scientists have attempted to find alternative solutions that are more sustainable and innovative. The combination of nanotechnology and green chemistry has provided a prospective solution in the form of BNMs. These are biosourced or produced *via* green technologies, through which they possess properties specifically applicable in the fight against plastic pollution. Their nanoscale size offers high surface-to-volume ratios that lead to increased reactivity and binding to plastic pollutants. Moreover, they have biocompatibility and biodegradability, decreasing environmental risks compared to traditional nanomaterials [10].

BNMs have functionalities capable of degrading plastics through physical, chemical, and biological processes. For instance, biopolymer-derived nanomaterials, such as chitosan nanoparticles and cellulose nanofibers, can improve catalytic performance by serving as carriers of plastic-degrading enzymes. The multifunctional BNMs, such as hybrid materials containing both organic and inorganic materials, also have photocatalytic and antimicrobial potential. Despite their potential, challenges remain in the extensive use of BNMs. The scalability, cost-efficiency, and stability under environmental conditions must be considered for large-scale implementation. The possible hazards caused by nanoparticle release should be prevented. Regulatory mechanisms must adapt to ensure safe and responsible use while promoting innovation and collaboration [11].

The main objective of this chapter is to provide a detailed overview of BNMs in the context of plastic pollution. It describes the state of knowledge regarding their synthesis, mechanisms of action, and applications.

By incorporating BNMs within circular economy models and policies, plastic waste could be successfully managed, resulting in a cleaner and more sustainable environment.

4.2 Overview of Plastic Pollution and Its Challenges

4.2.1 Scale and Scope of Plastic Pollution

4.2.1.1 Global Statistics on Plastic Production, Use, and Waste Accumulation

Since its invention, plastic manufacturing has grown exponentially and is currently of more than 400 million tons every year. This increase is indicative of the necessity of plastics in contemporary society, in areas such as packaging, construction, consumer electronics, healthcare, and agricultural areas. Plastics are lightweight, durable, and cost-effective and have the revolutionized industries, leading to economic and technological development. Nevertheless, this mass dependency has been at a great environmental price [12]. There has been recycling of only 9% of cumulative plastic waste, and about 12% incineration. Only 79% is in landfills, natural scenery, or aquatic habitats [13]. Since plastics are composed of synthetics and are non-degradable, they cause long-term pollution challenges lasting decades or centuries.

Globally, more than 8 million tons of plastic waste leak into oceans each year. This is roughly one garbage truck of plastic getting into the ocean every minute [14]. This has led to the development of large areas of accumulation zone known as the Great Pacific Garbage Patch which is approximately three times the size of France. This shows the dire necessity of sustainable and systematic interventions to the management of plastic waste.

4.2.1.2 Ecological and Health Impacts of Plastics

The infiltration of plastics into ecosystems has resulted in significant ecological disruption. In marine environments, plastics damage habitats, disrupt food chains, and harm over 800 species of wildlife. Marine mammals, birds, and fish often ingest plastics or become entangled in them, resulting in injury, malnutrition, or death. The presence of microplastics, defined as plastic pieces less than 5 millimeters in size, is found in marine life, ranging from the plankton to the top predators [15].

Microplastics are also dangerous to human health. These particles have the potential to transport toxic components, such as pesticides and heavy metals, which stick to their surfaces. When ingested by marine creatures, the toxins become bioaccumulative and biomagnifying *via* the food chain, and they eventually accumulate in humans. Research has found microplastic exposure to disrupt endocrines and cause inflammation and other health effects [15].

Terrestrial ecosystems are similarly vulnerable. Plastic debris changes the chemical composition of soil and decreases its permeability, which adversely affects the process of water infiltration and the growth of plant roots. Plastic contamination of soils can also reduce the fertility of soil and crop yields [16]. Moreover, the additives like phthalates and bisphenols are capable of leaching from plastics in the surrounding soil and groundwater to contaminate the ecosystem and pose a threat to biodiversity.

4.2.1.3 Economic Costs of Managing Plastic Pollution

Plastic pollution has a significant economic impact that encompasses the burden on public institutions, private businesses, and the economy of nations. Coastal countries, particularly the ones that depend on tourism and fisheries, suffer significant losses because of the deterioration in the ecosystems. Removal of plastics on the coastline and sand dunes costs billions of dollars every year.

The fisheries and aquaculture sector is also experiencing serious losses as a result of the depleting fish stocks and pollution of aquatic life. Tourism-reliant economies suffer because when beaches and coastal waters are polluted, fewer people want to travel to those areas, causing revenue losses and degrading local livelihoods.

4.2.2 The Need for Green Remediation

4.2.2.1 Limitations of Traditional Remediation Approaches

Traditional waste management strategies, such as landfilling, incineration, and recycling, have proven insufficient in mitigating the scale of plastic pollution.

- Landfilling leads to the prolonged accumulation of plastics. The accumulation of larger plastic materials gradually forms microplastics and gets into soil and groundwater. This aggravates environmental pollution and impacts agriculture [9].

- ➢ Incineration decreases the amount of waste but releases greenhouse gases such as carbon dioxide and methane. It also emits toxic chemicals, such as dioxins and furans, which are very dangerous to human and environmental health [10].
- ➢ Recycling, is also significant, but is limited in many ways. Pollution of waste, polymer multiplicity, and poor structure lower the effectiveness of recycling processes.

Moreover, the recycling rate of plastics in the world today is significantly less than the production rate [10].

The challenges reflect the urgency of alternative remediation technologies that target the root causes of plastic pollution and reduce detrimental effects on the environment.

4.2.2.2 *Introduction to Green Chemistry Principles and Sustainable Alternatives*

Green remediation is an emerging approach based on the principles of green chemistry and sustainable engineering. Unlike conventional methods, green remediation focuses on minimizing environmental impact while enhancing efficiency. Key principles include the following:

- Reduction of hazardous substances by eliminating toxic chemicals and promoting the use of environmentally safe alternatives [17].
- Utilization of renewable resources through biodegradable and biobased materials that prevent long-term waste accumulation.
- Energy efficiency by adopting low-energy processes that reduce carbon emissions.
- BNMs exemplify the potential of green remediation. By integrating the practical benefits of nanotechnology with the eco-friendliness of biological materials, they can be considered an effective approach toward plastic degradation in several ways [18, 19].
- Chemical degradation that breaks down complex plastic polymers into smaller, non-toxic compounds.
- Catalytic activity that accelerates degradation reactions, producing safer byproducts.

- Synergistic enzyme enhancement by encapsulating or modifying enzymes to improve their performance in breaking down plastics.

By integrating BNMs into waste management systems, these technologies can help overcome the drawbacks of conventional methods. Green remediation supports international sustainability goals, such as the United Nations Sustainable Development Goals (SDGs), by promoting innovation, resource recovery, and circular economy models.

4.3 What are BNMs?

BNMs include advanced materials that are made using bio-building blocks of nanoscale with various sources, such as biological sources or built-in biologically and environmentally friendly ways. These substances normally have at least one or two dimensions less than 100 nanometers. It is an integration of biocompatibility, sustainability, and functionality of the biological systems with the increased surface responsiveness and the physicochemical capabilities of the nanomaterials. Such a combination of properties enables BNMs to be employed efficiently in the context of environmental cleanup, biomedical, and energy systems [20].

4.3.1 Classification of BNMs

BNMs are generally divided into three major categories:

4.3.1.1 Biopolymer-Based Nanomaterials

These materials are derived from natural polymers such as cellulose, chitosan, starch, and alginate. They are naturally abundant, biodegradable, and have the ability to interact with different pollutants at the nanoscale. Common examples include the following:

- **Cellulose Nanocrystals (CNCs):** CNCs are derived by using plant biomass and valuable in composites and biofilms, owing to their high mechanical strength.
- **Chitosan Nanoparticles:** These nanoparticles are derived from chitin in crustacean shells and exhibit antimicrobial properties and are often utilized as an enzyme carrier in degradation of plastic [21].

4.3.1.2 Protein-Based Nanomaterials

These are composed of natural proteins such as collagen, silk fibroin, and keratin. Their advantages include excellent mechanical flexibility, biocompatibility, and the ability to self-assemble into nanostructures.

- **Silk Fibroin Nanoparticles:** These enhance enzymatic activity and are compatible with microbial systems, making them suitable for soil and marine plastic remediation [22].
- **Keratin Nanoparticles:** They are grown on human hair or the wool of animals and serve as adsorbents to toxic fragmentation byproducts [23].

4.3.1.3 Hybrid BNMs

These materials integrate organic (biopolymer-based) and inorganic (metal or carbon-based) components to acquire multifunctionality. Some common systems include the following:

- **Zinc Oxide (ZnO) in Biopolymer Matrices:** Utilized for photocatalytic degradation of plastics under light as an energy source [24].
- **Silver Nanoparticles in Chitosan:** Combine antimicrobial and catalytic properties for enhanced plastic breakdown [25].
- **Carbon-Based Hybrids:** Graphene oxide functionalized with biopolymers has demonstrated effectiveness in absorbance and gradual breakdown of persistent plastic contaminants [26].

4.3.2 Case Studies of BNMs in Plastic Remediation

Some examples of BNMs in practical uses in plastic degradation include several case studies:

- **Chitosan-Based Nanoparticles:** They have been employed in the production of lipase enzymes that have been encapsulated to enhance their stability against polyethylene terephthalate (PET) to hasten the degradation of this material [27].

- **Cellulose Nanofibers:** They are used to process biofilms in water, which entrap microplastics and release them to enzymatic degradation. It is a highly effective strategy in marine organisms [28].
- **Hybrid ZnO–Cellulose Systems:** Zinc oxide nanoparticles embedded in cellulose matrices have shown strong photocatalytic effects against polystyrene, degrading it into simpler and environmentally benign byproducts under light exposure [29].

4.3.3 Advantages in Plastic Remediation

4.3.3.1 Properties of BNMs

The BNMs have the following strengths that render them ideal for use in plastic remediation:

- **Eco-Friendliness:** They are biodegradable and sourced from renewable biological materials, which make them less harmful to the environment [30].
- **High Surface Reactivity:** Their nanoscale structure strengthens their interactions with plastic surfaces, which facilitates efficient degradation [31].
- **Synergistic Potential:** They can be used with enzymes or microbes to accelerate and intensify the degradation of plastics [30, 31].

4.3.3.2 Comparison with Conventional Nanomaterials

BNMs have various environmental and functional advantages over the standard synthetic nanomaterials:

- **Lower Environmental Impact:** BNMs are achieved through agricultural sources of renewable raw materials, forestry sources, or marine sources. They result in lower toxic byproducts attributed to their production and the harmful impacts to the ecology because they are less resources-intensive than conventional nanomaterials [31].

- **Higher Biodegradability:** These materials commonly break down into non-toxic products, such as CO_2, water, and simple organic molecules, within natural systems, and a large fraction of them will not accumulate in the environment over time.
- **Cost Efficiency:** They may be produced using inexpensive biowaste (e.g., crop residues, shells of seafood), which reduces the cost of raw materials and enables circular and sustainable production.

These characteristics render BNMs among the pillars of green remediation. They can be integrated into plastic waste management practices to expedite the degradation processes, achieve sustainability goals, and reduce the effects of environmental degradation.

4.4 Synthesis of BNMs

4.4.1 Green Synthesis Methods

4.4.1.1 Biogenic Routes Using Plants, Microorganisms, and Enzymes

Green synthesis of BNMs refers to the utilization of natural, renewable feedstock to form functional materials with a limited impact on the environment. Plants, microorganisms, and enzymes are biological components that largely contribute to the reduction of metal ions, stabilization of nanoparticles, and acquisition of functional properties (Figure 4.2).

a. **Plants**: Plant extracts are rich in secondary metabolites that include flavonoids, phenolics, terpenoids, alkaloids, and tannins. These bioactive compounds also serve as reducing agents, transforming metal ions into nanoparticles and capping agents that keep the particles stable (Figure 4.3). Examples:
    - **Aloe vera gel**: Promotes the production of silver nanoparticles (AgNPs), which have outstanding catalytic capacity to degrade polyethylene in the presence of UV light [32].
    - **Neem leaves (Azadirachta indica)**: Enhance the synthesis of gold and zinc oxide nanoparticles with controlled size and morphology [33].

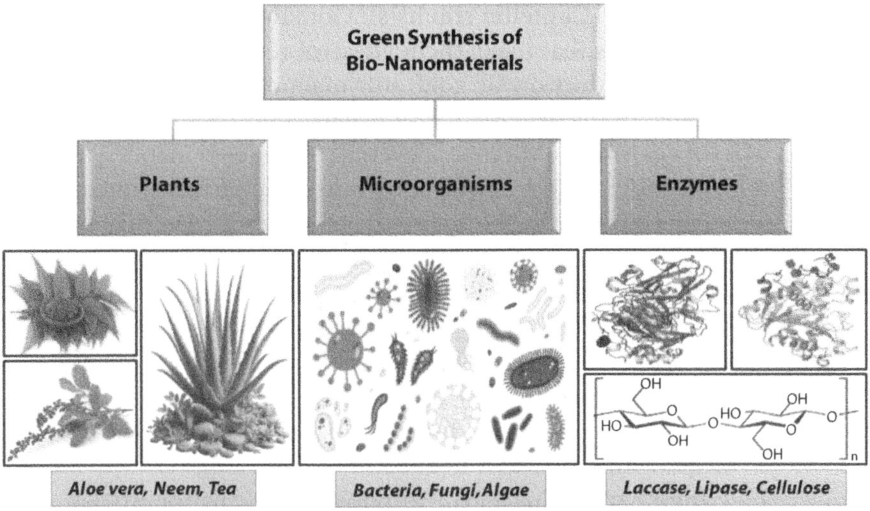

Figure 4.2 Biological routes for green synthesis of BNMs involving plants, microorganisms, and enzymes.

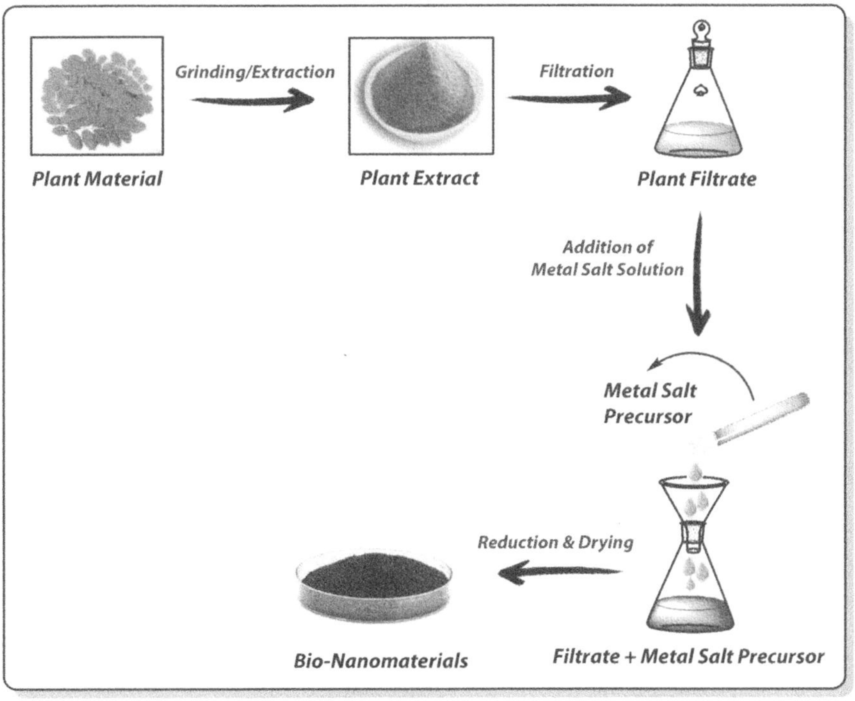

Figure 4.3 Biological routes for green synthesis of BNMs involving plants, microorganisms, and enzymes.

- **Tea leaves (Camellia sinensis)**: Contain tannin and catechin levels that make them effective reducing mediums in the production of gold and magnetic nanoparticles applied in the degradation of plastics [34].

b. **Microorganisms**: Microbes leverage their enzymatic systems to biotransform metal ions into nanoparticles under ambient conditions. The approach is beneficial because of the high specificity, low toxicity, and ease of cultivation. **Examples**:
   - **Bacteria**: *Bacillus subtilis* synthesizes polymer-coated nanoparticles with enzymatic properties that depolymerize polyethylene and polystyrene [35]. Pseudomonas aeruginosa generates functionalized nanoparticles to degrade microplastics in aquatic systems [36].
   - **Fungi**: Aspergillus niger forms metal oxide nanoparticles that improve the ability to trap microplastics [37].
   - **Algae**: Spirulina platensis and Chlorella vulgaris produce metal-based nanoparticles and eliminate plastic pollutants by bioadsorption [38].

c. **Enzymes**: Enzymatic synthesis involves a combination of catalytic accuracy of biological molecules and nanoscale engineering. Enzymes promote nanoparticle activity and promote site-specific reaction with plastic contaminant. **Examples**:
   - **Laccase**: Facilitates production of nanoparticles and breakdown of plastics [39].
   - **Lipase**: Hydrolysis of polyester-based plastics such as PET is accelerated [40].
   - **Cellulases**: Bioactive nanomaterials of cellulose nanocrystals incorporated in plastic depolymerization [41].

4.4.2 Sustainable Chemical Synthesis Methods

Sustainability can be attained through chemical synthesis, considering green chemistry principles, including utilization of renewable inputs, solvent-free reactions, and energy-efficient conditions.

- **Sol-Gel Processes**: Formation of a colloidal sol transformed into a gel to form nanoparticles. Chitosan nanocomposites produced *via* sol-gel methods exhibit biodegradability and catalytic efficiency [42].

- **Hydrothermal Synthesis**: Precursors dissolved in aqueous solutions under high pressure and temperature produce well-structured nanomaterials. Zinc oxide nanoparticles coated with biopolymers degrade polyethylene and PVC in wastewater [43].
- **Microwave-Assisted Methods**: Microwave heating provides equal heating, high reaction rates, and reduced energy use. Silver nanoparticles incorporated into cellulose matrices degrade plastics catalytically and enzymatically [44].
- **Ionic Liquids (ILs) and Supercritical Fluids**: Environmentally benign reaction media. PETase enzymes functionalized on cellulose nanofibers processed with ILs degrade PET plastics [45].

4.5 Case Studies of BNM Synthesis

4.5.1 Silver Nanoparticles from Plant Extracts

- **Synthesis:** In a green synthesis pathway, *Azadirachta indica* (neem) leaf extract is used, in which terpenoids and flavonoids found naturally serve as reducing and capping agents in the synthesis of silver nanoparticles (AgNPs).
- **Application:** The resultant AgNPs exhibit photocatalytic behavior and allow oxidative degradation of low-density polyethylene (LDPE) films in the presence of UV and natural sunlight [46].
- **Impact:** It is a promising solution to plastic waste management in water bodies through this eco-friendly and scalable approach using renewable sources of biological inputs and ambient energy.

4.5.2 Chitosan-Coated Zinc Oxide Nanoparticles

- **Synthesis:** ZnO nanoparticles are synthesized *via* a hydrothermal process and subsequently coated with biopolymeric chitosan to enhance surface reactivity and biocompatibility.
- **Application:** The composite nanomaterial is dual-action degradable, both through photocatalytic degradation under UV light and through possible biocatalytic degradation with

microbial enzymes, of common polymers such as polystyrene and polyethylene in soil and water environments [47].

- **Impact:** The synergistic properties of ZnO and chitosan broaden the degradation pathways, improving efficacy under natural conditions and supporting multifunctional remediation strategies.

4.5.3 Cellulose Nanocrystals Functionalized with PETase

- **Synthesis:** The CNCs are synthesized by subjecting agricultural residues (e.g., sugarcane bagasse or wheat straw) to acid hydrolysis, which gives rise to the cellulose, followed by a surface modification involving covalent attachment or adsorption of engineered enzymes, namely PETase.
- **Application:** This nanobiocatalyst specifically attaches itself to polyethylene terephthalate (PET), which it breaks down under mild conditions to biodegradable monomers, including terephthalic acid and ethylene glycol [48].
- **Impact:** The enzyme-functionalized CNCs have been introduced to facilitate selective plastic degradation, value addition to agrowaste, and the creation of models of a circular bioeconomy.

4.6 Mechanisms of Action

The mechanism of BNMs in plastic degradation is a complex process involving interactions with plastic polymers at the molecular level. Such interactions facilitate the degradation of plastics into smaller, less harmful substances using physical, chemical, and biological processes. Its key mechanisms are surface binding, oxidative degradation, synergy with biological systems, and nanoenzymes.

4.6.1 Interaction with Plastics

BNMs react with plastics in several ways, breaking them down into less hazardous byproducts. These interactions mainly occur at the molecular level and depend on the structure and properties of the BNM.

4.6.1.1 *Surface Binding and Polymer Chain Weakening*

BNMs have a high surface area-to-volume ratio, enhancing interaction with plastic surfaces:

- **Hydrophobic Interactions**: BNMs attach to non-polar surfaces of plastics, weakening polymer chains and making them more susceptible to degradation [49].
- **Electrostatic Interactions**: Charged BNMs interact with sections of plastic polymers, facilitating bond breaking [50].

4.6.1.2 *Oxidative Degradation*

BNMs generate reactive oxygen species (ROS), such as hydroxyl radicals (•OH) and superoxide anions ($O_2^{\bullet-}$), which oxidize plastic polymers and break long chains into smaller fragments [51].

- **Photocatalytic Degradation**: Certain BNMs, such as TiO_2 or ZnO, harness light energy to generate ROS, initiating a process known as photodegradation. When exposed to ultraviolet (UV) light, these materials produce ROS that attack the plastic structure, causing it to fragment [52].

4.6.2 Case Studies Highlighting these Mechanisms

4.6.2.1 *Chitosan-Silver Nanocomposites*

A novel method that shows potential is the preparation of chitosan, a biopolymer generated from chitin, and silver nanoparticles. These composites are efficient in fragmenting polyethylene under UV light. The silver nanoparticles also improve the photocatalytic performance of the chitosan, and the exposure of UV light stimulates the formation of ROS, which facilitates the degradation of the plastics [53].

4.6.2.2 *Enzyme-Loaded Cellulose Nanofibers*

Cellulose nanofiber carriers of plastic-degrading enzymes have been demonstrated to degrade polyethylene terephthalate (PET) by enzymatic degradation. These BNMs comprise enzymes that have a lower environmental impact than current plastic waste management systems, using the natural capacity of enzymes to degrade the complex forms of plastic [54].

4.6.2.3 Synergy with Biological Systems

BNMs have the potential to augment the naturally occurring degradation processes of plastics by providing an environment that allows the growth of microbes. These systems encourage the development of plastic-consuming microorganisms and enhances the inherent process of degradation of plastics. The interaction of BNMs with biological systems becomes a synergy in achieving effective remediation procedures.

4.6.2.4 Microbial Colonization on Plastic Surfaces

BNMs serve as scaffolds or the growth supporting surfaces to the proliferation of microbes capable of degrading plastic polymers. This increases the colonization and activity of plastic-degrading bacteria, fungi, or algae, thereby increasing the rate of biodegradation. Some of the common examples of BNMs to aid colonization by microbes are as follows:

- **Nanocellulose Films**: These films can immobilize the microbial colonies on their surfaces which result in a stable environment that maintains constant degradation of the plastics. Nanocellulose provides a viable surface that the microbes can easily attach to and come into contact with the plastic surface to initiate degradation [55].
- **Protein-Based Nanoparticles**: These nanoparticles act as the vehicle of the microorganisms and transmit them directly to plastic surfaces. The microbes are close to the plastic, which further promotes the degradation process through direct contact and enzyme attack of the plastic [56].

4.6.2.5 Activation of Biodegradative Enzymes

BNMs can also activate or stabilize enzymes that play key roles in plastic biodegradation. For example:

- **Metallic Nanoparticles**: Nanoparticles made of metals such as silver, copper, or gold can act as co-factors for enzymes, enhancing their catalytic activity. These nanoparticles can stabilize enzymes, improving their efficiency in breaking down complex plastic polymers, such as polystyrene or polyvinyl chloride (PVC) [57].

- **Hybrid BNMs**: These materials combine multiple degradation pathways, such as enzymatic and photocatalytic processes. For example, hybrid systems that integrate enzymes with photocatalytic nanomaterials can accelerate plastic degradation by combining the action of ROS with enzymatic hydrolysis, effectively doubling the efficiency of degradation [58].

4.7 Role of Nanoenzymes

Nanoenzymes are enzyme-mimicking nanostructures that exhibit catalytic properties similar to natural enzymes. They are more stable under harsh conditions, making them ideal for plastic degradation.

4.7.1 Mechanistic Insights into Nanoenzyme-Assisted Plastic Degradation

Nanoenzymes can target various bonds in plastic polymers, including ester, amide, and carbon-carbon bonds. By selectively cleaving these bonds, nanoenzymes accelerate the degradation of plastics. The key mechanisms involved include the following:

- **Catalytic Cleavage**: Break specific polymer bonds, reducing molecular weight and converting long chains into biodegradable fragments [59].
- **Selective Reactivity**: Functionalized nanoenzymes can target specific plastics like polyethylene or polystyrene [60].

4.7.2 Integration with BNM Systems

Nanoenzymes are commonly incorporated into BNM structures to enhance degradation of plastics. These are integrated systems that have catalytic and adsorptive ability which will enable the breakdown of plastics in various environments. Examples of nanoenzyme integration are as follows:

- **Iron Oxide Nanoparticles Coated with Chitosan**: Catalyze microplastic degradation in water. Chitosan enhances adsorption and stabilizes the process [61].

- **Lipase-Functionalized Nanoparticles**: Transport and attach lipase to degrade polyesters like PET in a targeted manner [10, 62].

4.8 Applications in Plastic Remediation

Plastic pollution threatens ecosystems and health worldwide. Traditional methods like landfilling, incineration, and recycling are often ineffective, especially for microplastics. BNMs have emerged as sustainable tools for plastic remediation through adsorption, catalytic degradation, and biological activation. They can be applied in soils, aquatic systems, and industrial wastewater treatment. Studies confirm their efficacy in enzyme-mediated degradation, microbial colonization, and microplastic trapping [63].

4.8.1 Soil Remediation

Plastic pollution of soils is becoming an issue of great concern, especially in agricultural and industrial fields. During the years, plastics, like mulch films, packaging, and industrial-discharge wastes, lie in accumulation in the soil. These plastics are not biodegradable and may last decades altering soil structure, changing microbial communities and lowering soil fertility.

4.8.1.1 Removal of Plastic Contaminants from Agricultural and Industrial Soils

BNMs provide various routes to plastic contaminant removal in the soil, including adsorption, degradation, and promotion of biodegradation reactions:

- **Bind and Sequester Microplastics:** They are technologically modified nanomaterials with a large surface area and functional groups that can help capture microplastics in the soils. The BNM surface chemistry such as hydrophobicity or electrostatic charge allows the adhesion to specific plastics, selectively binding to them, therefore aiding in removing them to the environment [64].
- **Promote Degradation:** Enzyme-loaded cellulose nanofibers can be used to degrade plastics into environmentally non-hazardous products. Such nanomaterials can then

speed up the breakdown of complex polymer structures, so that they can be broken down into smaller, more easily manageable fragments, thus easier to break down by the microbes [65].

- **Support Biodegradation:** They may also induce the activity of microorganisms in the soil, which naturally break down the polymers of plastic. These materials also enable the formation of microhabitats, fostering the colonization of microbes and plastic degradation by providing scaffolds or substrates of growing microbes, thus speeding up the process of biodegradation [66].

4.8.1.2 Enhancements in Soil Quality and Fertility through Remediation

BNMs provide various routes to plastic contaminant removal in the soil, including adsorption, degradation, and promotion of biodegradation reactions:

- **Reduction of Plastic Residues**: As plastics are broken down, the soil's porosity and water retention properties are improved. The fragmentation of plastics also reduces their ability to form harmful physical barriers in the soil, which can impede plant root growth.
- **Nutrient Enrichment**: The resulting degradation products of the plastic, which include oligomers and monomers, may serve as a carbon source of microorganisms in the soil. These byproducts encourage microbe growth and rendering the soil nutritious and fertile by increasing its supply of nutrients. The degradation of plastics by microorganisms makes important nutrients available in the soil, including nitrogen and phosphorus [67].
- **Rejuvenation of Microbial Communities**: The wastes made of plastics usually cause toxicity to the microorganism in the soil. Degrading plastics or eliminating them helps find the ecological balance between beneficial microbes that are needed to keep the health and growth of plants and the functionality of soil [67].

4.8.2 Marine Plastic Cleanup

Marine environments are among the most adversely affected areas by the plastic pollution. Millions of tons of plastic pollution are entering the oceans annually resulting in a significant burden on ocean life, biodiversity, and the marine ecosystem. Microplastics, usually smaller than 5 mm in diameter, present a particular challenge to decontaminate because their size and distribution in large areas makes it difficult to decontaminate them in combination with the vast magnitude of release. Potential solutions to overcome these challenges would be to use BNMs to enhance the removal and degradation of plastics in marine environments [68].

4.8.2.1 Strategies for Addressing Marine Microplastic Contamination

BNMs can be utilized to remove marine plastic pollution in many ways:

- **Adsorption Mechanisms**: BNMs, which are functionalized nanoparticles, are characterized by large surface area and have the ability to specifically adsorb microplastics to water. These surfaces could be developed with the desired surface properties (e.g., hydrophobic or electrostatic) so that they will efficiently capture the plastic particles in the marine.
- **Photocatalytic Degradation**: Hybrid BNMs, such as the integration of titanium dioxide (TiO_2) or zinc oxide (ZnO) nanoparticles, have been shown to disintegrate plastics under the scrutiny of sunlight. The photocatalytic materials transform any plastic polymer into smaller, less harmful products which are easily degraded biologically.
- **Biodegradation Enhancement**: BNMs would also be used to develop favorable microhabitats to encourage indigenous marine microorganisms to naturally degrade plastics. The BNMs will help contribute to the biological degradation of the plastics in the ocean through helping to encourage the growth and development of these microbes [68].

4.8.2.2 Deployment Strategies for BNMs in Aquatic Systems

The practical deployment of BNMs in marine environments requires effective strategies to target areas of high plastic contamination:

- **Nanoparticle Dispersal Systems**: One strategy will be to release BNMs in regions where there is a high concentration of plastic waste under controlled conditions. They are able to disperse the materials in a controlled fashion allowing them to target and extract the plastic in the water more effectively [69].
- **BNM Coatings on Filtration Devices**: The advanced filtration systems may be upgraded by coating the filtration membranes with BNMs. Such biofunctionalized filters are capable of capturing microplastics and other pollutants within seawater with high throughput and efficiency of the filtration system.
- **Integration with Booms and Barriers**: BNMs are also deployable to floating structures like booms and barriers to concentrate and break down plastics in high-density marine areas. These systems have the ability to be deployed to contain and clean up floating debris with a more focused remediation of marine plastic waste [63].

4.9 Industrial Applications

Plastic contamination is a major problem in industrial settings, especially in wastewater and recycling. Traditional methods for treating industrial effluents are often ineffective against microplastics and nanoplastics that escape conventional filtration. To a significant extent, BNMs have demonstrated potential in enhancing wastewater treatment, recycling, and recovery processes in industrial applications.

4.9.1 Incorporation of BNMs in Wastewater Treatment

Plastic contamination is a major problem in industrial settings, especially in wastewater and recycling. Traditional methods for treating industrial effluents are often ineffective against microplastics and nanoplastics that escape conventional filtration.

- **Microplastic Removal**: Microplastics can be captured using magnetically functionalized chitosan. These materials contain magnetic nanoparticles, allowing efficient extraction of microplastics from wastewater.

- **Degradation of Plastic Particles**: Enzyme-loaded BNMs can degrade plastic particles in industrial wastewater, ensuring effluents are environmentally safe before discharge [70].
- **Improved Filtration Systems**: BNMs can be incorporated into membrane filters to capture smaller plastic particles. Such advanced systems can effectively remove micro- and nanoplastics from sewage [71].

4.9.2 Role in Industrial Plastic Recycling and Recovery Processes

BNMs also serve to efficiently recycle and recover industrial plastics to make the circular economy more sustainable:

- **Depolymerization Catalysts**: BNMs possessing the capability of acting as catalysts, like enzyme-functionalized nanoparticles, would allow us to catalyze the process of depolymerizing plastic polymers into reusable monomers. This process simplifies the recycling of plastics and using them as raw materials to produce new products [72].
- **Hybrid Recycling Systems**: When BNMs are combined with conventional mechanical recycling routes, then more useful parts can be extracted through their recyclable process. Such mixed systems enhance the purity and quality of recyclable plastics to be reused in the production of new products [73].
- **Integration with Circular Economy Models**: BNMs can facilitate the circular economy by allowing high-capacity plastic recycling and recovery. By increasing recycling and limiting waste of plastic, such material contributes to more environmentally friendly industrial activities and helps to decrease the need to use non-recycled plastics [74].

BNMs provide greener alternatives for plastic remediation across soil, marine, and industrial systems by degrading, sequestering, and promoting natural biodegradation, supporting ecosystem health and sustainability [10, 75].

4.10 Case Studies

4.10.1 Chitosan-Based BNMs

4.10.1.1 Applications in Marine and Soil Plastic Degradation

Chitosan, derived from chitin in crustacean shells, is a promising BNM for addressing plastic pollution in marine and soil ecosystems.

- **Marine Applications**: Chitosan nanoparticles adsorb and degrade microplastics effectively. They are used in hybrid systems with photocatalysts such as titanium dioxide or zinc oxide to enhance degradation under light and promote microbial biodegradation [76].
- **Soil Applications**: The BNMs of chitosan can be applied to the soil to help in degrading plastic wastes, including the waste products of mulch film. They assist in the degradation *via* enzymes such as laccase and PETase and help to make soils healthy by providing a source of carbon to microbes [77].

4.10.1.2 Efficiency Metrics and Field Study Results

- **Laboratory Studies**: Chitosan nanoparticles achieved about 70% reduction in microplastic size within 60 days and enhanced enzymatic activity [78].
- **Field Studies**: In coastal regions, they reduced microplastic concentrations by 45% in 3 months; in agricultural soils, compost integration improved degradation by 30% [79].

4.10.2 Metal Nanoparticles for Plastic Degradation

Metal nanoparticles such as silver, gold, zinc oxide, and iron oxide are effective in plastic degradation due to their catalytic and photoreactive nature.

- **Silver Nanoparticles (AgNPs)**: Influence the degradation of polyethylene and polystyrene by producing reactive oxygen species (ROS) in the presence of light [80].
- **Gold Nanoparticles (AuNPs)**: Assist in the hydrolysis of polymers such as PET in the presence of enzymes to enhance their enzymatic degradation [81].

- **Zinc Oxide (ZnO) and Iron Oxide (Fe_3O_4)**: ZnO degrades plastics under UV light, while Fe_3O_4 facilitates the magnetic recovery of degraded plastic fragments [82].

4.10.3 Hybrid Systems

4.10.3.1 Combining BNMs with Catalytic/Biological Agents

Hybrid systems of BNMs with catalysts or microbes accelerate plastic degradation through synergistic effects:

- **Enzyme-Nanoparticle Hybrids**: Hybrid chitosan nanoparticles modified with such enzymes as PETase and laccase increase enzymatic degradation [83].
- **Photocatalyst-Biofilm Hybrids**: Biofilm structures with embedded photocatalytic nanoparticles catalyze the degradation of plastics in the presence of light [84].
- **Microbial-Nanoparticle Systems**: BNMs facilitate microbial colonization, which improves biodegradation [85].

4.10.3.2 Pilot Project and Successful Implementations

- **Marine Cleanup**: A Mediterranean pilot with integrative BNM-loaded booms decreased the quantity of microplastics by 60 percent in 6 months [86].
- **Soil Remediation**: Hybrid systems in agricultural fields improved plastic degradation by 40% [87].
- **Industrial Wastewater**: Hybrid systems could remove 80% of microplastics in wastewater treatment plants, proving to be scalable and cost-effective [88].

These case studies demonstrate the success of chitosan-based, metal, and hybrid BNMs in reducing plastic pollution across ecosystems. They confirm the scalability, flexibility, and environmental compatibility of bio-nano strategies and emphasize the need for translational research and regulatory support.

4.11 Challenges and Limitations

The conceptualization and use of BNMs in plastic remediation are currently confronted by numerous technical, environmental, economic, regulatory, and ethical objections. These constraints have to be considered in detail to maximize the potential of BNMs in addressing plastic pollution.

4.11.1 Technical Challenges

4.11.1.1 Issues of Scalability and Reproducibility in Synthesis

Among the most serious obstacles to the broad acceptance of BNMs is the difficulty of re-designing synthesis techniques that work well in laboratory environments to an industrial scale. Although laboratory-scale production has demonstrated potential in BNM-based plastic remediation, challenges of variability and inconsistency are important considerations when scaling up methods to industrial production.

- **Batch-to-Batch Variability**: Batch-to-batch variations experienced due to differences in the raw materials used, such as plant extracts or microbial strains, or variations in the synthesis conditions can result in differences in size, shape, and functional property of the BNMs. This discrepancy can be an impediment to their performance and dependability in large application [89].
- **High Energy and Resource Demands**: Although BNMs have the potential to be renamed environmentally friendly materials, large quantities of energy and raw materials are required to scale up their production. Such increased demands may offset some of the environmental advantages that are used by the BNMs, and therefore, their impact towards the environment is a factor of concern.
- **Process Optimization**: Developing standardized, reproducible methods for synthesizing BNMs at scale is a complex process. Ensuring consistent size, shape, and functionalization while maintaining cost-effectiveness remains a challenge for large-scale commercial production [90].

4.11.1.2 Stability of BNMs Under Real-World Conditions

The stability of BNMs in real-world circumstances, especially outdoor settings where the materials would be at risk of wide temperature, change in ultraviolet radiation, and pH changes, is another important issue.

- **Degradation of Functional Properties**: It is also possible that the BNM experiences changes in its functional capabilities after it is subjected to atmospheric conditions like high temperatures or UV radiation over a prolonged period. These may involve degradation of the catalytic performance or the physical structure, which has the effect of rendering the phase ineffective after some time [91].
- **Encapsulation and Delivery**: In order to minimize environmental degradation of BNMs, studies are underway to find suitable methods of encapsulation.

4.11.2 Environmental and Economic Concerns

4.11.2.1 Evaluating Potential Nanoparticle Toxicity

While BNMs are found to be more ecologically and biologically friendly as compared to traditional synthetic materials, there are doubts regarding long-term ecological and biological effects, particularly when the nanomaterials harbor metallic nanoparticles.

- **Bioaccumulation**: When BNMs are not effectively broken down or eradicated by the ecosystems, they accumulate over time and harm both terrestrial and aquatic life forms. Nanoparticles may enter food chains and cause unpredictable long-term effects on biological diversity [92].
- **Cytotoxic Effects**: Some BNMs, especially those that contain nanoparticles of metals like silver or gold, may cause harm to cells or organisms when large concentrations are exposed. This poses concern over the safety of such materials in the environment, especially when they come into contact with non-target species [93].
- **Monitoring and Assessment**: There exists a gap in acquiring standardized and reliable ways of measuring nanoparticle

toxicity and their presence in the environment. It is essential to comprehend the degradation and dispersion of these materials in the ecological interface to manage their environmental implications [81].

4.11.2.2 *Balancing Cost-Efficiency with Environmental Benefits*

The large-scale use of BNMs in plastic remediation has economic obstacles. To make BNMs a viable option on a large scale, it must be proven that they are financially competitive with current remediation practices:

- **Production Costs**: The production of BNMs may be expensive in terms of raw materials, energy, and specialized equipment needed to synthesize the material. Without a reduction in these costs, BNMs can end up being too expensive to apply on a large scale in managing plastic wastes [94].
- **Infrastructure Investment**: The production of BNMs may be expensive in terms of raw materials, energy, and specialized equipment needed to synthesize the material. Without a reduction in these costs, BNMs can end up being too expensive to apply on a large scale in managing plastic wastes [95].
- **Market Competitiveness**: To be globally adopted, BNMs need to possess major economic and environmental advantages over current plastic remediation methods. It might be a challenge to convince industries, governments, and other stakeholders to invest in such technologies [96].

4.11.3 Regulatory and Ethical Issues

4.11.3.1 *The Current Regulatory Landscape for BNM Deployment*

The implementation of BNMs is subjected to a complex regulatory environment. The current policies lack proper establishment, vary region to region, and may become an obstacle to their popularization:

- **Lack of Standardization**: There is no universally acknowledged definition or classification of BNMs, which adds to the inaccessibility of the regulatory environment. There is a need to develop uniformity in testing and approval, leading to the development and entry of BNMs into the world of plastic remediation [97].

- **Safety and Testing Requirements**: The new materials are usually subject to safety tests such as toxicity and environmental assessment by regulatory bodies. This may slow down product development and create higher expenses, especially in the case of BNMs used in environmental applications [98].
- **Cross-Border Regulations**: The new materials are usually subject to safety tests such as toxicity and environmental assessment by regulatory bodies. This may slow down product development and create higher expenses, especially in the case of BNMs used in environmental applications [99].

4.11.3.2 Ethical Concerns in Testing and Large-Scale Applications

The large-scale usage of BNMs has several ethical implications, most of them dependent on potentially adverse effects on vulnerable people and ecosystems:

- **Risk of Unintended Consequences:** BNMs that are not tested widely may harm non-target species or entire ecosystems accidentally. This can cause environmental imbalance and further deterioration instead of providing relief to the environment [100].
- **Transparency and Stakeholder Engagement**: The deployment of BNMs should be discussed with local citizens, governments, and other interested parties to make them participate in the decision-making process. Transparency in risks and benefits will enable stakeholders to understand potentials and work toward ethical use and improvement of such materials [101].
- **Equity and Accessibility**: The advantages of BNMs should reach all areas affected by plastic waste. It is also important to ensure that these technologies are available in less-developed or low-income regions for a balanced approach in environmental rehabilitation.

To harness the full potential of bio-nanomaterials in plastic remediation industry, several issues must be resolved, such as technical, environmental, economic, regulatory, and ethical issues of application. The key directions include improving scalability of synthesis procedures, establishing standardized testing and deployment protocols, and ensuring safety and environmental friendliness. Researchers, policymakers, industries, and local communities must collaborate to make BNMs an integral part of global plastic waste management for a greener future.

4.12 Future Perspectives

The potential of BNMs in plastic waste remediation is a promising avenue for future scientific, economic, and policy development. As we continue to seek sustainable solutions to the growing plastic pollution problem, enhancement of BNM research, integration into circular economy strategies, and effective policy interaction are important areas of progress (Table 4.1).

Table 4.1 Summary of emerging strategies and benefits associated with the future use of BNMs in sustainable plastic waste management.

Focus area	Key future direction	Expected impacts
Material design	Development of smarter, highly functional BNMs	Enhanced plastic degradation efficiency and broader environmental applicability
Hybrid system innovation	Integration with enzymes, photocatalysts, or responsive components	Synergistic degradation under varied environmental conditions
Recycling and circularity	Use in sorting technologies, biodegradable plastic production, and depolymerization	Reduction in non-recycled plastic demand and promotion of closed-loop recycling
Policy and regulation	Establishment of safety standards, incentives, and R&D funding	Safe deployment, increased industrial and societal adoption
Community engagement	Awareness campaigns, local remediation initiatives, stakeholder involvement	Grassroots participation and socially inclusive plastic waste management

4.12.1 Advancements in BNM Design

BNMs with the potential to degrade, recycle, and repurpose plastic waste have proven to be powerful solutions developed to be biologically based and mimic natural processes. The future of BNMs lies in enhancing their functionality and applicability in the context of plastic pollution (Figure 4.4).

4.12.1.1 Innovations in Functionalization and Hybridization Techniques

Functionalization involves altering the surface of BNMs to increase their capabilities for specific applications. Future advancements will focus on developing methods that allow better interaction with a wider range of plastics.

- **Enhanced Interaction with Plastics:** Researchers are studying how BNMs can be modified with chemical functional groups to bond with various plastic polymers.
- **Increased Efficiency in Degradation:** Developing specific enzymes or catalysts embedded within BNMs can enable them to degrade targeted plastics more effectively.

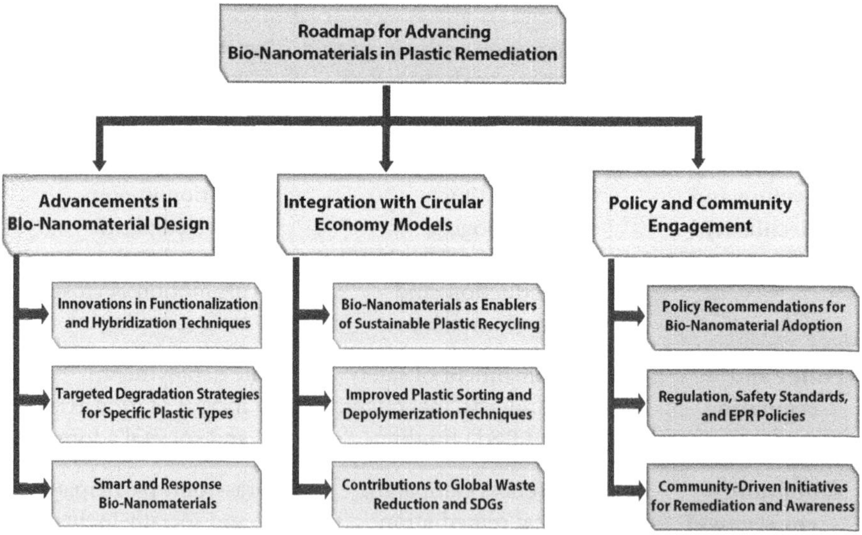

Figure 4.4 Roadmap outlining future perspectives for the development, deployment, and societal integration of BNMs in sustainable plastic remediation.

- **Nanocomposites and Hybrid Materials:** Combining the biodegradability of biological materials with the strength of synthetic polymers could lead to new applications in eco-friendly packaging.

4.12.1.2 *Targeted Degradation Strategies for Specific Plastic Types*

Different kinds of plastics (e.g., polyethylene, polypropylene, and polystyrene) have distinct structures that resist natural decomposition. Future research can focus on designing BNMs for specific degradation strategies.

- **Enzyme-Based Degradation:** Bio-nanocatalysts containing enzymes capable of breaking down plastic polymers.
- **Biodegradable Microbes:** Genetically engineered microbes or fungi can be integrated with BNM structures.
- **Smart Degradation:** BNMs with built-in sensors that identify the type of plastic and trigger degradation only when that plastic is present.

4.12.2 Integration with Circular Economy Models

The principles of the circular economy emphasize reuse, recycling, and re-creation of materials to reduce waste. BNMs can contribute by promoting efficient plastic decomposition and recycling in green economy models.

4.12.2.1 *Role in Sustainable Recycling Processes*

The adaptation of BNMs to recycling processes will be a central development towards limiting plastic waste and enhancing the reuse of valuable material. It is possible to use BNMs in the following ways:

- **Biobased Plastics for Recycling:** BNMs can serve as feedstocks for recyclable and biodegradable biobased plastics, minimizing fossil fuel-based usage.
- **Improved Plastic Sorting and Processing:** BNMs can aid in sorting plastics, improving separation and recycling efficiency.
- **Biodegradable Additives:** Incorporating BNMs into plastics can promote biodegradation after use, making single-use plastics more sustainable.

4.12.2.2 Contributions to Global Waste Reduction Goals

BNMs can support global waste reduction targets, including those under the United Nations Sustainable Development Goals (SDGs). The most significant contributions can be as follows:

- **Reduction of Landfill Waste:** BNMs have the potential to ease the recycling and biodegradation process, resulting in less plastic waste cluttering the landfill site.
- **Marine and Ocean Pollution Mitigation:** The development of BNMs can chemically degrade within marine conditions, preventing plastic pollution of the ocean.
- **Achieving Carbon Neutrality:** BNMs will assist in reducing reliance on fossil fuel and plastic refuse.

4.12.3 Policy and Community Engagement

BNMs that can remediate plastic waste will have to be developed, which will necessitate the use of policies, regulations, and community engagement strategies. The government and the industries along with local communities should collaborate and initiate the large-scale adoption of these technologies and strive to make their venture effective and reasonable.

4.12.3.1 Policy Recommendations for BNM Adoption

The initiatives of integrating BNM into the system of plastic waste management can be realized with the assistance of the policy and regulatory environment. The key policy recommendations are as follows:

- **Incentivizing Research and Development:** Governments are expected to allocate funding and incentives to research on BNMs.
- **Regulation and Safety Standards:** Policies should be issued to guide the security of the BNMs and their environmental impact.
- **Extended Producer Responsibility (EPR):** Governments may implement the EPR policies to encourage manufacturers to recycle or responsibly dispose of products.

4.12.3.2 Community-Driven Initiatives for Remediation and Awareness

The community involvements will be important in the attempts to realize the successful adoption and implementation of the BNM emerging technologies. The area of great change on the grassroots level can be attributed to the local initiatives and public awareness:

- **Public Education Campaigns:** Awareness campaigns could inform the people of the significance of reducing waste in plastics and how BNMs can help rectify the situation.
- **Local Recycling Programs:** Communities could initiate programs targeting plastics that are efficiently degraded by BNMs.
- **Collaborations with Local Governments and NGOs:** NGOs in collaboration with local governments can implement BNM-based technologies in waste management.

The plastic waste management with BNMs has massive potential in the future. BNMs can be developed as the building blocks of a sustainable future by targeting innovation in the properties of these materials, embedding them in a circular economy structure, and actively shaping policy and community engagement.

4.13 Conclusion

Plastics are one of the most prevalent and complex environmental challenges of the contemporary world and have depressing impacts on the environment, biodiversity, and human health. The vast use of plastics and the ways in which they ensure durability and resistance to natural decomposition have led to the accumulation of plastic waste in soils, water, and oceans. Microplastics are a relatively new threat, having found their way into food chains and ecosystems, carrying toxic pollutants, which further increases the issue. Traditional remediation techniques, like landfilling, incineration, and recycling, are found to be inadequate, as they have limited capabilities in accomplishing remediation.

BNMs present a revolutionary solution to plastic pollution. Due to the benefits of nanotechnology, such as the high reactivity of nanoparticles and high surface-to-volume ratio, as well as the sustainability and biocompatibility of the biological materials, BNMs can be regarded as a paradigm shift towards managing plastic wastes. BNMs, in contrast to traditional nanomaterials, are environmentally safe and biodegradable, contributing to sustainability and the circular economy.

In this chapter, the applicability of BNMs to the remediation of plastics has been discussed, including their synthesis through green methodologies and mechanisms of action. BNMs can react with plastics through adsorption to the surface or enzymatic degradation, breaking polymers into harmless byproducts. The synergy with enzymes and microbial communities also enhances the biodegradation of plastic pollutants in various ecosystems.

On a real-world level, the potential application of BNMs has been deliberated in soil treatment, marine plastic cleanup, and industrial wastewater treatment. Case studies indicate the feasibility of BNM-based systems to trap microplastics, depolymerize polymer chains, and enhance the performance of traditional recycling systems. The examples show the versatility of BNMs in managing plastic pollution in coastal land, agricultural soils, and cities. Despite the immense potential of BNMs, certain concerns need to be addressed to enable larger-scale application. Scalability, reproducibility, cost-effectiveness, and potential environmental hazards of nanoparticle release require further studies. The toxicity, environmental longevity, and ecological accumulation of nanoparticles should be well studied to avoid adverse effects. Moreover, regulatory frameworks guiding BNM deployment need to be clarified for safe application in plastic waste management.

BNMs can be seen as a revolutionary tool in the war against plastic pollution, promising scalability, sustainability, and efficacy. Through research, enhanced synthesis, and collaboration with researchers, industries, and policymakers, new solutions can reduce plastic waste and support environmental sustainability. BNMs have the potential to transform plastic pollution management and could become a key component in addressing one of the most pressing environmental challenges today.

AI disclosure statement: The author confirms that no artificial intelligence (AI)-generated content or AI-assisted writing tools were used in the preparation of this chapter (including summaries and the abstract). All text, analysis, and conclusions are the result of the author's original work.

References

1. Thompson, R.C., Moore, C.J., Vom Saal, F.S., Swan, S.H., Plastics, the environment and human health: current consensus and future trends. *Philos. Trans. R. Soc. Lond. B Biol. Sci.*, *364*, 1526, 2153–2166, 2009.
2. AlMaadeed, M.A.A., Ponnamma, D., El-Samak, A.A., Polymers to improve the world and lifestyle: physical, mechanical, and chemical needs, i*n: Polymer Science and Innovative Applications*, pp. 1–19, Elsevier, Amsterdam, 2020.
3. Gewert, B., Plassmann, M.M., MacLeod, M., Pathways for degradation of plastic polymers floating in the marine environment. *Environ. Sci. Process Impacts*, *17*, 9, 1513–1521, 2015.
4. Oberoi, G. and Garg, A., Single-use plastics: A roadmap for sustainability? *Supremo Amicus*, *24*, 585, 2021.
5. Kibria, M.G., Masuk, N.I., Safayet, R., Nguyen, H.Q., Mourshed, M., Plastic waste: challenges and opportunities to mitigate pollution and effective management. *Int. J. Environ. Res.*, *17*, 1, 20, 2023.
6. Lebreton, L. and Andrady, A., Future scenarios of global plastic waste generation and disposal. *Palgrave Commun.*, *5*, 1–11, 2019.
7. Zhang, K., Hamidian, A.H., Tubić, A., Zhang, Y., Fang, J.K.H., Wu, C., Lam, P.K.S., Understanding plastic degradation and microplastic formation in the environment: A review. *Environ. Pollut.*, *274*, 116554, 2021.
8. Kassab, A., Al Nabhani, D., Mohanty, P., Pannier, C., Ayoub, G.Y., Advancing plastic recycling: Challenges and opportunities in the integration of 3D printing and distributed recycling for a circular economy. *Polymers*, *15*, 19, 3881, 2023.
9. Geyer, R., Jambeck, J.R., Law, K.L., Production, use, and fate of all plastics ever made. *Sci. Adv.*, *3*, 7, e1700782, 2017.
10. Bhattacharjee, J. and Roy, S., Green remediation of microplastics using bionanomaterials, in: *Remediation of Plastic and Microplastic Waste*, pp. 240–260, CRC Press, Boca Raton, FL, 2024.
11. Srinivas, A., Chandran, K.R., Nehra, S., Kumar, D., Future Aspects in the Field of Bio-Nanomaterials Toward Environmental Assessment, in: *Bio-Nanomaterials in Environmental Remediation: Industrial Applications*, pp. 295–317, 2025, https://doi.org/10.1002/9783527848546.ch12.
12. Gautam, B.P.S., Qureshi, A., Gwasikoti, A., Kumar, V., Gondwal, M., Global scenario of plastic production, consumption, and waste generation and their impacts on environment and human health, in: *Advanced Strategies for Biodegradation of Plastic Polymers*, pp. 1–34, Springer, Cham, 2024.
13. Organisation for Economic Co-operation and Development (OECD), *Global Plastics Outlook: Economic Drivers, Environmental Impacts and Policy Options*, OECD Publishing, Paris, 2022, https://doi.org/10.1787/de747aef-en.
14. Williams, A.T. and Rangel-Buitrago, N., The past, present, and future of plastic pollution. *Mar. Pollut. Bull.*, *176*, 113429, 2022.

15. Chatterjee, S. and Sharma, S., Microplastics in our oceans and marine health. *Field Actions Sci. Rep.*, *19*, 54–61, 2019.
16. Wang, F., Wang, Q., Adams, C.A., Sun, Y., Zhang, S., Effects of microplastics on soil properties: current knowledge and future perspectives. *J. Hazard. Mater.*, *424*, 127531, 2022.
17. Anastas, P.T. and Warner, J.C., *Green Chemistry: Theory and Practice*, *135*, 1–135, Oxford Universty Press, 2000, https://doi.org/10.1093/oso/9780198506980.001.0001.
18. Shah, A.A., Hasan, F., Hameed, A., Ahmed, S., Biological degradation of plastics: a comprehensive review. *Biotechnol. Adv.*, *26*, 3, 246–265, 2008.
19. Sun, S., Enzyme-embedded biodegradable plastic for sustainable applications: advances, challenges, and perspectives, *ACS Appl. Bio Mater.*, *8*, 3, 1785–1796, 2025.
20. Sharma, N.K., Sharma, R., Dakal, T.C., *Bio-Nanomaterials in Environmental Remediationin: Industrial Applications*, 1–46, John Wiley & Sons, 2025, 10.1002/9783527848546.
21. Omoregie, I.P., Osagie, A.D., Oritseweyinmi, O.O., Crustacean nanochitosan-based bioremediation of nanoplastic-polluted aquatic habitat: A review pursuant to SDG 6. *Sci. Afr.*, *21*, e01881, 2023.
22. Miditana, S.R., Tirukkovalluri, S.R., Raju, I.M., Synthesis and antibacterial activity of transition metal (Ni/Mn) co-doped TiO_2 nanophotocatalyst on different pathogens under visible light irradiation. *Nanosyst.: Phys. Chem. Math.*, *13*, 1, 104–114, 2022, https://doi.org/10.17586/2220-8054-2022-13-1-104-114.
23. Ramakrishnan, N., Sharma, S., Gupta, A., Alashwal, B.Y., Keratin-based bioplastic film from chicken feathers and its characterization. *Int. J. Biol. Macromol.*, *111*, 352–358, 2018.
24. Kamalian, P., Khorasani, S.N., Abdolmaleki, A., Neisiany, R.E., Grafted ZnO nanoparticles used for development in photocatalytic degradation performance of polyethylene. *Polym. Bull.*, *76*, 3593–3606, 2019.
25. Ediyilyam, S., George, B., Shankar, S.S., Dennis, T.T., Wacławek, S., Černík, M., Padil, V.V., Chitosan/gelatin/silver nanoparticles composites films for biodegradable food packaging applications. *Polymers*, *13*, 11, 1680, 2021.
26. Rouf, T.B. and Kokini, J.L., Biodegradable biopolymer–graphene nanocomposites. *J. Mater. Sci.*, *51*, 9915–9945, 2016.
27. Ullah, A., Lee, G.J., Kim, H., Kwon, H.T., Lim, S.I., Development and evaluation of bioinspired pH-responsive sericin–chitosan-based hydrogel for controlled colonic delivery of PETase: Harnessing PETase-triggered degradation of microplastics. *Int. J. Biol. Macromol.*, *242*, 124698, 2023.
28. Hamid, I.D.H.A., Soni, R., Hsu, Y.I., Uyama, H., TEMPO-oxidized cellulose nanofiber reinforced starch film with pH-responsive weakening and rapid marine-degradability. *Polym. Degrad. Stab.*, *219*, 110618, 2024.
29. Lou, Y., Xi, J., Meng, L., Yan, Z., Deng, W., Bian, H., Wu, W., High-permeance nanocellulose/ZnO hybrid membranes with photo-induced anti-fouling

performance for wastewater purification. *Carbohydr. Polym.*, *348*, 122807, 2025.

30. Khedulkar, A.P., Bobade, R.G., Doong, R., Pandit, B., Ky, N.M., Ambare, R.C., Hoang, T.D., Kumar, K.J., Bio-based nanomaterials as effective, friendly solutions and their applications for protecting water, soil, and air. *Mater. Today Chem.*, *46*, 102688, 2025.
31. Abady, M.M., Mohammed, D.M., Soliman, T.N., Shalaby, R.A., Sakr, F.A., Sustainable synthesis of nanomaterials using different renewable sources. *Bull. Natl. Res. Cent.*, *49*, 1, 24, 2025.
32. Kumar, D., Salvi, A., Thakur, V., Kharbanda, S., Sangwan, S., Thakur, P., Thakur, A., Aloe vera assisted green synthesis of silver nanoparticles: structural characterization, electrochemical behaviour, and antimicrobial efficiency. *Discov. Appl. Sci.*, *7*, 6, 1–11, 2025.
33. Yadeta Gemachu, L. and Lealem Birhanu, A., Green synthesis of ZnO, CuO and NiO nanoparticles using Neem leaf extract and comparing their photocatalytic activity under solar irradiation. *Green Chem. Lett. Rev.*, *17*, 1, 2293841, 2024.
34. Abbasi, I.A., Aurangzeb, J., Du Ri, P., Soomro, A.F., Tae, Y.I., Green Synthesis of Magnetic Nanoparticles Using Black Tea Extract (BTE) to Remove Polyethylene (PE) Microplastics from Water. *Pol. J. Environ. Stud.*, 2025.
35. Sial, A., Zafar, I., Shafiq, S., Sustainable Microbial Approaches for Plastic Biodegradation: Impact on Human Health, Research Advancements, and Challenges. *Front. Health Inform.*, 14, 1, 2025.
36. El-Kurdi, N., ElBaghdady, K., El-Shatoury, S., Hammad, S., Ghazy, M., Sustainable Removal of Nanoplastics: Exploiting the Lipolytic Activity of Pseudomonas aeruginosa O6 Isolated from Mariout Wetland, Egypt. *Egypt. J. Aquat. Biol. Fish.*, 28, 2, 2024.
37. Ullah, H., Hayat, M., Raza, B., Alam, K., Ziad, M., Afridi, J., Khan, F., Biodegradation of Microplastic: The Role of Aspergillus Species in Sustainable Plastic Waste Management. *Res. Med. Sci. Rev.*, *3*, 1, 314–329, 2025.
38. Shalaby, S.M., Madkour, F.F., El-Kassas, H.Y., Mohamed, A.A., Elgarahy, A.M., Green synthesis of recyclable iron oxide nanoparticles using Spirulina platensis microalgae for adsorptive removal of cationic and anionic dyes. *Environ. Sci. Pollut. Res.*, *28*, 65549–65572, 2021.
39. Ramamurthy, K., Thomas, N.P., Gopi, S., Sudhakaran, G., Haridevamuthu, B., Namasivayam, K.R., Arockiaraj, J., Is Laccase derived from Pleurotus ostreatus effective in microplastic degradation? A critical review of current progress, challenges, and future prospects. *Int. J. Biol. Macromol.*, 133971, 2024.
40. Urbanek, A.K., Mirończuk, A.M., García-Martín, A., Saborido, A., de la Mata, I., Arroyo, M., Biochemical properties and biotechnological applications of microbial enzymes involved in the degradation of polyester-type plastics. *Biochim. Biophys. Acta Proteins Proteom.*, *1868*, 2, 140315, 2020.

41. Rovera, C., Fiori, F., Trabattoni, S., Romano, D., Farris, S., Enzymatic hydrolysis of bacterial cellulose for the production of nanocrystals for the food packaging industry. *Nanomaterials, 10*, 4, 735, 2020.
42. Wrońska, N., Katir, N., Nowak-Lange, M., El Kadib, A., Lisowska, K., Biodegradable chitosan-based films as an alternative to plastic packaging. *Foods, 12*, 18, 3519, 2023.
43. Raghavendra, V.B., Shankar, S., Govindappa, M., Pugazhendhi, A., Sharma, M., Nayaka, S.C., Green synthesis of zinc oxide nanoparticles (ZnO NPs) for effective degradation of dye, polyethylene and antibacterial performance in waste water treatment. *J. Inorg. Organomet. Polym. Mater.*, 1–17, 2022.
44. Kamal, T., Ahmad, I., Khan, S.B., Ul-Islam, M., Asiri, A.M., Microwave assisted synthesis and carboxymethyl cellulose stabilized copper nanoparticles on bacterial cellulose nanofibers support for pollutants degradation. *J. Polym. Environ.*, *27*, 2867–2877, 2019.
45. Zara, Z., Mishra, D., Pandey, S.K., Csefalvay, E., Fadaei, F., Minofar, B., Řeha, D., Surface interaction of ionic liquids: stabilization of polyethylene terephthalate-degrading enzymes in solution. *Mol.*, *27*, 1, 119, 2021.
46. Manik, U.P., Nande, A., Raut, S., Dhoble, S.J., Green synthesis of silver nanoparticles using plant leaf extraction of Artocarpus heterophyllus and Azadirachta indica. *Results Mater.*, *6*, 100086, 2020.
47. Faizan, M., Fatima, M., Javed, M., Iqbal, S., Fiaz, M., Shoaib, M., Bahadur, A., Enhancing environmental remediation: sustainable photocatalytic degradation of water pollutants with biodegradable chitosan–ZnO nanocomposites. *Res. Chem. Intermed.*, 1–21, 2025.
48. Thompson, L., Azadmanjiri, J., Nikzad, M., Sbarski, I., Wang, J., Yu, A., Cellulose nanocrystals: production, functionalization and advanced applications. *Rev. Adv. Mater. Sci.*, *58*, 1, 1–16, 2019.
49. Lin, W., Hu, X., You, X., Sun, Y., Wen, Y., Yang, W., Zhang, X., Li, Y., Chen, H., Hydrophobic modification of nanocellulose *via* a two-step silanation method. *Polymers, 10*, 9, 1035, 2018.
50. Shen, X., Zhang, Y., He, H., Yi, C., Dong, W., Ye, S., Nie, Z., Electrostatic adsorption behaviors of charged polymer-tethered nanoparticles on oppositely charged surfaces. *Macromol. Rapid Commun.*, *43*, 14, 2200171, 2022.
51. Oh, S. and Stache, E.E., Recent advances in oxidative degradation of plastics. *Chem. Soc. Rev.*, *53*, 7309–7327, 2024.
52. He, Y., Rehman, A.U., Xu, M., Not, C.A., Ng, A.M., Djurišić, A.B., Photocatalytic degradation of different types of microplastics by TiOx/ZnO tetrapod photocatalysts. *Heliyon*, *9*, 11, 22562, 2023.
53. Yu, D., Basumatary, I.B., Liu, Y., Zhang, X., Kumar, S., Ye, F., Dutta, J., Chitosan-photocatalyst nanocomposite on polyethylene films as antimicrobial coating for food packaging. *Prog. Org. Coat.*, *186*, 108069, 2024.
54. Thomsen, T.B., Radmer, T.S., Meyer, A.S., Enzymatic degradation of poly(ethylene terephthalate) (PET): Identifying the cause of the hypersensitive

enzyme kinetic response to increased PET crystallinity. *Enzyme Microb. Technol.*, *173*, 110353, 2024.

55. Si, Y., Lin, Q., Zhou, F., Qing, J., Luo, H., Zhang, C., Cha, R., The interaction between nanocellulose and microorganisms for new degradable packaging: A review. *Carbohydr. Polym.*, *295*, 119899, 2022.
56. Naidu, G., Nagar, N., Poluri, K.M., Mechanistic insights into cellular and molecular basis of protein–nanoplastic interactions. *Small*, *20*, 5, 2305094, 2024.
57. Kumar, N. and Upadhyay, L.S.B., Enzyme immobilization over polystyrene surface using cysteine functionalized copper nanoparticle as a linker molecule. *Appl. Biochem. Biotechnol.*, *191*, 1247–1257, 2020.
58. Bora, L.V., Bhatt, M., Patel, A., Bora, N.V., Plastic degradation by photocatalysis: Basic concepts and general mechanisms, *ACS Symposium Series*, 1, 1–22, 2024, https://doi.org/10.1021/BK-2024-1489.CH001.
59. Zandieh, M. and Liu, J., Removal and degradation of microplastics using the magnetic and nanozyme activities of bare iron oxide nanoaggregates. *Angew. Chem.*, *134*, 47, e202212013, 2022.
60. El-Khawaga, A.M., Farrag, A.A., Elsayed, M.A., El-Sayyad, G.S., El-Batal, A.I., Antimicrobial and photocatalytic degradation activities of chitosan-coated magnetite nanocomposite. *J. Clust. Sci.*, *32*, 1107–1119, 2021.
61. Milovanović, J., Banjanac, K., Nikolić, J., Nikodinović-Runić, J., Prlainović, N.Ž., The organic-functionalized silica nanoparticles as lipase carriers for biocatalytic application: Future perspective in biodegradation. *Catalysts*, *15*, 1, 54, 2025.
62. Rizvi, N.B., Sarwar, A., Waheed, S., Iqbal, Z.F., Imran, M., Javaid, A., Kim, T.H., Khan, M.S., Nano-based remediation strategies for micro and nanoplastic pollution: A review. *J. Contam. Hydrol.*, 104380, 2024.
63. Miditana, S., Korlam, S., Babu, A., Agarwal, N., Solanki, V., Nanotechnology for Sustainable Environment, 2025, 10.1201/9781003484189-7.
64. Bhandari, G., Gangola, S., Dhasmana, A., Rajput, V., Gupta, S., Malik, S., Slama, P., Nano-biochar:Recent progress, challenges, and opportunities for sustainable environmental remediation. *Front. Microbiol.*, *14*, 1214870, 2023.
65. Rillig, M.C., Ziersch, L., Hempel, S., Microplastic transport in soil by earthworms. *Sci. Rep.*, *7*, 1, 1362, 2017.
66. Aralappanavar, V.K., Mukhopadhyay, R., Yu, Y., Liu, J., Bhatnagar, A., Praveena, S.M., Li, Y., Effects of microplastics on soil microorganisms and microbial functions in nutrients and carbon cycling–A review. *Sci. Total Environ.*, 171435, 2024.
67. Mitra, A.K., Dey, S., Solanki, V.S., Agarwal, N., Nanosensors to Monitor Wastewater Pollutants, in: *Nanomaterials in Wastewater Research*, Advances in Wastewater Research. N. Agarwal, M.P. Shah, V.S. Solanki, N. Singh (Eds.), Springer, Singapore, 2025, https://doi.org/10.1007/978-981-96-3908-3_3.

68. Goh, P.S., Kang, H.S., Ismail, A.F., Khor, W.H., Quen, L.K., Higgins, D., Nanomaterials for microplastic remediation from aquatic environment: Why nano matters? *Chemosphere, 299,* 134418, 2022.
69. Anjana, K., Hinduja, M., Sujitha, K., Dharani, G., Review on plastic wastes in marine environment – Biodegradation and biotechnological solutions. *Mar. Pollut. Bull., 150,* 110733, 2020.
70. Schwaminger, S.P., Fehn, S., Steegmüller, T., Rauwolf, S., Löwe, H., Pflüger-Grau, K., Berensmeier, S., Immobilization of PETase enzymes on magnetic iron oxide nanoparticles for the decomposition of microplastic PET. *Nanoscale Adv., 3,* 15, 4395–4399, 2021.
71. de Rosset, A., Torres-Mendieta, R., Pasternak, G., Yalcinkaya, F., Synergistic effects of natural biosurfactant and metal oxides modification on PVDF nanofiber filters for efficient microplastic and oil removal. *Process Saf. Environ. Prot., 194,* 997–1009, 2025.
72. Kotnis, S., Gulati, S., Sun, Q., High-Efficiency PET Degradation With a Duo-Enzyme System Immobilized on Magnetic Nanoparticles. *Biotechnol. Bioeng., 122,* 6, 1397–1401, 2025.
73. Nikolaivits, E., Pantelic, B., Azeem, M., Taxeidis, G., Babu, R., Topakas, E., Brennan Fournet, M., Nikodinovic-Runic, J., Progressing plastics circularity: A review of mechano-biocatalytic approaches for waste plastic (re)valorization. *Front. Bioeng. Biotechnol., 9,* 696040, 2021.
74. Akram, M.A., Savitha, R., Kinsella, G.K., Nolan, K., Ryan, B.J., Henehan, G.T., Microbial and Enzymatic Biodegradation of Plastic Waste for a Circular Economy. *Appl. Sci., 14,* 24, 11942, 2024.
75. Zimmermann, W., Biocatalytic recycling of plastics: facts and fiction. *Chem. Sci., 16,* 16, 6573–6582, 2025.
76. Kumar, S., Ye, F., Mazinani, B., Dobretsov, S., Dutta, J., Chitosan nanocomposite coatings containing chemically resistant ZnO–SnOx core–shell nanoparticles for photocatalytic antifouling. *Int. J. Mol. Sci., 22,* 9, 4513, 2021.
77. Koyani, R.D. and Vazquez-Duhalt, R., Laccase encapsulation in chitosan nanoparticles enhances the protein stability against microbial degradation. *Environ. Sci. Pollut. Res., 23,* 18850–18857, 2016.
78. Djajadi, D.T., Müller, S., Fiutowski, J., Rubahn, H.G., Thygesen, L.G., Posth, N.R., Interaction of chitosan with nanoplastic in water: the effect of environmental conditions, particle properties, and potential for *in situ* remediation. *Sci. Total Environ., 907,* 167918, 2024.
79. Sharifi, M.J., Nouralishahi, A., Hallajisani, A., Fe_3O_4-chitosan nanocomposite as a magnetic biosorbent for removal of nickel and cobalt heavy metals from polluted water. *Int. J. Biol. Macromol., 248,* 125984, 2023.
80. Jayaprakash, V. and Palempalli, U.M.D., Studying the effect of biosilver nanoparticles on polyethylene degradation. *Appl. Nanosci., 9,* 4, 491–504, 2019.

81. Agarwal, N., Solanki, V., Jonnalagadda, S., Ameta, K., Singh, N., Singh, A., Bind, V., Emerging Pollutants in Aquatic Environment: Critical Risk Assessment and Treatment Options, 2024, 10.2174/97898152389691240 10004.
82. Quy, B.M., Thu, N.T.N., Xuan, V.T., Hoa, N.T.H., Linh, N.T.N., Tung, V.Q., Le, V.T.T., Photocatalytic degradation performance of a chitosan/ZnO–Fe_3O_4 nanocomposite over cationic and anionic dyes under visible-light irradiation. *RSC Adv.*, *15*, 3, 1590–1603, 2025.
83. Sihtmäe, M., Laanoja, J., Blinova, I., Kahru, A., Kasemets, K., Toxicity of Silver–Chitosan Nanocomposites to Aquatic Species. *Proceedings*, *92*, 1, 17, 2023.
84. Dave, D., Chauhan, K., Khimani, A., Soni, K., Vaidya, Y., Photocatalytic degradation of low-density polythene using protein-coated titania nanoparticles and Lactobacillus plantarum. *Environ. Technol.*, *44*, 5, 619–630, 2023.
85. Yoshida, S., Hiraga, K., Takehana, T., Taniguchi, I., Yamaji, H., Maeda, Y., Toyohara, K., Miyamoto, K., Kimura, Y., Oda, K.A., bacterium that degrades and assimilates poly(ethylene terephthalate). *Science*, *351*, 6278, 1196–1199, 2016.
86. Gkanasos, A., Tsiaras, K., Triantaphyllidis, G., Panagopoulos, A., Pantazakos, G., Owens, T., Karametsis, C., Stopping macroplastic and microplastic pollution at source by installing novel technologies in river estuaries and wastewater treatment plants: the CLAIM project. *Front. Mar. Sci.*, *8*, 738876, 2021.
87. Hao, Y., Min, J., Ju, S., Zeng, X., Xu, J., Li, J., Wang, H., Possible hazards from biodegradation of soil plastic mulch: Increases in microplastics and CO_2 emissions. *J. Hazard. Mater.*, *467*, 133680, 2024.
88. Thathsarani, N., Khiadani, M., Azhar, M.R., Zargar, M., Enhanced removal of microplastic fibres using aluminium and chitosan-based coagulants assisted with microbubble technology. *J. Environ. Chem. Eng.*, *13*, 116780, 2025.
89. Su, S. and Kang, P.M., Systemic review of biodegradable nanomaterials in nanomedicine. *Nanomaterials*, *10*, 4, 656, 2020.
90. Ahmed, S., Ahmad, M., Swami, B.L., Ikram, S., A review on plants extract mediated synthesis of silver nanoparticles for antimicrobial applications: a green expertise. *J. Adv. Res.*, *7*, 1, 17–28, 2016.
91. Długa, A., Bajer, D., Kaczmarek, H., Photochemical and Thermal Stability of Bionanocellulose/Poly(Vinyl Alcohol) Blends. *Polymers*, *14*, 20, 4364, 2022.
92. Banu, A.N., Kudesia, N., Raut, A.M., Pakrudheen, I., Wahengbam, J., Toxicity, bioaccumulation, and transformation of silver nanoparticles in aqua biota: A review. *Environ. Chem. Lett.*, *19*, 6, 4275–4296, 2021.
93. Kus-Liśkiewicz, M., Fickers, P., Ben, T.I., Biocompatibility and cytotoxicity of gold nanoparticles: recent advances in methodologies and regulations. *Int. J. Mol. Sci.*, *22*, 20, 10952, 2021.
94. Adeleye, A.S., Conway, J.R., Garner, K., Huang, Y., Su, Y., Keller, A.A., Engineered nanomaterials for water treatment and remediation: Costs, benefits, and applicability. *Chem. Eng. J.*, *286*, 640–662, 2016.

95. Augusto, P.A., Castelo-Grande, T., Vargas, D., Pascual, A., Hernández, L., Estevez, A.M., Barbosa, D., Upscale design, process development, and economic analysis of industrial plants for nanomagnetic particle production for environmental and biomedical use. *Materials, 13*, 11, 2477, 2020.
96. Al-Gethami, W., Qamar, M.A., Shariq, M., Alaghaz, A.N.M., Farhan, A., Areshi, A.A., Alnasir, M.H., Emerging environmentally friendly bio-based nanocomposites for the efficient removal of dyes and micropollutants from wastewater by adsorption: a comprehensive review. *RSC Adv., 14*, 4, 2804–2834, 2024.
97. Allan, J., Belz, S., Hoeveler, A., Hugas, M., Okuda, H., Patri, A., Rauscher, H., Regulatory landscape of nanotechnology and nanoplastics from a global perspective. *Regul. Toxicol. Pharmacol., 122*, 104885, 2021.
98. Miditana, S., Babu, A., Korlam, S., Ampolu, S., Agarwal, N., Solanki, V., Photocatalytic Activity and Potential of NMs in Environmental Remediation, 2025, 10.2174/9789815322941125010014.
99. Kumar, C.M.V., Karthick, V., Ganesh Kumar, V., Inbakandan, D., Rene, E.R., Uma Suganya, K.S., Embrandiri, A., The impact of engineered nanomaterials on the environment: Release mechanism, toxicity, transformation, and remediation. *Environ. Res., 212*, 113202, 2022.
100. Isigonis, P., Hristozov, D., Benighaus, C., Giubilato, E., Grieger, K., Pizzol, L., Semenzin, E., Risk governance of nanomaterials: review of criteria and tools for risk communication, evaluation, and mitigation. *Nanomaterials, 9*, 5, 696, 2019.
101. Asghar, N., Hussain, A., Nguyen, D.A., Ali, S., Hussain, I., Junejo, A., Ali, A., Advancement in nanomaterials for environmental pollutants remediation: a systematic review on bibliometrics analysis, material types, synthesis pathways, and related mechanisms. *J. Nanobiotechnol., 22*, 1, 26, 2024.

5

Biological Approaches for the Degradation of Plastics: Current Developments and Future Prospects

Nagraj[1], Pankaj Kumar Chaurasia[1*], Shashi Lata Bharati[2], Rima Kumari[3], Apoorv Saraswat[4], Naveen Kumar[5] and Rahul D. Kamble[6]

[1]P.G. Department of Chemistry, L.S. College, B.R. Ambedkar Bihar University, Muzaffarpur, Bihar, India
[2]Department of Chemistry, North Eastern Regional Institute of Science and Technology, Itanagar, Arunachal Pradesh, India
[3]Department of Botany, L.S. College, B.R. Ambedkar Bihar University, Muzaffarpur, Bihar, India
[4]Department of Chemistry, Ramashray Baleshwar College, Dalsingsarai (A Constituent College of Lalit Narayan Mithila University, Darbhanga) Samastipur, Bihar, India
[5]Department of Chemistry, Sanatan Dharm College, Muzaffarnagar, Uttar Pradesh, India
[6]Amruteshwar Arts, Commerce and Science College, Pune (MS), India

Abstract
Plastic is an organic polymer with a high molecular weight, made from several hydrocarbons and petroleum derivatives. Excessive consumption of plastics poses a significant threat to the environment and human health. In the present time, plastic pollution is a serious global concern, making its proper management, handling, processing, and treatment an essential task. Among the various physical, chemical, and biological methods of plastic degradation, biological techniques are well known for their high efficiency, cost-effectiveness, environmental friendliness, and the easy availability of biological sources. Biodegradation of plastic polymers has been studied through various biological approaches using fungi, bacteria, algae, microbial enzymes, and combined technologies (integration of

**Corresponding author*: pankaj.chaurasia31@gmail.com

Neha Agarwal, Brijesh Pare, Vijendra Singh Solanki and Sankara Rao Miditana (eds.) Plastic Remediation: Transformative Pathways for a Rejuvenated Sustainable Environment, (123–146)

biological methods with chemical, physical, and/or other methods). Among these, fungi, bacteria, and microbial enzyme–based biodegradation methods are dominant over other biological approaches. Microorganisms play a major role in the degradation of plastic. This review discusses the efficiency of various biological sources and methods in degrading plastic polymers, the factors influencing the rate of biodegradation, and future prospects. In addition to a critical assessment of the biological efficiency of plastic degradation, there is also a need to raise awareness and foster greater interest in the use of these methods for large-scale plastic degradation.

Keywords: Plastics, polymer, plastic pollution, biodegradation, microbial, enzymes, fungi, bacteria

5.1 Introduction

The automotive, agricultural, construction, packaging, and textile industries all depend heavily on plastics, malleable materials composed predominantly of long polymer chains with high mechanical and chemical adaptability [1–3]. Packaging is the leading market for plastics, which is responsible for the rising demand for disposable containers [4]. Based on their ability to degrade in nature, plastics are classified into two groups: degradable and non-degradable [5, 6]. Typical plastics such as polyethylene (PE), polystyrene (PS), polyurethane (PUR), polyvinyl chloride (PVC), polyethylene terephthalate (PET), polypropylene (PP), and other polymer compounds are broken down extremely slowly in the environment [7, 8].

The unrestrained usage of plastic has led to significant environmental issues due to its resistance to degradation [9]. This resistance causes plastic to accumulate in natural environments. As living organisms consume the accumulating plastic, it disrupts population balance and harms ecosystems. Plastic debris endangers marine life by preventing or restricting its movement when it accumulates in oceans. Furthermore, the combustion of plastics releases dioxins, carbon dioxide, and other harmful substances, which can lead to lung cancer and other respiratory illnesses [10]. These hazardous substances contaminate air, land, and aquatic ecosystems [11]. To mitigate this risk, plastic waste should be properly treated and managed [9, 12].

According to a recent study, organic polymers with high molecular weight are synthesized from various petroleum derivatives and hydrocarbons [13]. Ahmed *et al.* identified these polymers as plastic [14]. The word "plastic" is derived from the Greek word plastikos, which means something that can be molded into different shapes. Plastics are polymers that can be molded and become pliable when heated. Apart from biodegradable bioplastics, the majority of plastic products are derived from petrochemicals [15, 16]. Plastics may have several elements like nitrogen, silicon, carbon, hydrogen, oxygen, and chloride. Polyethylene, which accounts for 64% of all plastics, has the general formula C_nH_{2n} [10, 13]. Plastic has become a necessity and impacts nearly every aspect of our daily lives [17]. Its diverse properties enable a wide range of applications, from medical implants to food packaging. However, due to their low cost and rapid production, plastics are often used for single-use items, which contribute significantly to their misuse [13].

The growing usage of synthetic plastic has raised concerns about the polymerization process [18]. Between 1940 and 1950, a significant expansion in industrialization led to a sharp increase in plastic production. By 2050, the annual production of plastics is projected to rise to 590 million metric tons, up from an estimated 367 million metric tons in 2021 [19]. Out of 76% of all plastics that are termed as plastic garbage, 9% are reprocessed, 12% are burned, and 79% are discarded into the surroundings [20]. As a result of these factors, plastic waste poses an increasing risk to the environment and human health [18]. A recent study reported the continuous breakdown of waste plastics into microplastics (MPs, < 5 mm) and nanoplastics (NPs, < 100 nm) [21]. These particles have become persistent environmental pollutants [22–26]. Polystyrene (PS), which is produced in excess of 20 million tons per year worldwide and has the formula $[\text{-CH-}(C_6H_5)\ CH_2\text{-}]_n$, is one of the most common plastic polymers found in environmental litter. Microplastics (MPs) have been found in aquatic environments such as rivers, lakes, sludge, and municipal wastewater, largely due to expanded polystyrene (EPS), a byproduct of PS used in packaging materials, food containers, and building insulation [27–30]. Very little work has been done in the field of soil pollution caused by plastic waste. This has sparked curiosity among researchers to closely examine the potential of biodegradation by *Achatina fulica*, a specific soil snail, for the removal of plastic pollution from soil, including all forms of PS. A gut microbiome analysis and an antibiotic suppression test were performed in order to determine the role of gut microbiota in the snails' PS depolymerization [21].

When microorganisms like bacteria and fungi eat plastic materials, biological degradation takes place, turning them into biomass, carbon dioxide, and water. The chemical composition of plastic, environmental factors, and the presence of degradation initiators all have a significant impact on the rate of degradation of plastic [31]. The properties and structures of polymers have been described, highlighting them as valuable chemicals composed of multiple repeating units [32, 33]. The term "-mer" designates the fundamental repeating unit of a polymer, while "poly-mer" designates a molecule made up of several repeating units. Depending on the chemical properties of the monomers, polymers can be synthesized through various chemical methods to achieve the desired properties. Numerous examples of polymers can be found in nature, which can be utilized directly or converted into materials that meet specific social needs. Common polymers are typically composed of carbon and hydrogen, often with additional functional groups containing elements such as oxygen, nitrogen, and chlorine. The qualities that continue to attract people to polymers include their chemical resistance, electrical and thermal insulation, strength combined with light weight, and their suitability for a wide range of applications where other materials may not be effective. Polymers are extensively used in the automobile, building and construction, and packaging sectors [32, 34].

Landfill, incineration, regenerative granulation, and pyrolysis are the four primary methods that have been employed, globally, for the disposal of plastic waste [35, 36]. These methods handled and decomposed the majority of plastic waste, but they also had certain disadvantages. Compaction and soil leaching are the risks associated with landfills that can contaminate groundwater. When plastic waste passes through treatment systems, it may generate organic contaminants that can have several negative health effects on human beings [37, 38]. The previous recycling process for plastics, which often involved energy-intensive treatment stages, is one example of how improper disposal of plastic scraps can result in the loss of their economic value. Furthermore, environmental contamination has led to increased costs for treatment and remediation [39]. Pyrolysis and solvent decomposition processes are efficient ways to recycle chemicals by converting polymers into monomers and oligomers. However, chemical recycling is not economically viable due to the low cost of virgin monomers derived from fossil carbon-based feedstock [40]. Biodegradation and sustainable conversion techniques for plastic waste are gaining increased attention due to their potential to address the limitations of conventional disposal systems. Enzymatic and microbiological degradation are two biological methods used for the biodegradation of plastic waste. Enzymatic degradation has proven to be a gentler, more energy-efficient approach with lower environmental impact compared to chemical recycling. Furthermore, biological techniques such as enzyme immobilization can reduce costs and facilitate enzyme reuse. Additionally, the transformation of trash into valuable molecules made it possible to recycle and reuse resources, which decreased the need for virgin resources [35].

Upcycling is an emerging concept that aims to reduce waste and minimize the use of raw materials by converting plastic monomers into valuable products [41, 42]. Microbial or enzymatic decomposition offers a viable alternative to traditional recycling and upcycling technologies, as it uses microorganisms capable of breaking down plastics in mild conditions without harmful chemicals. This makes it possible for the waste to be decomposed without the need of any earlier sorting or cleaning [43, 44]. Combining the bio-upcycling and plastic waste degradation in a single bioprocess presents an innovative approach to managing plastic waste, which offers opportunities for a more sustainable and ecological friendly plastic industry [42]. This chapter discusses the potential of various biological techniques in the biodegradation of plastic polymer waste. It also aims to encourage further research in this field while addressing current challenges and gaps in knowledge.

5.2 Methods of Plastic Degradation

Polymers are materials that undergo physical or chemical changes due to natural processes [45]. Those physical or chemical conversions are caused by environmental factors such as chemical conditions, or biological activity, light, heat, and moisture. These chemical transformations alter the properties of polymers, leading to a loss of functionality. Degradation may involve phase separation, erosion, discoloration, cracking, crazing or delamination as well. Those changes are associated with the generation of novel functional group, bond scission, and chemical transformations. The term polymer degradation refers to the biological, chemical, or physical methods that lead to bond breaking and subsequent structural changes [46].

There are four stages of plastic biodegradation: biodeterioration, depolymerization, assimilation, and mineralization [9]. In biodeterioration, polymers are disintegrated into simple molecules by the combined action of microbes and abiotic elements. During depolymerization, the polymers are gradually broken down by catalytic substances. The microorganisms, such as enzymes and free radicals, give rise to these catalytic materials [47]. Mechanical and physiochemical properties of plastic polymers change due to biodeterioration [48, 49]. Biodeterioration is accelerated by microbial biofilms, while depolymerases catalyze the depolymerization reactions and produce dimers, monomers, or oligomers. The presence of oxygen molecules plays a key role in determining the fate of smaller molecules. For plastic biodegradation, both extracellular and intracellular depolymerase enzymes produced by microbes are required. These enzymes break down polymers into smaller intermediates that can be further metabolized. These intermediates, once dissolved in water, are readily taken up by microbial cells. Recent research indicates that assimilation occurs in the cytoplasm of microbial cells, where metabolic processes produce biomass, primary or secondary metabolites, and energy. Anaerobic metabolism results in the production of CO_2, H_2O, CH_4, and H_2S, whereas aerobic degradation yields CO_2 and H_2O [47, 50]. The final step in the biodegradation of plastic is mineralization. During this phase, toxic and hazardous compounds are converted into safer, non-toxic substances [9, 51].

5.3 The Role of Microorganisms in Plastic Biodegradation

Microbes are excellent biological agents for the remediation of various organic and inorganic pollutants [52–55]. Bacteria and fungi are among the microorganisms involved in the breakdown of plastics [56]. Each of the microorganisms involved in biodegradation has unique growth requirements and characteristics. Both, non-enzymatic and enzyme-catalyzed hydrolysis, are part of the biodegradation process [57]. Simple molecules like oligomers, dimers, and monomers are derived from complex polymers that are degraded by the exoenzymes of microorganisms. These molecules are soluble in aqueous media and small enough to pass through semi-permeable external bacterial membranes [58]. The source of energy and carbon (C) is the exoenzymes of microorganism. This initial stage of polymer breakdown is known as depolymerization. The degradation mechanisms involved in polymer breakdown are influenced by environmental factors. Complete degradation of polymer gives H_2O, organic acids, CH_4, and CO_2 [59, 60].

During the complex process of biological degradation of polymeric substances, enzymes play an important role in initiating a series of subsequent reactions. The most significant type of enzymatic polymer cleavage is hydrolysis, as confirmed by Schink *et al.* [61]. Bonds such as glycosidic linkages, peptide bonds, and ester bonds are subjected to hydrolysis through a nucleophilic attack on the carbonyl carbon. According to Rajendran *et al.* [56], in the presence of enzymes and microbes, both, the biotic availability of the polymeric structure and the correct abiotic conditions, have an impact on the biological degradation of polymeric substances.

Saritha *et al.* [62] state that the microorganisms are widespread in the biosphere, and their presence influences the environments in which they thrive. The impact of bacteria on their surroundings can be either beneficial or detrimental. The ability of bacteria to degrade organic matter is one of their most important functions on Earth. They play a crucial role in recycling the three major elements that constitute all living systems: nitrogen (N), carbon (C), and oxygen (O). Plastic biodegradation is the breakdown of complex polymers into simpler monomers, and it depends on several factors such as polymer's molecular weight, substrate availability, surface properties, and shape [63–65]. Different microorganisms break down different types of plastic, and microbes play a vital role in this process. According to a scientific investigation, certain bacteria in mangrove

Table 5.1 Efficiency of microorganisms in polymer degradation [62].

Microorganisms	Types of polymer	Microbial sources	Effectiveness of degradation	Ref.
Bacillus cereus	Polythene (PE)	Discarding soil	2.4–7.2%	[72]
Pseudomonas putida	Milk coat	Garden earth	75.3%	[73]
Streptomyces sp.	LDPE	Garbage soil	46.7%	[71]
Pseudomonas sp.	Natural and synthetic polythene	Household garbage Dumpsite Textile Effluent Drainage site	31.4 and 16.3% 39.7 and 19.6%	[74]
Aspergillus glacus	Plastic and polythene Plastic and polythene	Mangrove soil Mangrove soil	20.54 and 8.16% 20.8 and 7.26%	[66]
Micrococcus luteus	Synthetic dishes	Soil of forest	38.00%	[75]
Masoniella sp.	Plastic cups	Soils of forest	27.4%	[75]

soil have the capacity to break down plastic [66]. The fungal isolates of *Aspergillus* sp. and bacterial isolates of *Bacillus* are found to be the finest degraders after analyzing their works on LDPE, low-density polyethylene [67]. Other researchers [51, 68–75] have also presented similar studies which are summarized in Table 5.1 [62].

Microbes such as bacteria, fungus, actinomycetes, and saccharomonospora are associated with the degradation of both natural and manmade plastics. It has been discovered that nine fungal species and seventeen bacterial genera are capable of breaking down plastics [62, 76–78]. It is assumed that natural polymers biodegrade more efficiently than do synthetic polymer backbone chains. Their degradation leaves voids in the polymer matrix, increasing the surface-to-volume ratio and enhancing microbial activity within the plastic material [79–81].

The synthesis of biosurfactants leads to the emulsification of the hydrophobic hydrocarbons in aquatic environments, thereby increasing their bioavailability to the microbial cells. This process is typically associated with microbial growth on hydrocarbons [82]. For instance, rhamnolipids were produced by *Renibacterium salmoninarum* 27BN during its growth on *n*-hexadecane [83]. *Dietzia maris* As-13-3, an oil-degrading bacterium, was isolated from a deep marine hydrothermal area [84]. This oil-degrading bacterium is able to generate di-rhamnolipid as a biosurfactant. On the other hand, tetradecane, n-hexadecane, and pristine were used as the sole carbon sources [84]. Thus, such types of bacteria may utilize the pyrolysis hydrocarbons derived from polyethylene, polyvinyl chloride, and other plastics as feedstock to produce biosurfactants [82]. Figure 5.1 illustrates the potential steps involved in the biodegradation of plastic polymers into various end products.

5.3.1 Bacteria in Plastic Degradation

It has been found that polyethylene (PE) is an extremely inert and resistant polymer that is difficult to break down, even after being buried in a landfill for a number of years [85]. After 12 to 32 years in moist soil, a polyethylene sheet is only partially degraded and has lost very little weight [86]. Studies have demonstrated that over 20 bacterial genera are capable of breaking down various forms of polyethylene (PE) [87–89]. These species include Gram-negative and Gram-positive species from the genera *Streptomyces*, *Bacillus Pseudomonas*, *Rhodococcus*, *Staphylococcus*, *Streptococcus*, *Klebsiella* and *Ralstonia*, *Stenotrophomonas*, *Acinetobactor*, etc. According to a recent study [85], most of these bacterial strains have the potential to degrade polyethylene (PE) surfaces or form biofilms on them.

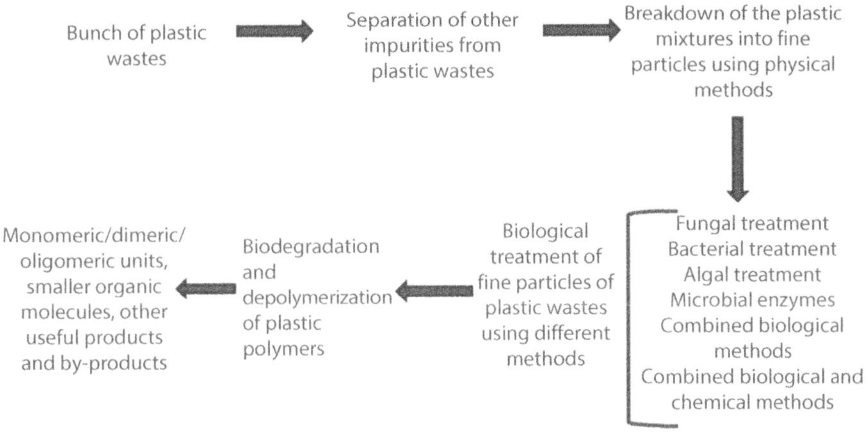

Figure 5.1 Steps involved in the microbial degradation of plastic waste.

According to a recent study, the degradation of plastic is carried out by the isolates of bacteria [82]. The bacterial isolates involved in plastic degradation include *Ralstonia eutropha*, *Azotobacter* sp., *Pseudomonas* sp., *Bacillus megaterium*, *Halomonas* sp. [77], *Bacillus brevis* [90], *Acidovorax delafieldii* [91], *B. petrii* [92], *P. amyloticus* [93], *B. pumilus* [94], *P. aeruginosa* [95], and *Shewanella* sp. [96]. Apart from this, a thermophilic bacterium, *Bacillus brevis*, has been isolated from soil that has the ability to degrade polylactic acid. Research on bacterial degradation of polyethylene have revealed the potential of *Acinetobacter baumannii*, *Rahnella aquatills*, *Xylosus* sp., *Bacillus amyloliquefaciens*, *Pseudomonas* sp., *Arthrobacter* sp., *B. thuringiensis* and *B. pumilus*, *Mycoides*, *Staphylococcus cohnii*, *Cereus*, *Pseudomonas fluorescens*, *A. Viscosus* sp., *Micrococcus luteus*, *Arthrobacter* sp., *M. lylae*, *Flavobacterium* sp., *Paenibacillus macerans*, *Ralstonia* sp., *Delftia acidovorans*, *R. erythropolis*, and *P. aeruginosa* [97] species in plastic biodegradation. *Bacillus* species, *Diplococcus* species, *Pseudomonas* species, *Streptococcus* species, *Staphylococcus* species, *Micrococcus* species, and *Moraxella* species were selected from the polythene bag samples [82].

5.3.2 Fungi in Plastic Degradation

Lignin is an aromatic polymer found in nature [10]. It is a component of woody plant biomass and is highly resistant to natural degradation. Among fungi, white rot fungi are the most efficient at breaking down lignin. Microorganisms that break down lignin have been extensively researched for their potential to degrade plastics. *Phanerochaete chrysosporium*,

IZU-154, a lignin-degrading fungus, is used to degrade polyethylene membrane. Under different dietary circumstances, *Trametes versicolor* revealed that manganese peroxidase (MnP) (a primary enzyme) is involved in the degradation of polyethylene. Significant degradation occurred when partially pure MnP was applied to polyethylene membranes in the presence of Mn (II) and Mn (III) chelators and Tween 80 [98].

Aspergillus flavus is one of several fungi that are known to produce enzymes such as laccase, lignin peroxidase, amylase, and manganese peroxidase [99]. *Aspergillus niger* produces lignin peroxidase, laccase, and manganese peroxidase. Amylase, laccase, lignin peroxidase, and manganese peroxidase are also produced by *Fusarium graminearum*. These four types of enzymes are crucial in the breaking down of carbon bonds in polyethylene polymers [100]. *Pestalotiopsis microspora* produce the serine hydrolase enzyme, which is used as a catalyst in the degradation of PUR (polyurethane) polymers [101, 102].

Fungi, which belong to ten genera of the Ascomycota and Mucor, break down the polyethylene more quickly than bacteria [103]. This difference is mostly due to the fact that fungus may stick to hydrophobic polymer surfaces. It might be feasible to cultivate *Penicillium*, *Fusarium oxysporum*, and *Aspergillus* for three months in order to examine their potential for degradation when PP and PE are the only carbon sources. The research also reveals surface changes in polypropylene (PP) and polyethylene (PE), confirming the presence of biofilms. Further examination of the incubated fungi shows that they can survive for nearly three months without any additional carbon sources [104].

5.3.3 Algae in Plastic Degradation

Microorganisms are essential to the natural breakdown of polymers. Algae, in addition to bacteria and fungi, participate in this process in a number of ways. These organisms have the ability to colonize plastic garbage and use biochemical mechanisms to break it down. The capacity of microorganisms to release enzymes allows them to target specific types of plastic. Those enzymes interact with the chemical structure of plastics. Several variables, including temperature, pH, and the availability of nutrients and oxygen, affect how well these microbes break down the plastics. In an effort to accelerate the degradation process and reduce environmental pollution, scientists are actively researching the ways to identify and develop microbes with enhanced plastic-degrading abilities [31].

5.3.4 Plastic-Degrading Microbial Symbionts of Invertebrates

Based upon their studies of plastic degradation, a recent study concludes that some invertebrates can consume plastics [105]. This has drawn increased attention to the gut microbiomes of these organisms, as they appear to break down plastics more efficiently. Those gut microbiomes attack on plastic more efficiently. In the preliminary stage, it is observed that certain insects are able to create voids in PE film. This observation was first reported in 2007 by Riudavets *et al.* [106]. Subsequently, strains of *Bacillus* sp. and *Enterobacter asburiae* are separated from the gut of *Plodia interpunctella* (Indian meal moth). Those strains of *Bacillus* sp. and *Enterobacter asburiae* were found to degrade PE film by 6 and 10%, over a period of 60 days, accompanied by a reduction in molecular weight of the remaining PE [107]. The lesser wax moth *Achatina fulica* [21], the mealworm beetle *Tenebrio molitor* [108], *Achroia grisella* [109], the giant mealworm beetle *Zophobas atratus* [110], and the dark mealworm beetle *Tenebrio obscurus* [111, 112] have all been shown to biodegrade PE and other plastics. Strains of *Bacillus* sp. and *Acinetobacter* sp., isolated from the abdomen of *T. molitor* larvae and grown in co-culture, were found to grow on LDPE and reduce its mass by 18% after 30 days of incubation [113]. The larvae of the mealworm beetle (*Tenebrio molitor*) were fed polyvinyl chloride (PVC) powder, resulting in promising outcomes [105, 110].

5.3.5 Microbiological Enzymes That Contribute to Plastic Breakdown

Microbial enzymes also participate in the breakdown of plastic. These microbial enzymes are divided into intracellular and extracellular categories [58, 114]. Extracellular enzymes are the more extensively studied of the two groups due to their higher reactivity and ability to facilitate both oxidative and hydrolytic functions [32]. As seen below, their primary function is to depolymerize the lengthy carbon chains of polymeric polymers into a combination of oligomers, dimers, and monomers.

The combination of plastic polymer and microbiological enzymes can lead to the formation of monomers.

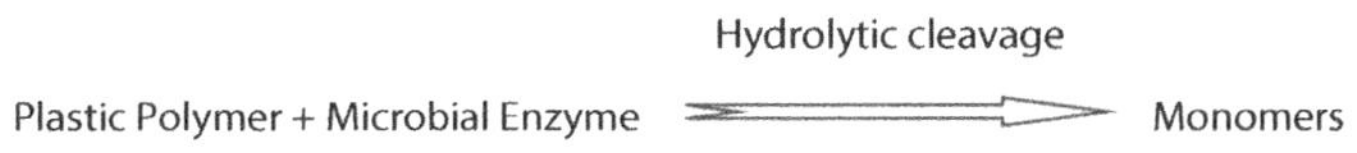

The breakdown of plastic polymers is catalyzed by enzymes that are produced from many microorganisms, such as actinomycetes fungi, and bacteria [115].

5.4 Factors Influencing the Biodegradation of Plastic

According to Ahmed *et al.* [14], biodegradation is influenced by many factors such as characteristics of enzymes, properties of polymer, and the environmental condition. Moisture influences polymer biodegradation in several ways, as water is essential for microbial growth and reproduction. Therefore, adequate moisture promotes rapid microbial activity, which in turn accelerates the rate of polymer degradation [116]. Moreover, plenty of moisture boosts hydrolysis by escalating the number of chain scission proceedings. The pH can change the rate of hydrolysis by varying the acidic or basic conditions. For instance, PLA (polylactic acid or polylactide) capsules are hydrolyzed at the optimal rate of pH 5 [117, 118]. Temperature also plays a crucial role in the plastic degradation. With the increase in temperature, enzymatic degradation diminishes potentially. For example, polyester like polycaprolactone with low melting point has been effectively hydrolyzed by pure lipase of *Rhizopus delemar* [119, 120].

Different kinds of polymers are biodegraded by different enzymes, and each of them has its own active site. For instance, polyesters synthesized from diacid monomers containing six to twelve carbon atoms have been rapidly degraded by enzymes secreted by the fungal species *Aspergillus flavus* and *Aspergillus niger*, compared to straight-chain polyesters derived from other monomers [121]. The degradation of PHB (polyhydroxybutyrate) by extracellular enzymes occurs through specific mechanisms that rely on microbially produced depolymerases [122]. Hydrophobic properties of PE stop bacteria from forming a biofilm, which decrease the rate of biodegradation [14]. A number of enzymes from bacterial and fungal sources, such as cutinases, esterases, lipases, laccases, peroxidases, proteases, and ureases, have been shown to be able to degrade PE, PET, and PP [103]. The role of a two-enzyme system in the depolymerization of plastics has also been studied by recently [123]. Fungal enzymes, particularly the depolymerases, have the ability to break down a variety of polymers like PET (polyethylene terephthalate) and PE [103]. There are several other studies that demonstrate the microbial degradation of different types of plastics under various conditions [124–126].

5.5 Future Prospects

The role of fungi in the biodegradation of plastics is significant [99]. Fungal biodegradation is considered an environmentally friendly solution to the harmful effects of plastic waste, offering a safer and more sustainable alternative. Fungi produce a variety of extracellular and intracellular enzymes that break down plastic polymers. Fungi use plastic polymers as the source of energy and carbon for their metabolism. Implementing this technique could play a vital role in reducing the accumulation of plastic waste. Abiotic factors such as pH, temperature, and moisture influence the rate of plastic biodegradation. Biotic factors, including enzyme activity and the hydrophobicity of the plastic, also play a key role in determining the degradation rate. In addition to this, biodegradation of plastic polymer is affected by the properties like size and shape, molecular weight, additives, and biosurfactants [14]. Each fungal species is capable of breaking down only specific types of plastic waste [99].

The environment is currently under threat from non-biodegradable petrochemical materials used in plastic production, particularly due to inadequate waste management infrastructure and lack of effective control measures. In some applications, there is a steady rise in demand for environmentally friendly polymers. Using biodegradable plastics for sterile products, fishing nets, agricultural films, and bio-absorbable materials in surgical frameworks and therapies is beneficial for the environment. However, to fully benefit society from such polymers, litter reduction and efficient waste management are essential. To build a more sustainable future, the use of next-generation bio-based biodegradable polymers is imperative. To enable their reuse, these polymers should also undergo balanced biodegradation and recycling. Hence, it is important to produce new biodegradable plastic polymers for different purposes. Microbiologists, process engineers, synthetic chemists, and biomass researchers should all apply their expertise to develop environmentally friendly materials that contribute to a more sustainable society.

5.6 Conclusion

The use and demand for plastics are continuously increasing. For long-term environmental safety, this trend must be responsibly aligned with effective litter prevention and waste management practices. The use of fossil-based biodegradable and bio-based materials is strongly recommended for specific applications. Certain plastic polymers can be biologically broken down into simpler compounds by diverse microbial communities through aerobic and anaerobic processes. The development of innovative bio-based biodegradable polymers and/or the investigation of newer microbial strains and their processes for breaking down fossil-based polymers require more attention of scientific communities.

AI disclosure statement: The author confirms that no artificial intelligence (AI)-generated content or AI-assisted writing tools were used in the preparation of this chapter (including summaries and the abstract). All text, analysis, and conclusions are the result of the author's original work.

References

1. Andrady, A.L. and Neal, M.A., Applications and societal benefits of plastics. *Philos. Trans. R. Soc. B.*, 364, 1977–1984, 2009, https://doi.org/10.1098/rstb.2008.0304.
2. Andrady, A.L., Microplastics in the marine environment. *Mar. Pollut. Bull.*, 62, 1596–1605, 2011, https://doi.org/10.1016/j.marpolbul.2011.05.030.
3. Dris, R., Imhof, H., Sanchez, W., Gasperi, J., Galgani, F., Tassin, B., Laforsch, C., Beyond the ocean: contamination of freshwater ecosystems with (micro-) plastic particles. *Environ. Chem.*, 12, 539–550, 2015, https://doi.org/10.1071/EN14172.
4. Geyer, R., Jambeck, J., Law, K., Production, use, and fate of all plastics ever made. *Sci. Adv.*, 3, e1700782, 2017, https://doi.org/10.1126/sciadv.1700782.
5. Major, I., Fuenmayor, E., Mcconville, C., The production of solid dosage forms from non-degradable polymers. *Curr. Pharm. Des.*, 22, 2738–2760, 2016, https://doi.org/10.2174/1381612822666160217141049.
6. Brodhagen, M., Peyron, M., Miles, C., Inglis, D.A., Biodegradable plastic agricultural mulches and key features of microbial degradation. *Appl. Microbiol. Biotechnol.*, 99, 1039–1056, 2015, https://doi.org/10.1007/s00253-014-6267-5.
7. Wang, J., Tan, Z., Peng, J., Qiu, Q., Li, M., The behaviors of microplastics in the marine environment. *Mar. Environ. Res.*, 113, 7–17, 2016, https://doi.org/10.1016/j.marenvres.2015.10.014.
8. Yang, X.G., Wen, P.-P., Yang, Y.F., Jia, P.P., Li, W.G., Pei, D.S., Plastic degradation by *in vitro* environmental microorganism and *in vivo* gut microorganism of insects. *Front. Microbiol.*, 13, 1001750, 2023, https://doi.org/10.3389/fmicb.2022.1001750.
9. Heris, Y.S., Bacterial biodegradation of synthetic plastic: a review. *Bull. Natl. Res. Cent.*, 48, 87, 2024, https://doi.org/10.1186/s42269-024-01241-y.
10. Kale, S.K., Deshmukh, A.G., Dudhare, M.S., Patil, V.B., Microbial degradation of plastic: a review. *J. Biochem. Technol.*, 6, 2, 952–961, 2015.
11. Soud, S.A., Biodegradation of polyethylene LDPE plastic waste using locally isolated. *Streptomyces* sp. *J. Pharm. Sci. Res.*, 11, 4, 1333–1339, 2019.

12. Sowmya, H.V., Ramalingappa, V., Krishnappa, T.B., Low density polyethylene degrading fungi isolated from local dumpsite of Shivamogga District. *Int. J. Biolog. Res.*, 2, 2, 39–43, 2014, https://doi.org/10.14419/ijbr.v2i2.2877.
13. Zeenat, Elahi, A., Bukhari, D.A., Shamim, S., Rehman, A., Plastic degradation by microbes: A sustainable approach. *J. King Saud Univ. Sci.*, 33, 6, 101538, 2021, https://doi.org/10.1016/j.jksus.2021.101538.
14. Ahmed, T., Shahid, M., Azeem, F., Rasul, I., Shah, A.A., Noman, M., Hameed, A., Manzoor, N., Manzoor, I., Muhammad, S., Biodegradation of plastics: current scenario and future prospects for environmental safety. *Environ. Sci. Pollut. Res.*, *25*, 8, 7287–7298, 2018, https://doi.org/10.1007/s11356-018-1234-9.
15. Getachew, A. and Woldesenbet, F., Production of biodrgradable plastic by polyhydroxybutyrate(PHB) accumulating bacteria using low cost agricultural waste material. *BMC Res. Notes*, 9, 1, 509, 2016, https://doi.org/10.1186/s13104-016-2321-y.
16. Akmal, D., Asiska, P., Wangi, Q., Rivai, H., Agustein, A., Biosynthesis of copolymer poly (3-hydroxybutyrate-CO-3-hydroxyvalerate) from palm oil and n-pentanol in a 10L bioreactor. *Rasayan J. Chem.*, 8, 389–395, 2015.
17. Bergeson, A.R., Silvera, A.J., Alper, H.S., Bottlenecks in biobased approaches to plastic degradation. *Nat. Commun.*, 15, 4715, 2024, https://doi.org/10.1038/s41467-024-49146-8.
18. Gilani, I.E., Sayadi, S., Zouari, N., Al-Ghouti, M.A., Plastic waste impact and biotechnology: Exploring polymer degradation microbial role and sustainable development implications. *Bioresour. Technol. Rep.*, 24, 101606, 2023, https://doi.org/10.1016/j.biteb.2023.101606.
19. Dimassi, S.N., Hahladakis, J.N., Yahia, M.N.D., Ahmad, M.I., Sayadi, S., Al-Ghouti, M.A., Degradation-fragmentation of marine plastic waste and their environmental implications: a critical review. *Arab. J. Chem.*, 15, 11, 104262, 2022, https://doi.org/10.1016/j.arabjc.2022.104262.
20. Kumar., R., Verma., A., Shome., A., Sinha., R., Sinha., S., Jha., P.K., Kumar., R., Kumar., P., Das, S.S., Sharma, P., Vara Prasad, P.V., Impacts of plastic pollution on ecosystem services, sustainable development goals, and need to focus on circular economy and policy interventions. *Sustainability*, 13, 17, 9963, 2021, https://doi.org/10.3390/su13179963.
21. Song, Y., Qui, R., Hu, J., Li, X., Zhang, X., Chen, Y., Wu, W.-M., He, D., Biodegradation and disintegration of expanded polystyrene by land snails *Achatina fulica*. *Sci. Total Environ.*, 746, 141289, 2020, https://doi.org/10.1016/j.scitotenv.2020.141289.
22. Lambert, S. and Wagner, M., Formation of microscopic particles during the degradation of different polymers. *Chemosphere*, 161, 510–517, 2016, https://doi.org/10.1016/j.chemosphere.2016.07.042.
23. Cole, M., Lindeque, P.K., Fileman, E., Clark, J., Lewis, C., Halsband, C., Galloway, T.S., Microplastics alter the properties and sinking rates of

zooplankton faecal pellet. *Environ. Sci. Technol.*, 50, 3239–3246, 2016, https://doi.org/10.1021/acs.est.5b05905.

24. Peters, C.A. and Bratton, S.P., Urbanization is a major influence on microplastic ingestion by sunfish in the Brazos River basin, Central Texas. *USA. Environ. Pollut.*, 210, 380–387, 2016, https://doi.org/10.1016/j.envpol.2016.01.018.
25. Wang, L., Wu, W.-M., Bolan, N.S., Tsang, D.C.W., Li, Y., Qin, M., Hou, D., Environmental fate, toxicity and risk management strategies of nanoplastics in the environment: current status and future perspectives. *J. Hazar. Mat.*, 401, 123415, 2020, https://doi.org/10.1016/j.jhazmat.2020.123415.
26. Wu, W.M., Yang, J., Criddle, C.S., Microplastics pollution and reduction strategies. *Front. Environ. Sci. Eng.*, 11, 6, 2017, https://doi.org/10.1007/s11783-017-0897-7.
27. Lv, X., Dong, Q., Zuo, Z., Liu, Y., Huang, X., Wu, W.M., Microplastics in a municipal wastewater treatment plant: fate, dynamic distribution, removal efficiencies, and control strategies. *J. Clean. Prod.*, 225, 579–586, 2019, https://doi.org/10.1016/j.jclepro.2019.03.321.
28. Zhang, K., Xiong, X., Hu, H., Wu, C., Bi, Y., Wu, Y., Zhou, B., Lam, P.K.S., Liu, J., Occurrence and characteristics of microplastic pollution in Xiangxi Bay of three Gorges reservoir, China. *Environ. Sci. Technol.*, 51, 3794–3801, 2017, https://doi.org/10.1021/acs.est.7b00369.
29. Zhou, Q., Zhang, H., Fu, C., Zhou, Y., Dai, Z., Li, Y., Tu, C., Luo, Y., The distribution and morphology of microplastics in coastal soils adjacent to the Bohai Sea and the Yellow Sea. *Geoderma*, 322, 201–208, 2018, https://doi.org/10.1016/j.geoderma.2018.02.015.
30. Mani, T., Hauk, A., Walter, U., Burkhardt-Holm, P., Microplastics profile along the Rhine River. *Sci. Rep.*, 5, 17988, 2015, https://doi.org/10.1038/srep17988.
31. Vaia, Plastic Degradation: Causes, Effects | Vaia, 2024, [online] Available at: https://www.vaia.com/en-us/explanations/environmental science/pollution/plastic-degradation/.
32. Glaser, J.A., Biological Degradation of Polymers in the Environment, in: *Plastics in the Environment*, A. Gomiero (Ed.), IntechOpen, London, U.K., 2019, Available from: http://dx.doi.org/10.5772/intechopen.85124.
33. Sperling, L.H., *Introduction to Physical Polymer Science*, John Wiley & Sons, New York, 2005.
34. Elias, H.G., Macromolecules, Volume 4: Applications of Polymers. *J. Am. Chem. Soc.*, 131, 48, 17720–17720, Wiley, New York, 2009, doi:https://doi.org/10.1021/ja909401b.
35. Li, M., Chen, Z., Huo, Y., Advancement in the Bio-degradation of Plastic waste into value-added chemicals: A recent perspective. *Synth. Biol. Eng.*, 2, 2, 10009, 2024, https://doi.org/10.35534/sbe.2024.10009.
36. Jatoi, A.S., Ahmed, J., Mubarak, N.M., Shaikh, M.S., Kumar, S., Ahmed, M.I., Karri, R.R., A Comprehensive Review on Handling of Plastic Waste For

Energy Generation—Current Status and Future Challenges. *Water Air Soil Pollu.*, *234*, 773, 2023, https://doi.org/10.1007/s11270-023-06784-w.
37. Huang, Q., Chen, G., Wang, Y., Chen, S., Xu, L., Wang, R., Modelling the global impact of China's ban on plastic waste imports. *Res. Conser. Recycl.*, *154*, 104607, 2020, https://doi.org/10.1016/j.resconrec.2019.104607.
38. Jung, Y.S., Sampath, V., Prunicki, M., Aguilera, J., Allen, H., LaBeaud, D., *et al.*, Characterization and regulation of microplastic pollution for protecting planetary and human health. *Environ. Pollu.*, *315*, 120442, 2022, https://doi.org/10.1016/j.envpol.2022.120442.
39. Banu, J.R. and Sharmila, V.G., A systematic review on plastic waste conversion for a circular economy: recent trends and emerging technologies. *Catal. Sci. Tec.*, *13*, 2291–2302, 2023, https://doi.org/10.1039/D2CY02066A.
40. Weckhuysen, B.M., Creating value from plastic waste. *Science*, *370*, 400–401, 2020, https://doi.org/10.1126/science.abe3873.
41. Rezaei, Z., Dinani, A.S., Moghimi, H., Cutting-edge developments in plastic biodegradation and upcycling *via* engineering approaches. *Metab. Eng. Commun.*, 19, e00256, 2024, https://doi.org/10.1016/j.mec.2024.e00256.
42. Lee, S., Lee, Y.R., Kim, S.J., Lee, J.S., Min, K., Recent advances and challenges in the biotechnological upcycling of plastic wastes for constructing a circular bioeconomy. *Chem. Eng. J.*, 454, P4, 140470, 2023, https://doi.org/10.1016/j.cej.2022.140470.
43. Kumar, R., Sadeghi, K., Jang, J., Seo, J., Mechanical, chemical, and biorecycling of biodegradable plastics: a review. *Sci. Total Environ.*, 882, 163446, 2023, https://doi.org/10.1016/j.scitotenv.2023.163446.
44. Rad, M.M., Moghimi., H., Azin., E., Biodegradation of thermo-oxidative pretreated low-density polyethylene (LDPE) and polyvinyl chloride (PVC) microplastics by *Achromobacter denitrificans Ebl13. Mar. Pollut. Bull.*, 181, 113830, 2022, https://doi.org/10.1016/j.marpolbul.2022.113830.
45. Shah, A.A., Hasan, F., Hameed, A., Ahmed, S., Biological degradation of Plastic: A Comprehensive review. *Biotechnol. Adv.*, 26, 3, 246–265, 2008, https://doi.org/10.1016/j.biotechadv.2007.12.005.
46. Pospisil, J. and Nespurek, S., Highlights in chemistry and physics of polymer stabilization. *Macromol. Symp.*, 115, 143–63, 1997, https://doi.org/10.1002/masy.19971150110.
47. Puja Asiandu, A., Wahyudi, A., Widiya, Sari, S., A review: plastics waste biodegradation using plastics-degrading bacteria. *J. Environ. Treat. Tech.*, 9, 1, 148–157, 2021, http://dx.doi.org/10.47277/JETT/9(1)157.
48. Arutchlevi, J., Sudhakar, M., Arjatkar, A., Doble, M., Bhaduri, S., Uppara, P., Biodegradation of polyethylene and polypropylene. *Indian J. Biotechnol.*, 7, 1, 9–22, 2008.
49. Kumar, V.R., Kanna, G.R., Elumalai, S., Biodegradation of polyethylene by green photosynthetic microalgae. *J. Bioremediat. Biodegrad.*, 8, 1, 1–8, 2017.

50. Tiwari, A.K., Gautam, M., Maurya, H.K., Recent development of biodegradation techniques of polymer. *Int<./j> J. Res. Granthaalayah*, 6, 6, 414–452, 2018.
51. Alshehrei, F., Biodegradation of low density polyethylene by fungi isolated from red sea water. *Int. J. Curr. Microbiol. App. Sci.*, 6, 8, 1703–1709, 2017, http://dx.doi.org/10.20546/ijcmas.2017.602.204.
52. Ayilara, M.S. and Babalola, O.O., Bioremediation of environmental wastes: the role of microorganisms. *Front. Agron.*, 5, 1183691, 2023, https://doi.org/10.3389/fagro.2023.1183691.
53. Chaurasia, P.K., Sharma, N., Nagraj, Rudakia, D.M., Singh, S., Bharati, S.L., Fungal-Assisted Bioremediation of Agricultural Organic Pollutants (Pesticides and Herbicides). *Curr. Green Chem.*, 9, 1, 14–25, 2022, http://dx.doi.org/10.2174/2213346109666220927121948.
54. Chaurasia, P.K., Nagraj, Sharma, N., Kumari, S., Yadav, M., Singh, S., Mani, A., Yadava, S., Bharati, S.L., Fungal Assisted Bio-Treatment of Environmental Pollutants with Comprehensive Emphasis on Noxious Heavy Metals: Recent Updates. *Biotechnol. Bioeng.*, 120, 1, 57–81, 2023, https://doi.org/10.1002/bit.28268.
55. Nagraj, Chaurasia, P.K., Bharati, S.L., Sharma, N., Kumar, J., Sivalingam, A.M., Degradation of Dyes by Fungi: An Overview on Recent Updates. *Microbe*, 6, 100232, 2025, https://doi.org/10.1016/j.microb.2024.100232.
56. Devi, R.S., Kannan, V.R., Natarajan, K., Nivas, D., Kannan, K., Chandru, S., Antony, A.R., The role of Microbes in Plastic Degradation, in: *Environmental Waste Management*, R. Chandra (Ed.), pp. 341–370, 2015, doi:https://doi.org/10.1201/B19243-16.
57. Wackett, L.P. and Hershberger, C.D., *Biocatalysis and Biodegradation, Microbial Transformation of Organic Compounds*, ASM Press, Washington, 2001.
58. Gu, J.D., Microbiological deterioration and degradation of synthetic polymeric materials: Recent research advances. *Int. Biodeterior. Biodegrad., 52*, 69–91, 2003, https://doi.org/10.1016/S0964-8305(02)00177-4.
59. Alexander, M., *Introduction to Soil Microbiology*, 2nd edn., Wiley, New York, 1977.
60. Narayan, R., Biodegradation of polymeric materials (anthropogenic macromolecules) during composting, in: *Science and Engineering of Composting: Design, Environmental, Microbiological and Utilization Aspects*, H.A.J. Hoitink and H.M. Keener (Eds.), pp. 339–362, Renaissance Publishers, Washington, 1993.
61. Brune, A. and Schink, B., Anaerobic degradation of hydroaromatic compounds by newly isolated fermenting bacteria. *Arch. Microbiol.*, 158, 5, 320–327, 1992, doi:https://doi.org/10.1007/bf00245360.
62. Saritha, B., Sindgi, S.A., Remadevi, O.K., Plastic Degrading Microbe: A Review. *Microbiol. Res. J. Int.*, 31, 6, 22–28, 2021.

63. Ammala, A., Bateman, S., Deana, K., Petinakis, E., Sangwan, P., Wong, S., Yuan, Q., Yu, L., Patrick, C., Leong, K.H., An overview of degradable and biodegradable polyolefins. *Prog. Polym. Sci.*, 36, 1015–1049, 2011, https://doi.org/10.1016/j.progpolymsci.2010.12.002.
64. Albertsson, A.C., Anderson, S.O., Karlsson, S., Mechanism of biodegradation of polyethylene. *Polym. Degrad. Stab.*, 18, 73–87, 1987, https://doi.org/10.1016/0141-3910(87)90084-X.
65. Harrison, J.P., Boardman, C., O'Callaghan, K., Delort, A.M., Song, J., Biodegradability standards for carrier bags and plastic films in aquatic environments: a critical review. *R. Soc. Open Sci.*, 5, 171792, 2018, https://doi.org/10.1098/rsos.171792.
66. Kathiresan, K., Polythene and Plastics degrading microbes from the mangrove soil. *Rev. Biol. Trop.*, 51, 3, 629–633, 2003.
67. Muhonja, C.N., Makonde, H., Magoma, G., Imbuga, M., Biodegradability of polyethylene by bacteria and fungi from Dandora dumpsite Nairobi-Kenya. *PLoS One*, 13, 7, e0198446, 2018, https://doi.org/10.1371/journal.pone.0198446.
68. Sharma, B., Rawat, H., Pooja, Sharma, R., Bioremediation - A progressive approach toward reducing plastic wastes. *Int. J. Curr. Microbiol. App. Sci.*, 6, 12, 1116–1131, 2017, http://dx.doi.org/10.20546/ijcmas.2017.612.126.
69. El-Morsy, E.M., Hassan, H.M., Ahmed, E., Biodegradative activities of fungal isolates from plastic contaminated soils. *Mycosphere.*, 8, 8, 1071–1087, 2017.
70. Singh, G., Singh, A.K., Bhatt, K., Biodegradation of polythenes by bacteria isolated from soil. *Int. J. Res. Dev. Pharm. L Sci.*, 5, 2, 2056–2062, 2016.
71. Deepika, S. and Madhuri, R.J., Biodegradation of low density polyethylene by microorganisms from garbage soil. *J. Exp. Bio. Agri. Sci.*, 3, 1, 15–21, 2015.
72. Sowmya, H., Ramalingappa, M.K., Thippeswamy, B., Biodegradation of polyethylene by *bacillus cereus*. *Adv. Polym. Sci. Technol. An Int. J.*, 4, 2, 28–32, 2014.
73. Saminathan, P., Sripriya, A., Nalini, K., Sivakumar, T., Thangapandian, V., Biodegradation of plastics by *Pseudomonas putida* isolated from garden soil samples. *J. Adv. Bot. Zool.*, 1, 3, 1–4, 2014.
74. Nanda, S., Sahu, S.S., Abraham, J., Studies on the biodegradation of natural and synthetic polyethylene by Pseudomonas spp. *J. Appl. Sci. Environ. Manage.*, 14, 2, 57–60, 2010.
75. Sivasankari, S. and Vinotha, T., *In Vitro* Degradation of plastics (Plastic Cup) Using micrococcus luteus and Masoniella Sp. *Sch. Acad. J. Biosci.*, 2, 2, 85–89, 2014, https://www.saspublishers.com/article/9017/10.36347/sajb.2014.v02i02.004.
76. Gu, J.D., Mitton, B., TE, F., Mitchell, R., Microbial degradation and deterioration of polymeric materials, in: *Uhlig Corrosion Handbook*, 2nd ed, W. Revie (Ed.), Wiley, New York, 2000.
77. Chee, J.Y., Yoga, S.S., Lau, N.S., Ling, S.C., Abed, R.M.M., Sudesh, K.L., Bacterially produced polyhydroxyalkanoate (PHA): Converting renewable

resources into bioplastics, in: *Current Research, Technology and Education Topics in Applied Microbiology and Microbial Biotechnology*, A. Méndez-Vilas (Ed.), Formatex Research Center, Spain, 2010.
78. Hugenholtz, P., Goebel, B., Pace, N., Impact of culture-independent studies on the emerging phylogenetic view of bacterial diversity. *J. Bacteriol.*, 180, 4765–4774, 1998, https://doi.org/10.1128/jb.180.18.4765-4774.1998.
79. Erlandsson, B., Karlsson, S., Albertsson, A.-C., The mode of action of corn starch and a pro-oxidant system in LDPE: influence of thermo-oxidation and UV-irradiation on the molecular weight changes. *Polym. Degrad. Stab.*, 55, 2, 237–245, 1997, https://doi.org/10.1016/S0141-3910(96)00139-5.
80. Thew, C.X.E., Lee, Z.S., Srinophakun, P., Ooi, C.W., Recent Advances and Challenges in Sustainable management of plastic waste using biodegradation approach. *Bioresour. Technol.*, 374, 128772, 2023, https://doi.org/10.1016/j.biortech.2023.128772.
81. Crystal Thew, X.E., Lo, S.C., Ramanan, R.N., Tey, B.T., Huy, N.D., Chien Wei, O., Enhancing plastic biodegradation process: strategies and opportunities. *Crit. Rev. Biotechnol.*, 44, 3, 1–18, 2023, doi:https://doi.org/10.1080/07388551.2023.2170861.
82. Ru, J., Huo, Y., Yang, Y., Microbial Degradation and valorization of Plastic Wastes. *Front. Microbial.*, 11, 442, 2020, https://doi.org/10.3389/fmicb.2020.00442.
83. Christova, N., Tuleva, B., Lalchev, Z., Jordanova, A., Jordanov, B., Rhamnolipid biosurfactants produced by *Renibacterium salmoninarum* 27BN during growth on n-hexadecane. *Z. Naturforsch. C. J. Biosci.*, 59, 1–2, 70–74, 2004, https://doi.org/10.1515/znc-2004-1-215.
84. Wang, W., Cai, B., Shao, Z., Oil degradation and biosurfactant production by the deep sea bacterium Dietzia maris As-13-3. *Front. Microbiol.*, 5, 711, 2014, https://doi.org/10.3389/fmicb.2014.00711.
85. Ghatge, S., Yang, Y., Ahn, J.-H., Hur, H.-G., Biodegradation of polyethylene: a brief review. *Appl. Biol. Chem.*, 63, 27, 2020, https://doi.org/10.1186/s13765-020-00511-3.
86. Otake, Y., Kobayashi, T., Asabe, H., Murakam, I.N., Ono, K., Biodegradation of low density polyethylene, polystyrene, polyvinyl chloride, and urea formaldehyde resin buried under soil for over 32 years. *J. Appl. Polym. Sci.*, 56, 1789–1796, 1995, https://doi.org/10.1002/app.1995.070561309.
87. Danso, D., Chow, J., Streit, W.R., Plastics: environmental and biotechnological perspectives on microbial degradation. *Appl. Environ. Microbiol.*, 85, 1–14, 2019, https://doi.org/10.1128/AEM.01095-19.
88. Restrepo-Florez, J.M., Bassi, A., Thompson, M.R., Microbial degradation and deterioration of polyethylene-a review. *Int. Biodeterior. Biodegrad.*, 88, 83–90, 2014, https://doi.org/10.1016/j.ibiod.2013.12.014.
89. Sen, S.K. and Raut, S., Microbial degradation of low density polyethylene (LDPE): a review. *J. Environ. Chem. Eng.*, 3, 462–473, 2015, https://doi.org/10.1016/j.jece.2015.01.003.

90. Tomita, K., Kuroki, Y., Nagai, K., Isolation of thermophiles degrading poly (L-lactic acid). *J. Biosci. Bioeng.*, 87, 752–755, 1999, https://doi.org/10.1016/S1389-1723(99)80148-0.
91. Uchida, H., Nakajima-Kambe, T., ShigenoAkutsu, Y., Nomura, N., Tokiwa, Y., Nakahara, T., Properties of a bacterium which degrades solid poly (tetramethylene succinate)-coadipate, a biodegradable plastic. *FEMS Microbiol. Lett.*, 189, 25–29, 2000, https://doi.org/10.1111/j.1574-6968.2000.tb09201.x.
92. Kim, M. and Park, S., Degradation of poly (L-lactide) by a mesophilic bacterium. *J. Appl. Polym. Sci.*, 117, 67–74, 2010, https://doi.org/10.1002/app.31950.
93. Teeraphatpornchai, T., Nakajima-Kambe, T., Shigeno-Akutsu, Y., Nakayama, M., Nomura, N., Nakahara, T., Uchiyama, H., Isolation and characterization of a bacterium that degrades various polyester-based biodegradable plastics. *Biotechnol. Lett*, 25, 23–28, 2003, https://doi.org/10.1023/a:1021713711160.
94. Hayase, N., Yano, H., Kudoh, E., Tsutsumi, C., Ushio, K., Miyahara, Y., Tanaka, S., Nakagawa, K., Isolation and characterization of poly (butylene succinate-co-butylene adipate)- degrading microorganism. *J. Biosci. Bioeng.*, 97, 131–133, 2004, https://doi.org/10.1016/S1389-1723(04)70180-2.
95. Lee, S. and Kim, M., Isolation of bacteria degrading poly (butylenesuccinate-co butylene adipate) and their lip A gene. *Int. Biodeterior. Biodegrad.*, 64, 184–190, 2010, https://doi.org/10.1016/j.ibiod.2010.01.002.
96. Sekiguchi, T., Sato, T., Enoki, M., Kanehiro, H., Uematsu, K., Kato, C., Isolation and characterization of bidegradable plastic degrading bacteria from deep-sea environments. *Rep. Res. Dev.*, 11, 3341, 2010.
97. Koutny, M., Amato, P., Muchova, M., Ruzicka, J., Delort, A.M., Soil bacterial strains ableto grow on the surface of oxidized polyethylene film containing prooxidant additives. *Int. Biodeterior. Biodegrad.*, 63, 1, 354–357, 2009, https://doi.org/10.1016/j.ibiod.2008.11.003.
98. Iiyoshi, Y., Tsutsumi, Y., Nishida, T., Polyethylene degradation by lignin-degrading fungi and manganese peroxidise. *J. Wood Sci.*, 44, 3, 222–229, 1998, https://doi.org/10.1007/BF00521967.
99. Asiandu, A.P., Wahyudi, A., Sari, S.W., Biodegradation of Plastic Waste using Fungi: A Review. *Curr. Res. Environ. Appl. Mycol. (J. Fungal Biol.)*, 11, 1, 1–15, 2021.
100. Ganesh, P., Dineshraj, D., Yoganathan, K., Production and screening of depolymerizing enzymes by potential bacteria and fungi isolated from plastic waste dump yard sites. *Int. J. Appl. Res.*, 3, 693–695, 2017.
101. Al-Obaidi, J.R., Ability of endophytic fungi isolated from *Nephenthes ampullaria* to degrade polyurethane. *Malays. J. Microbiol.*, 13, 172–179, 2017.
102. Russell, J.R., Huang, J., Anand, P., Kucera, K., *et al.*, Biodegradation of polyester polyurethane by endophytic fungi. *Appl. Environ. Microbiol.*, 77, 6076–6084, 2011, https://doi.org/10.1128/AEM.00521-11.

103. Cai, Z., Li, M., Zhu, Z., Wang, X., Huang, Y., Li, T., Gong, H., Yan, M., Biological Degradation of Plastics and Microplastics: A recent perspective on associated mechanism and influencing factors. *Microorganism*, 11, 7, 1661, 2023, https://doi.org/10.3390/microorganisms11071661.
104. Wróbel, M., Szymańska, S., Kowalkowski, T., Hrynkiewicz, K., Selection of microorganisms capable of polyethylene (PE) and polypropylene (PP) degradation. *Microbiol. Res.*, 267, 127251, 2023, https://doi.org/10.1016/j.micres.2022.127251.
105. Kotova, I.B., Taktarova, Y.V., Tsarkelova, E.A., Egorova, M.A., Bubhnor, I.A., Malakhova, D.V., Shivinkina, L.I., Sokolova, T.G., Osmolovskaya, E.A.M., Microbial Degradation of Plastic and Approach to make it more efficient. *Microbiology*, 90, 671–701, 2021, https://doi.org/10.1134/S0026261721060084.
106. Riudavets, J., Salas, I., Pons, M.J., Damage characteristics produced by insect pests in packaging film. *J. Stored Prod. Res.*, 43, 564–570, 2007, https://doi.org/10.1016/j.jspr.2007.03.006.
107. Yang, J., Yang, Y., Wu, W.M., Zhao, J., Jiang, L., Evidence of polyethylene biodegradation by bacterial strains from the guts of plastic-eating waxworms. *Environ. Sci. Technol.*, 48, 13776–13784, 2014, https://doi.org/10.1021/es504038a.
108. Billen, P., Khalifa, L., Van Gerven, F., Tavernier, S., Spatari, S., Technological application potential of polyethylene and polystyrene biodegradation by macro-organisms such as mealworms and wax moth larvae. *Sci. Total Environ.*, 735, 139521, 2020, https://doi.org/10.1016/j.scitotenv.2020.139521.
109. Kundungal, H., Gangarapu, M., Sarangapani, S., Patchaiyappan, A., Devipriya, S.P., Efficient biodegradation of polyethylene (HDPE) waste by the plastic-eating lesser waxworm (*Achroia grisella*). *Environ. Sci. Pollut. Res. Int.*, 26, 18509–18519, 2019, https://doi.org/10.1007/s11356-019-05038-9.
110. Peng, B.-Y., Chen, Z., Chen, J., Yu, H., Zhou, X., Criddle, C.S., Wu, W.-M., Zhang, Y., Biodegradation of polyvinyl chloride (PVC) in Tenebrio molitor (Coleoptera: Tenebrionidae) larvae. *Environ. Int.*, 145, 106106, 2020, https://doi.org/10.1016/j.envint.2020.106106.
111. Brandon, A.M., Gao, S.H., Tian, R., Ning, D., Yang, S.S., Zhou, J., Wu, W.-M., Criddle, C.S., Biodegradation of polyethylene and plastic mixtures in mealworms (larvae of Tenebrio molitor) and effects on the gut microbiome. *Environ. Sci. Technol.*, 52, 6526–6533, 2018, https://doi.org/10.1021/acs.est.8b02301.
112. Peng, B.-Y., Su, Y., Chen, Z., Chen, J., Zhou, X., Benbow, M.E., Criddle, C.S., Wu, W.-M., Zhang, Y., Biodegradation of polystyrene by dark (Tenebrio obscurus) and yellow (Tenebrio molitor) mealworms (Coleoptera: Tenebrionidae). *Environ. Sci. Technol.*, 53, 5256–5265, 2019.
113. Yin, C.-F., Xu, Y., Zhou, N.-Y., Biodegradation of polyethylene mulching films by a co-culture of *Acinetobacter sp.* strain NyZ450 and Bacillus sp. strain

NyZ451 isolated from T*enebrio molito*r larvae. *Int. Biodeterior. Biodegrad.*, 155, 105089, 2020, https://doi.org/10.1016/j.ibiod.2020.105089.
114. Gu, J.D., Microbiological deterioration and degradation of synthetic polymeric materials: recent research advances. *Int. Biodeterior. Biodegrad.*, 52, 69–91, 2003. https://doi.org/10.1016/S0964-8305(02)00177-4.
115. Saritha, B. and Ramadevi, O.K., Role of Microbes and their enzymes in biodegradation of Plastic waste, in: *Innovations in Microbiology and Biotechnology*, vol. 4, pp. 24–39, BP International, West Bengal, 2022.
116. Jyoti, K., Khatri, M., Arya, S.K., Recent insight into enzymatic degradation of plastics prevalent in the environment: A mini – review. *Chem. Eng. Technol.*, 2, 100083, 2021, https://doi.org/10.1016/j.clet.2021.100083.
117. Ho, K.-L.G., Pometto III, A.L., Hinz, P.N., Effects of temperature and relative humidity on polylactic acid plastic degradation. *J. Environ. Polym. Degrad.*, 7, 2, 83–92, 1999, https://doi.org/10.1023/A:1021808317416.
118. Auras, R., Harte, B., Selke, S., An overview of polylactides as packaging materials. *Macromol. Biosci.*, 4, 9, 835–864, 2004, https://doi.org/10.1002/mabi.200400043.
119. Henton, D.E., Gruber, P., Lunt, J., Randall, J., Polylactic acid technology, in: *Natural fibers, biopolymers, and biocomposites*, K.A. Mohanty, M. Misra, L.T. Drzal (Eds.), CRC press, Boca Raton, 2005.
120. Tokiwa, Y. and Calabia, B.P., Review degradation of microbial polyesters. *Biotechnol. Lett*, 26, 15, 1181–1189, 2004, https://doi.org/10.1023/B:-BILE.0000036599.15302.e5.
121. Tokiwa, Y., Calabia, B.P., Ugwu, C.U., Aiba, S., Biodegradability of plastics. *Int. J. Mol. Sci.*, 10, 9, 3722–3742, 2009, https://doi.org/10.3390/ijms10093722.
122. Kale, G., Kijchavengkul, T., Auras, R., Rubino, M., Selke, S.E., Singh, S.P., Compostability of bioplastic packaging materials: an overview. *Macromol. Biosci.*, 7, 3, 255–277, 2007, https://doi.org/10.1002/mabi.200600168.
123. Yamada-Onodera, K., Mukumoto, H., Katsuyaya, Y., Saiganji, A., Tani, Y., Degradation of polyethylene by a fungus, *Penicillium simplicissimum YK. Polym. Degrad. Stab.*, 72, 2, 323–327, 2001, https://doi.org/10.1016/S0141-3910(01)00027-1.
124. Hadad, D., Geresh, S., Sivan, A., Biodegradation of polyethylene by the thermophilic bacterium Brevibacillus borstelensis. *J. Appl. Microbio.*, 98, 1093–1100, 2005. https://doi.org/10.1111/j.1365-2672.2005.02553.x.
125. Amobonye, A., Bhagwat, P., Singh, S., Pillai, S., Plastic biodegradation: Frontline microbes and their enzymes. *Sci. Total Environ.*, 759, 143536, 2021. https://doi.org/10.1016/j.scitotenv.2020.143536.
126. Knott, B.C., Erickson, E., Allen, M.D., Gado, J.E., Graham, R., Kearns, F.L., Pardo, I., Topuzlu, E., Anderson, J.J., Austin, H.P., *et al.*, Characterization and engineering of a two-enzyme system for plastics depolymerisation. *Proc. Natl. Acad. Sci. U.S.A.*, 117, 25476–25485, 2020, https://doi.org/10.1073/pnas.2006753117.

6

Green Remediation of Plastic Pollution through Bio-Nanomaterials

Prabhat Kumar Patel

School of Science, Harbin Institute of Technology (Shenzhen), Shenzhen, China

Abstract
A class of substances known as plastics has come to dominate every aspect of contemporary life. Due to their long-chain hydrocarbon architecture, which is resilient and resistant to biological degradation, plastics are difficult to eliminate from the environment. Globally, plastic waste has been found in our surroundings, endangering aquatic organisms, contaminating the food chain, and posing risks to human health. The long-chain hydrocarbon structure of plastics makes them resilient and resistant to biological degradation, rendering them difficult to eliminate from the environment. Worldwide, plastic waste is pervasive, threatening aquatic life, contaminating the food chain, and posing serious risks to human health. We can mitigate the negative effects of pollution from plastics on the environment by developing innovative solutions for the management of plastic waste. For this purpose, this chapter will provide in-depth insights into the applications of bio-nanomaterials (BNMs) in plastic eradication, the safety and challenges associated with their use, the environmental impacts of plastics, and the recyclability of plastic materials.

Keywords: Green nanomaterials, plastics, sustainability, treatment, toxic pollutants

6.1 Introduction

As of 2015, approximately 6.3 billion tons of plastic waste had been generated worldwide, with about 79% of it resulting from dumping, disposal,

Email: pprabhat1991@gmail.com

Neha Agarwal, Brijesh Pare, Vijendra Singh Solanki and Sankara Rao Miditana (eds.) Plastic Remediation: Transformative Pathways for a Rejuvenated Sustainable Environment, (147–170)

residential effluents, and manufacturing activities [1]. The immense quantity of plastic pollution poses an imminent threat to human health and food security as well as to marine, land, and climatic ecosystems [2]. Rivers, one of the primary sources of aquatic pollution, transport between 0.8 and 2.7 million metric tons of plastic from land to the oceans each year [3]. The decomposition of plastics occurs in an unregulated manner in the environment, resulting in the formation of microplastics (MPs) (with a diameter of less than 5 mm) and nanoplastics (with a diameter of less than 1 mm) [4]. These plastic fragments can carry a variety of biological and chemical contaminants to other areas, potentially threatening ecosystem services and functions [5–8]. Consequently, nano-remediation is gaining popularity as an innovative approach that could revolutionize the management and remediation of hazardous and polluting substances [8]. The unique physicochemical, catalytic, electrical, optical, mechanical, and environmental properties of nanomaterials (NMs) are the focus of extensive research. They therefore have a great deal of promise and provide a lot of opportunities for a variety of potential uses in the areas of biomedicine, biosensors, pharmaceuticals, catalysis, drug administration, medical services, beauty products, home, ophthalmic, chemical, and antimicrobials [9].

As shown in Figure 6.1a, NMs are tiny particles ranging in size from 1 to 100 nm; however, size alone is not sufficient to define them. These materials can be incidental, engineered, or naturally occurring [10]. NMs exhibit various structural configurations, including 0D, 1D, 2D, and 3D architectures. In 0D structures, all three dimensions are confined to the nanometer scale. In 1D NMs, two dimensions are at the nanoscale, while the third extends beyond the nanoscale. Similarly, 2D nanostructures exhibit nanometer-scale characteristics only within a single plane. Interestingly, this category also includes 3D particles, even though they exceed 100 nm in size (Figure 6.1b). NMs can be synthesized through three primary methods: chemical, biological, and environmental [11]. Recently, there has been an increased use of NMs in ecological cleanup due to their potential as sorbents and catalysts [12]. They are effective in breaking down redox-sensitive contaminants due to their unique redox properties and key characteristics such as size, solubility, dimensions, surface charge, and interfacial chemical structure [13]. Various types of nanostructures have been developed for air purification and wastewater treatment. This approach is considered reliable and cost-effective, as NMs offer significantly larger surface areas and higher adsorption capacities [14]. Furthermore, NMs with well-engineered surface modifications and tunable properties offer enhanced functionalities and notable advantages in addressing environmental contaminants. While some NMs sequester pollutants, others

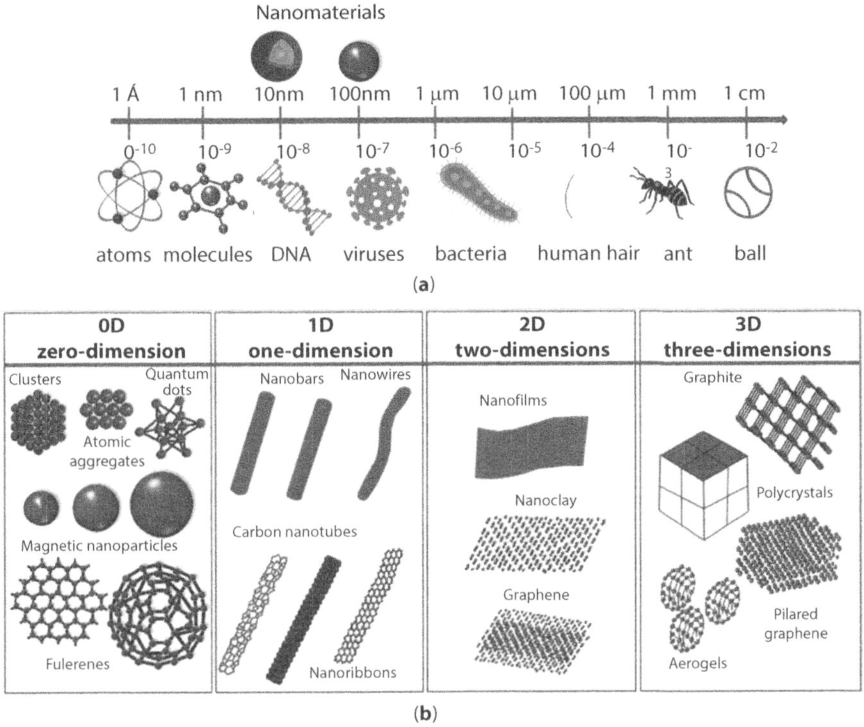

Figure 6.1 (a) NP size in contrast to conventional substances. (b) Simplified representation of distinct dimensioned nanocrystals [18].

actively degrade or remove them [15]. Unfortunately, there are certain drawbacks associated with NMs, including high costs, potential toxicity, challenges in reuse, and interactions with other media. Moreover, unintended exposure or accidental release of NMs poses significant health risks [16]. A variety of reliable, cost-effective, and eco-friendly NMs have been developed for the purification of air and water contaminants, offering a wide range of applications [17]. Nevertheless, NMs must be safe to use and exhibit strong sorption capacity and selectivity, particularly to ensure the efficient removal and recycling of contaminants at low concentrations [18]. Recent studies have demonstrated that NMs are capable of meeting most of these criteria. Additionally, their significance, types, sources, applications, and mechanisms of action have been extensively explored. Given the extensive research on innovative approaches combining biological remediation and nanoparticles (NPs) for MP removal, the primary objective of this chapter is to present bioremediation strategies that utilize s to enhance the elimination of MPs from the environment [19].

6.2 Biobased NMs

Nanotechnology has facilitated the utilization of NMs, which typically range from 1 to 100 nm in size. NMs serve as effective adsorbents for achieving desired goals due to their small size, large surface area, and high reactivity. Various strategies, including chemical, physical, and biological methods, are employed to produce NMs. Owing to their unique physicochemical characteristics, NMs attract significant interest from researchers in several fields such as drug delivery, agriculture, and environmental remediation. As a material is scaled down to the nanoscale, its surface area per unit mass increases significantly. Consequently, it can interact with a greater number of surrounding substances, thereby enhancing its reactivity. Owing to their complex morphologies and topologies, which facilitate the removal of diverse pollutants, NMs are widely employed in environmental remediation. Moreover, NMs have the unique ability to penetrate contaminated areas, an advantage not achievable with microparticles. In the early stages, NMs synthesized through physicochemical processes were primarily used in the electronics industry due to their exceptional electrical and thermal properties [20]. Nevertheless, these NMs proved highly effective in removing both inorganic and organic pollutants, owing to their increased surface area, ease of modification, and strong microbial disinfection capabilities. As a result, NMs and their different derivatives hold significant potential for emerging applications in water treatment [21]. However, chemically synthesized NMs, such as titanium oxide and graphene, have also been utilized in a variety of applications. Some metallic NMs are known to be persistent and resistant to degradation, potentially leading to several environmental concerns [22]. However, numerous drawbacks are also associated with the chemical processes used to fabricate NMs, including the risk of contamination from precursor compounds and the release of hazardous byproducts. Thus, there is a great demand for environmentally friendly, economically viable, and highly efficient processes for the synthesis of nanoscale materials.

There is a growing need to develop biologically synthesized substances (biobased NMs) capable of removing pollutants from environmental settings. Such NMs can be derived from viruses, plants, fungi, and bacteria and are often referred to as "bio-nano factories." Due to their microscopic size and suitability for heavy metal remediation, biobased NMs exhibit high effectiveness [23]. A wide range of biological sources can be used to produce these NMs, including microorganisms, plants and their derivatives, farm and agro-industrial waste, macromolecules, and aquatic organisms such as microalgae, micro-seaweeds, phytoplankton, zooplankton, and diatoms. Substances derived from biological sources are known as biobased NMs. Both organic and inorganic NMs can be naturally synthesized by living organisms. The interaction between biological species and inorganic materials has led to the synthesis of NMs through various intracellular and extracellular pathways [24]. When NMs interact with small biological linkers, bioactive compounds, cell membranes, DNA, and proteins, they generate a variety of surface properties influenced by dynamic biophysicochemical interactions and colloidal forces.

Biomaterial monomers are biopolymers that, upon polymerization, form biobased materials. In this context, the term "bio" refers to imparting life to polymers by incorporating living organisms or biodegradable substances derived from natural sources such as bacteria and crops. Several types of bacteria are involved in the production of biopolymers. Many researchers have explored the functional approaches for producing microbial biopolymers as part of understanding or enhancing their resistance mechanisms [25]. Natural processes such as microbial and enzymatic activities are known to degrade these types of compounds, ensuring their eventual reabsorption into the environment. NMs derived from microbe-based fabrication are produced at a relatively slower rate compared to those synthesized from plants, which may be considered a disadvantage. Moreover, considering their availability, high medicinal value, safe management, and existence of metabolites that can support the reduction process can be advantages for plant-derived NMs.

6.3 Micro/Nanoplastics' Detrimental Effects on the Environment

According to recent studies, micro/nanoplastics are emerging plastic contaminants in different environmental matrices. Numerous creatures may be impacted by these plastic particles directly and indirectly. The macromolecular architecture of plastics contributes to their widespread non-reactivity in biochemical processes [26]. However, due to their considerable surface-area-to-volume ratio and hydrophilic properties, micro- and nanoplastics (MNPs) may act as carriers, transporting additional environmental contaminants such as hydrophobic persistent organic pollutants (POPs) and infectious microorganisms from the surrounding media into the body [27]. Plastics are more persistent in aquatic environments than on land and are difficult to eradicate. Larger fragments break down into micro- and nanoplastics, increasing their accessibility and causing detrimental effects on ecosystems [28]. These small fragments can also irritate, abrade, and clog the gastrointestinal tract. There is ongoing research on the potential cellular absorption and toxic effects of MNPs. Nanosized polystyrene (PS) particles may readily penetrate lipid layers through various mechanisms, including pinocytosis, the caveolin-mediated pathway, clathrin-mediated endocytosis, and phagocytosis. In addition to disrupting membrane protein integrity, these penetrated PS fragments also impair cellular function. Ultimately, PS causes significant alterations in membrane structure, reducing molecular diffusion and membrane flexibility. It has recently been discovered that PS NPs exist in the cerebral cortex of several different animals. According to Almutairi *et al.*, these findings imply that nanoplastics can penetrate the blood/brain barrier, which is an exceptionally specific permeation layer necessary to shield the brain from peripheral toxic substances and to maintain the equilibrium for neural function. MNPs can enter the human body through ingestion, direct skin contact, inhalation, and the food chain [29].

6.3.1 Through Skin Contact

Although MPs are generally considered unable to penetrate the skin's protective barrier, their accumulation on the surface can still increase the risk of contamination. For example, using cosmetics such as moisturizers and cleansers containing MPs can elevate the risk of polyethylene (PE) exposure. Improper use or disposal of mobile screen protectors can release MPs into the environment, which may subsequently come into contact with human skin. Children, in particular, may be exposed to MPs on surfaces while crawling, playing, or moving around. Certain common plastic ingredients, such as phthalates, triclosan (TCS), bisphenols (BPs), and brominated flame retardants (BFRs), can be absorbed when MPs are exposed topically [30]. MNPs can enter the human body through three primary routes and may lead to various harmful effects, including oxidative stress, carcinogenesis, translocation within the body, and neurodegeneration. Limited uptake of 0.2% poly (lactic-co-glycolic acid) (PLGA) MPs with a diameter of 3 μm within human colonic mucous membranes is currently being assessed *in vitro*. In patients with severe conditions such as inflammatory bowel disease (IBD), the colonic mucosal membrane showed increased intestinal permeability and faster transit (0.45% vs. 0.2%)

compared to healthy controls. The capacity of NPs to pierce the intestinal membrane and cells is a unique problem. Researchers have been examining the consequences of oral intake of polystyrene NPs (NPs) for many years. Furthermore, both *in vitro* and *in vivo*, they have investigated the movement of intestinal absorption of generated NPs (such as ZnO, TiO_2, and Ag) [31]. In addition to altering oxidative damage, fatty acids, and energy utilization, MPs also have neurotoxicological consequences. These actions exacerbate the state of cell toxicity just like they do for human liver cells. Fragments of nanopolystyrene can slow down development and lower the amount of chlorophyll. MNPs may have detrimental effects on cholinergic neurons. It is likely that NPs (NPs) exert toxic effects on nematodes and interact with neurons [32].

6.3.2 Through Inhalation

Most airborne PE, PS, and PET fibers and fragments range in size from 8000 and 10 μm. The majority of airborne MPs originate from roads, which contribute approximately 84% of the total. In Paris, the average concentration of MP fibers in the air has been reported to be 0.9 fibers/m^3 indoors and 5.4 fibers/m^3 outdoors. In Shanghai, outdoor air contains approximately 1.42 particles/m^3, with particle sizes ranging from 23 to 5000 μm. Considering both respiratory and oral exposure, the estimated annual ingestion of MNPs ranges from 74,000 to 121,000 particles. The MPs detected in lung tissue are generally smaller than those found in the external environment. This highlights the potential damage to our bodies and provides further evidence that people may inhale MPs [33]. NP contamination can occur through spontaneous inhalation, dermal absorption from plastic objects, or *via* ingestion and assimilation [34]. Inhalation is likely the most critical route of exposure to aerosols containing NPs in workplace settings, although skin exposure is also possible through contaminated air, water, and products such as cosmetics, cleaning solutions, and materials with pore spaces that contain NPs [30].

6.3.3 Through Oral Intake

The human body absorbs between 39,000 and 52,000 MP fragments per person per year, depending on food intake, while seafood consumption exposes Europeans to approximately 11,000 MP fragments per person per year [35]. It is likely that MNPs are widely distributed in soil, particularly in agricultural areas. Particularly when negatively charged, MPs can enter a plant's water transport system and reach the stems, roots, leaves, and flowers.

When MPs enter agricultural systems through fertilizers, sewage effluent, or plastic mulching, they can contaminate food, thereby increasing the risk of human exposure [36]. NPs are most likely to be absorbed through ingestion, typically by consuming seafood or contaminated water. The compelling scientific evidence of nanoplastic (NP) intake, accumulation, and trophic transfer in marine organisms suggests that NPs may enter the human alimentary tract and ultimately cause harm [30]. The following are additional ways in which micro- and nano-sized polystyrene (PS) fragments impact ecological systems: the accumulation of latex micro- and nanoparticles (39–500 nm) in the testes, gills, intestinal tract, bloodstream, and liver of *Oryzias latipes* was found to reduce its lifespan. 20-nm PS particles disrupted metabolism and increased the formation of reactive oxygen species (ROS) in Chlorella and Scenedesmus species. Canesi *et al.* found that the presence of PS with a size of 200 nm led to the induction of proapoptotic pathways in *Mytilus galloprovincialis* species [37]. According to Wegner *et al.*, *Mytilus edulis* has been observed to produce pseudo feces and exhibit impaired purification operation when exposed to 30-nm-sized PS fragments [38]. The gastrointestinal tracts of *Mytilus edulis* and *Crassostrea virginica* have been found to accumulate 100-nm PS fragments. In *Artemia franciscana* larvae, the presence of 40-nm-sized PS hindered mobility and affected the maturation phenomena [39]. Researchers have shown that 64-nm PS causes a massive influx of neutrophils into the lungs of rats. According to recent studies, zebrafish exposed to low-density polyethylene (LDPE) particulates for 21 days did not exhibit any lasting effects from the exposure [40]. Furthermore, organisms may experience adverse effects due to the unintentional release of monomers and other chemical additives during polymer degradation. Numerous severe consequences, including early mortality, inflammation, slow or impeded developmental progress, decreased vitality, impaired nourishing operation, oxidative stress, compromised immunity, neural suppression, and behavioral malformation, can result from microorganisms' susceptibility to MNP particulates [41]. Due to their impacts on both human health and the environment, MNPs in agricultural soil are a major cause for concern. Recent studies have shown that polyethylene MP residues, as well as fragments from biodegradable plastics, negatively affect plant growth and crop fertilization both above and below the soil surface. Excessive leftover plastics also damage soil water opportunities, nutritional movement, and soil integrity, which lowers agricultural yields [42]. MNPs have the potential to linger in the soil, build up, and ultimately impact the viability and diversification of soil creatures, posing an environmental danger to the soil environment [43]. Moreover, plastic particles accelerate the breakdown of carbon and nitrogen, which reduces soil nutrients, increases greenhouse gas emissions, and causes soil water absorption [44]. MNPs have been shown to have potentially harmful effects on the gastrointestinal systems of invertebrates and can be transferred through the alimentary tract. However, it remains challenging to characterize these small particles and to understand how MNPs interact with living organisms [45]. Therefore, further research is needed to fully understand the effects of MNP contaminants on human health, aquatic ecosystems, and agricultural environments. One of the most intriguing aspects of this field is the relationship between MPs' persistence, interaction mechanisms, and degradation processes.

6.4 Source of BNMs

6.4.1 Chitin

The exoskeletons of crustaceans, fungi, yeast, and plants can be utilized as a source of chitin. Its lack of solubility and chemical reactivity is comparable to those of cellulose. Chitin is distinguished from chitosan by the presence of N-acetyl-glucosamine units. When the degree of acetylation is greater than 50%, the material is considered chitin; when it is less than 50%, it is classified as chitosan. Chitin and chitosan have a wide range of applications, including in medicine, food safety, agriculture, and environmental remediation (such as water purification) [46]. Chitosan is better suited for biological applications due to its solubility in water and various organic solvents. Chitin- and chitosan-based NMs have been synthesized using various methods, including cross-linking, hydrothermal processing, precipitation, and emulsification. In recent years, chitin-based NMs have been explored for a variety of applications, showing significant potential in environmental settings. Additionally, numerous biobased NMs derived from alginates, casein, starch, and soy protein have been studied for their diverse uses in biotechnology [20]. Nanocarriers are also employed in the efficient delivery of medications, including curcumin to the intended sites. Several protein polymers, such as silk, rice bran, albumin, and others, as well as polysaccharide-based BNMs, such as starch, chitosan, cellulose, and alginate, are frequently utilized for drug delivery applications [47].

6.4.2 Cellulose

Cellulose, a major organic compound and biopolymer, is distinct from other polymers. To produce microfibrillated cellulose (MFC), which ranges in size from 20 to 50 nanometers, 36 cellulose molecules come together to form structures known as elementary fibril units. The cellulosic chains are compacted in a way that resembles nanocrystalline rods, providing structural stability. Generally, nanocellulose is classified into two main types: (i) cellulose nanocrystals (CNCs) and (ii) cellulose microfibrils. Additionally, cellulose-based particles can be categorized into nine distinct types: algal cellulose nanocrystals (AC), microfibrillated cellulose (MFC), bacterial cellulose particles (BC), tunicate cellulose nanoparticles (t-CNCs), microcrystalline cellulose (MCC), plant fibers (PFs), and wood fibers (WFs).

The crystalline nature, hydrophilic characteristics, large surface area, and other properties associated with cellulose are also present in nanocellulose [48]. Nanocellulose can be extracted using a top-down method, a physical method (ultrasound, microwaves, and gamma rays), a mechanical method (crushing and cryo-crushing), or a chemical method (ionic solutions, alkaline and acidic hydrolysis, and so on). The isolation process involves two main stages: the first stage focuses on the purification and preparation of the source material, while the second stage involves extracting the pure cellulosic components, such as microfibrillar and/or crystalline elements.

6.4.3 Algae

Greener production methods or the use of BNMs are considered the optimal choice due to the importance of environmental protection. Algae are used in the fabrication of NMs in combination with other techniques [49]. Algae have been used as living cell production facilities, and their extracts are being utilized as natural reducing agents to produce NMs in more sustainable ways. Using algae as live cell factories is considered a clean and green method for the effective synthesis of various metallic NMs. Nowadays, marine polysaccharides are employed in several nanobiotechnological domains, such as tissue engineering, gene and medication delivery, wastewater treatment, and biosensors. Numerous marine creatures, especially algae, contain polysaccharides such as carrageenan, fucoidan, agarose, fucoidan, and alginate. In nature, these polysaccharides are abundant, harmless, non-toxic, and biodegradable [50]. These characteristics make them crucial for the synthesis of NMs. Aquatic polysaccharides can be efficiently transformed into various nanoforms such as beads, nanofibers, sponges, and nanoparticles, for use in a wide range of nanobiotechnology applications [51]. While traditionally valued as sources of food, additives, and medications, marine algae are now receiving increased attention for their role in nanoparticle (NP) production processes. A recent study suggests that using algae to cap and stabilize NPs is a less hazardous, safer, more cost-effective, and easier-to-handle approach. According to the study, natural polysaccharides derived from algae function were effective as capping and stabilizing agents during NP synthesis [52]. Furthermore, phytochemicals, dyes, and proteins found in algae are excellent bio-fabricators for the synthesis of NPs [53]. Algae are a suitable option for environmentally friendly NP production, as they can accumulate metals and convert them into nanoparticles.

6.4.4 Bacteria

Currently, various mechanisms have been developed for water remediation, including the use of UV radiation and chlorination. However, compared to other methods, membrane-based approaches are significantly simpler and more straightforward. They are also considered environmentally friendly, generate fewer contaminants, and can be scaled up. Additionally, they tend to disinfect water and effluents while consuming less energy [54]. Several biological sources contribute to the production of cellulose membranes or NMs, and one of these sources is BC. BC is a type of cellulose that is easily processed and produced through fermentation or bacterial metabolism, similar to plant-derived cellulose [55]. Due to their advantageous properties, they are a suitable alternative for the development and practical application of biobased membranes and NMs for environmental remediation. Due to its high rate of glucose molecule production, nearly 200,000 molecules per second *Gluconacetobacter xylinum* is commonly used in cellulose production. Magnetotactic bacteria produce various magnetic NMs using mineralization proteins. This biological method yields uniformly sized and shaped nanocrystals and highlights the potential of using proteins for the synthesis of additional magnetic NPs [56].

6.5 Applications of Bio-Nanomaterials for MNP Removal

The term BNMs refers to NPs that have biological origins or inorganic particles combined with organic substances. Due to their negligible adverse effects, low ecological impact, biocompatibility, and biodegradability, BNMs are currently being employed for the effective remediation of MNPs [57, 58]. BNMs are compounds derived from biological sources or engineered to mimic biological processes at the microscopic level.

A wide range of organisms including bacteria, algae, plants, and animals are capable of producing them. Bio-nanorobots have demonstrated superior effectiveness and specificity toward NPs due to the active moieties present on their surfaces [59]. Silver NPs produced *via* biosynthetic procedures are additionally employed for cleaning out MPs from the environment because they agglomerate into tiny fragments that are readily captured by filtering. Two types of bacteria *Pseudomonas* and *Bacillus* are known to produce lipases and proteases, which can break down various plastic materials such as polyethylene, polystyrene, and polypropylene, resulting in less hazardous byproducts. Modern science is opening new possibilities by enhancing bionanostructures, computational tools, and machine learning (ML) techniques to enable the controlled degradation of MP contaminants from the environment, food products, and skincare formulations. Cellulose nanocrystals (CNCs) generated from wood convert MPs into valuable products such as power storage devices, reused plastic layers, and therapeutic instruments. MPs composed of polyethylene terephthalate (PET) can be converted into biofuel using green BNMs, which serve to reduce MPs and act as a fuel replacement substrate through processes such as hydrothermal liquefaction. Researchers developed an aquatic purification membrane based on a non-permeable cellulose mesh [59]. Using a mold-encasing technique, they incorporated various polysaccharide nanocrystals, such as chitin nanocrystals (ChNCs), cellulose nanocrystals (CNCs), and 2,2,6,6-tetramethylpiperidine-1-oxyl (TEMPO)-oxidized cellulose nanofibers (T-CNF). The nano-crystal significantly increased the fabric's durability, modulus of elasticity, and hydrophilic properties. It also made the surface potential more adverse (favorable) for cellulose-based (chitin-based) alterations. While CNC and ChNC modifications enhanced the porousness of the membranes, T-CNF immobilization enhanced the rigidity in dry circumstances. All of the mentioned biological materials had the ability to eradicate MPs (<500 nm), whereas T-CNF was only competent to isolate fragments (<2 μm) [60]. Compared to other films found in retail outlets, lignin-zeolite hybrid microfiber films made by electrospinning offer superior flux and puncture speeds. Additionally, the uniformly dispersed zeolite NPs and a thermal treatment improve the mechanical attributes of the NP. The inclusion of 1 weight percent zeolite NPs drastically improved the modulus and durability. These BNMs have been effectively implemented to pre-filter for an array of MPs [61]. The hydrophobic magnetite NPs were reportedly generated through a substance derived from the aerial portions of the *Anthemis pseudocotula* herb. These NPs may eradicate 84% of MPs from clean water, 78% from soils, and 92–93% from saltwater with barely negative environmental effects. Lysozyme amyloid fibrils were one of the recently developed biological bio-flocculants. Because of its nanoscale dimension, the flocculant substances may be categorized as BNMs. This flocculant additive provides a substantially positively charged surface throughout a very broad pH span and an excellent turbidity mitigation effectiveness of 98.2% for distributed polystyrene MP and 97.9% for humic acid, respectively [62]. Because of its ease of the procedure, enhanced execution, reduced electricity expenses, and extended longevity, bionanoremediation has shown to be very effective and beneficial at the institutional stage. Although still a relatively new field, the biological cleanup of MPs using BNMs holds great promise, as numerous scientists are working toward commercial-scale implementation [60]. Due to their small size, high mobility, toxicity, and tendency to accumulate, the sustainable elimination of nanoplastics remains a significant challenge. Therefore, further research is needed to identify the key parameters and modifications that could enhance the effectiveness of nanoplastic removal.

6.6 Improved Recycled Plastic Using NMs

Recycling is one of the most important methods for reducing waste generated by everyday human activities. Plastic recycling involves converting discarded plastics into new products with distinct properties. However, recycled plastics often have limited market value, which restricts their widespread applications [63]. A report by the Organization for Economic Co-operation and Development (OECD) stated that global plastic production has risen significantly from 2 million tons in 1950 to 460 million tons in 2019. A striking fact reveals that only 8% of plastic waste is recovered through recycling, while approximately 50% ends up in landfills—underscoring the urgent need for improved waste management practices [64]. Plastic recycling has emerged as a promising solution, attracting attention for its potential to contribute to a sustainable future by reducing energy consumption, conserving valuable raw materials, and mitigating the environmental impact of plastic production [65]. As it was indicated previously, the properties of recycled plastic are inferior to those of original plastics. This is because various substances are added to the resins to obtain enhanced stability, strength, and durability [63]. Nanotechnology has thus offered an improved approach by incorporating nanofillers into polymeric matrices to create nanocomposites, which can help mitigate the negative effects of recycling on the physical and mechanical properties of materials. Incorporating nanoparticles (NPs) into a polymer matrix offers numerous advantages. Primarily, the strategic addition of nanofillers enhances the performance of the polymer by strengthening interfacial interactions, owing to the high surface area of the NMs. Moreover, even a small amount of these microscopic materials can significantly enhance the properties of polymers—without increasing cost, compromising their lightweight nature, or diminishing optical clarity, a combination of benefits that expands the potential of polymer innovation [66].

NPs are highly effective as additives, as only a small quantity is needed due to their exceptionally high surface area-to-volume (or mass) ratio. This property can help address several key challenges associated with plastic recycling. Therefore, it is essential to introduce additional functionalities to make plastic recycling more efficient and attractive. In this context, nano-additives offer two primary types of value addition during the recycling process. First, they can enhance or introduce specific electromagnetic properties such as electrical conductivity, thermal resistance, magnetic behavior, or radiation shielding. Second, they can improve mechanical properties, including hardness and tensile strength.

6.7 Deployment in Farming

The growth of nanotechnology and its applications in a variety of arenas have been observed in recent years. The agriculture system has been improved by nanotechnology through the production of new nano-agrochemicals (NACs). The use of NACs allows for precise application, minimizing undesirable biological effects and reducing excessive use. However, their use may still pose risks to both human health and the environment. BNMs, on the other hand, are generally considered to be less toxic and therefore receive greater attention [67]. They utilize waste crops as a substrate for biosynthesis and integrate into the circular bioeconomy. Besides this, nanobiochar is a comparatively new method for sustainable farming and soil usage. Biochar is commonly used in agroindustries, environmental remediation, and soil amendment. Nanobiochar, produced from animal-derived biomass, benefits from nano-enabled properties, making it a promising option for environmental management [68]. Agricultural wastes are also used to synthesize biobased NMs. Biobased nanofilms and nanopapers were fabricated using cellulose nanofibers (CNFs) derived from banana tree pseudostem (BTPT) waste and Eucalyptus kraft cellulose (EKC). These materials were evaluated for a variety of qualities, including their capacity to transmit light, durability, and their degradable nature [69]. A similar study was conducted using spent mushroom substrate (SMS), which contains both chitin and cellulose nanofibers. To explore its biological activity and potential applications in crop production, a chitin–cellulose nanofiber complex (CCNFC) was developed. When dispersed in water at low amounts, CCNFC significantly increased plant growth capabilities and resistance to plant infections. This study offers insight into the applicability of SMS in producing economically effective agricultural NMs [70]. Thus, it can be concluded that NMs synthesized from agricultural waste hold significant potential for use in agriculture, particularly as remedies against various biotic stresses.

6.8 Security, Safety Considerations, and Difficulties with BNMs

Using environmentally friendly methods and less hazardous substances without causing adverse effects on the surroundings is an important consideration in the manufacturing of NMs. Numerous microorganisms including yeasts, bacteria, microorganisms, and fungi contribute to the synthesis of highly active NMs [71, 72]. BNMs are often considered less hazardous due to their biological origin. However, during the early part of this era, some textile workers were reported to suffer from lung ailments caused by exposure to cotton spores associated with the use of cellulosic materials. Investigations suggest that cellulose fibers may lead to mesothelioma and other severe lung diseases. Due to their nanoscale size, these particles can penetrate biological barriers, potentially accumulating within cellular tissues. The increased surface area allows for additional interactions between the cell and its environment [73]. Moreover, a high surface area enhances contact with the biological surroundings, potentially leading to an increase in reactive oxygen species [74]. This highlights the need to address both environmental and individual safety concerns. Before the application of BNMs, several factors such as cytotoxicity, genotoxicity, immune response evaluation, oxidative stress, bacterial lipopolysaccharides (LPS), and endotoxins are considered; which can trigger inflammatory processes. Research focused on hazard evaluation throughout the nano-life cycle has been conducted to investigate the safety implications of using lignin-coated cellulose nanofibrils (L-CNF) and lignin-coated cellulose nanocrystals (L-CNCs). Both types of nanofibrils have been shown to be relatively less hazardous in several studies [75]. However, the production of NMs faces challenges related to cultural acceptance, economic feasibility, ecological sustainability, and the availability of local resources [76].

As shown in Figure 6.2, the key aspects of BNMs include (1) efficient extraction methods from biological sources, (2) commercialization, (3) toxicity concerns, (4) appropriate manufacturing processes, (5) accessibility, and (6) purification, among others. However, beneficial and efficient methods are also associated with certain challenges. Moreover, two major obstacles to their broader adoption are cost-effectiveness and energy consumption [77]. At the commercial level, collaboration between experts and industry professionals is essential to achieve high output with precise performance. Manufacturing companies must take proactive measures to ensure safety, preventing potential safety concerns from obstructing commercialization. Government agencies in many countries have established regulations regarding the safe use of substances, the development of commercial products, and testing methods for biologically produced NMs [73]. Additionally, a key objective in developing efficient NMs is to ensure their stability. The resilience of NMs can be influenced by various factors, depending on the type being produced. For instance, CNCs synthesized using sulfuric acid has been found to exhibit high thermal stability [78]. Alkaloids, polyphenols, acidic compounds, and other plant-derived molecules help stabilize biologically synthesized metallic NPs. Research has shown that the stabilization of nanoparticles such as AuNPs and AgNPs in *Ocimum* leaf extract is supported by a thin film of polymers. The impact of nanocellulose (NC) on the demand for textiles, paper, purification systems, coatings, finishes, and other products by 2027 was discussed in a global market report by Future Markets, which publishes articles on NC [79]. Sulfuric acid digestion, due to its low cost and reusability, currently serves as the foundation for most CNC production. Additionally, several

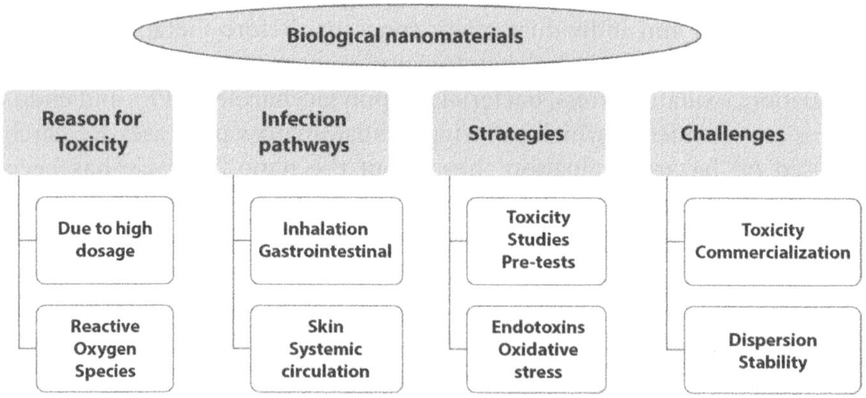

Figure 6.2 Problems and solutions for creating effective BNMs.

cost-effective manufacturing methods, such as oxidation-based techniques and pre-extraction treatments involving boiling water, are recognized as efficient steps within the biorefinery process. Research has also proposed dicarboxylic acid hydrolysis as an economical approach for the combined production of CNFs and CNCs [79]. Similarly, the production of bacterial cellulose is both costly and time-consuming. To reduce expenses, it is essential to use more affordable substrates, such as agricultural waste materials [80].

6.9 Conclusion

Environmental pollution caused by MPs is an increasingly serious issue that requires thorough assessment and mitigation. Rapid technological advancements worldwide have opened up possibilities for comprehensive solutions, and scientists have developed advanced methods for the removal of MPs. One promising approach is biological remediation using nanotechnology. The combination of nanotechnology and BNMs has been extensively applied in the cleanup of MPs. As there have been no reported adverse effects, BNMs are considered environmentally safe. Additionally, bio-NPs possess several properties, such as sorption capacity and the ability to facilitate photodegradation that can be harnessed to remove MPs from the environment. Advances in nanotechnology have made smart MP degradation feasible, and significant efforts are currently underway in this field. To effectively reduce plastic pollution, it is essential to consider the environmental impacts of MPs, technological advancements, social behaviors, and regulatory measures, despite the current lack of comprehensive scientific understanding needed to fully resolve this complex issue. Additionally, since MPs pose a new threat to human health, a range of mitigation measures should be progressively implemented to effectively reduce and manage MP formation, particularly in critical areas such as sewage systems and water treatment facilities. This chapter highlights that the green synthesis NPs provides highly efficient tools and a safer approach for environmental remediation. To gain consumer trust and acceptance as a sustainable alternative for the future, this method of NP fabrication continues to evolve. Nevertheless, despite the many advantages and proven efficacy of green NPs, several challenges and concerns still require further investigation and validation through continued research. Moreover, to benefit both society and the environment, it is essential to advance the application of green NMs and nanotechnology beyond the laboratory and into larger-scale, real-world research and implementation. Before the application of NPs, it is essential to evaluate their ecotoxicological effects. Accordingly, addressing the potential impact of NPs on human health is of critical importance. To support the continued development and long-term potential of nanotechnologies for promoting healthy environments, ongoing research validation, along with government support, is essential. The implementation of clear rules and regulations would help complete the framework, enabling the environmentally responsible removal of pollutants.

AI disclosure statement: The author confirms that no artificial intelligence (AI)-generated content or AI-assisted writing tools were used in the preparation of this chapter (including summaries and the abstract). All text, analysis, and conclusions are the result of the author's original work.

References

1. Geyer, R., Jambeck, J.R., Law, K.L., Production, use, and fate of all plastics ever made. *Sci. Adv.*, 3, 7, e1700782, 2017.
2. Borrelle, S.B., Ringma, J., Law, K.L., Monnahan, C.C., Lebreton, L., Mcgivern, A., Murphy, E., Jambeck, J., Leonard, G.H., Hilleary, M.A., Eriksen, M., Possingham, H.P., De Frond, H., Gerber, L.R., Polidoro, B., Tahir, A., Bernard, M., Mallos, N., Barnes, M., Rochman, C.M., Predicted growth in plastic waste exceeds efforts to mitigate plastic pollution. *Science*, 369, 6510, 1515–1518, 2020.
3. Meijer, L.J.J., Van Emmerik, T., Van Der Ent, R., Schmidt, C., Lebreton, L., More than 1000 rivers account for 80% of global riverine plastic emissions into the ocean. *Sci. Adv.*, 7, 18, eaaz5803, 2021.
4. da Costa, J.P., Santos, P.S.M., Duarte, A.C., Rocha-Santos, T., (Nano)plastics in the environment – Sources, fates and effects. *Sci. Total Environ.*, 566–567, 15–26, 2016.
5. Yuan, W., Xu, E.G., Shabaka, S., Chen, P., Yang, Y., The power of green: Harnessing phytoremediation to combat micro/nanoplastics. *Eco-environ. Health*, 3, 3, 260–265, 2024.
6. Patel, P.K. and Uppaluri, R.V.S., Adsorption of emerging pollutants utilizing chitosan derivatives: Recent advances and future perspective. *Int. J. Biol. Macromol.*, 299, 139–155, 2025.
7. Patel, P.K., Pandey, L.M., Uppaluri, R., Multi-metal adsorption and cyclic desorption characteristics of Zn^{2+} and Cu^{2+} constituting multi-component synthetic wastewater system using commercial resins. *Sustain. Environ.*, 3, 3–27, 2023.

8. Corsi, I., Venditti, I., Trotta, F., Punta, C., Environmental safety of nanotechnologies: The eco-design of manufactured NMs for environmental remediation. *Sci. Total Environ.*, 864, 160991, 2023.
9. Arsenov, D., Beljin, J., Jović, D., Maletić, S., Borišev, M., Borišev, I., NMs as endorsed environmental remediation tools for the next generation: Eco-safety and sustainability. *J. Geochem. Explor.*, 253, 107076, 2023.
10. Cheriyamundath, S. and Vavilala, S.L., Nanotechnology-based wastewater treatment. *Water Environ. J.*, 35, 1, 123–132, 2021.
11. González-Martín, J., Kraakman, N.J.R., Pérez, C., Lebrero, R., Muñoz, R., A state-of-the-art review on indoor air pollution and strategies for indoor air pollution control. *Chemosphere*, 262, 128376, 2021.
12. Najafpoor, A., Norouzian-Ostad, R., Alidadi, H., Rohani-Bastami, T., Davoudi, M., Barjasteh-Askari, F., Zanganeh, J., Effect of magnetic NPs and silver-loaded magnetic NPs on advanced wastewater treatment and disinfection. *J. Mol. Liq.*, 303, 112645, 2020.
13. Xue, S., Xiao, Y., Wang, G., Fan, J., Wan, K., He, Q., Gao, M., Miao, Z., Adsorption of heavy metals in water by modifying Fe_3O_4 NPs with oxidized humic acid. *Colloids Surf. a Physicochem. Eng. Asp.*, 616, 126333, 2021.
14. Abouzeid, R.E., Khiari, R., El-Wakil, N., Dufresne, A., Current state and new trends in the use of cellulose NMs for wastewater treatment. *Biomacromolecules*, 20, 2, 573–597, 2019.
15. Magudieshwaran, R., Ishii, J., Raja, K.C.N., Terashima, C., Venkatachalam, R., Fujishima, A., Pitchaimuthu, S., Green and chemical synthesized CeO_2 NPs for photocatalytic indoor air pollutant degradation. *Mater. Lett.*, 239, 40–44, 2019.
16. Bourdrel, T., Annesi-Maesano, I., Alahmad, B., Maesano, C.N., Bind, M.A., The impact of outdoor air pollution on COVID-19: A review of evidence from *in vitro*, animal, and human studies. *Eur. Respir. Rev.*, 30, 159, 200242, 2021.
17. Lu, F. and Astruc, D., Nanocatalysts and other NMs for water remediation from organic pollutants. *Coord. Chem. Rev.*, 408, 213180, 2020.
18. Asghar, N., Hussain, A., Nguyen, D.A., Ali, S., Hussain, I., Junejo, A., Ali, A., Advancement in NMs for environmental pollutants remediation: A systematic review on bibliometrics analysis, material types, synthesis pathways, and related mechanisms. *J. Nanobiotechnol.*, 22, 1, 149, 2024.
19. Chellasamy, G., Kiriyanthan, R.M., Maharajan, T., Radha, A., Yun, K., Remediation of MPs using bioNMs: A review. *Environ. Res.*, 208, 112719, 2022.
20. Saraswat, P., Singh, S., Prasad, M., Misra, R., Rajput, V.D., Ranjan, R., Applications of bio-based NMs in environment and agriculture: A review on recent progresses. *Hybrid Adv.*, 4, 100097, 2023.
21. Smith, S.C. and Rodrigues, D.F., Carbon-based NMs for removal of chemical and biological contaminants from water: A review of mechanisms and applications. *Carbon N. Y.*, 91, 122–143, 2015.

22. Khan, I., Saeed, K., Khan, I., NPs: Properties, applications and toxicities. *Arab. J. Chem.*, 12, 7, 908–931, 2019.
23. Zou, L., Luo, Y., Hooper, M., Hu, E., Removal of VOCs by photocatalysis process using adsorption enhanced TiO_2–SiO_2 catalyst. *Chem. Eng. Process. Process. Intensif.*, 45, 11, 959–964, 2006.
24. Li, X., Xu, H., Chen, Z.S., Chen, G., Biosynthesis of NPs by microorganisms and their applications. *J. Nanomater.*, 2011, 1–12, 2011.
25. Mokhtarzadeh, A., Alibakhshi, A., Hejazi, M., Omidi, Y., Ezzati Nazhad Dolatabadi, J., Bacterial-derived biopolymers: Advanced natural NMs for drug delivery and tissue engineering. *Trac Trends Anal. Chem.*, 82, 367–384, 2016.
26. Hammer, J., Kraak, M.H.S., Parsons, J.R., Plastics in the marine environment: The dark side of a modern gift. *Rev. Environ. Contam. Toxicol.*, 220, 1–44, 2012.
27. Xu, Y., He, Q., Liu, C., Huangfu, X., Are micro- or nanoplastics leached from drinking water distribution systems? *Environ. Sci. Technol.*, 53, 16, 9339–9340, 2019.
28. Chamas, A., Moon, H., Zheng, J., Qiu, Y., Tabassum, T., Jang, J.H., Abu-Omar, M., Scott, S.L., Suh, S., Degradation rates of plastics in the environment. *ACS Sustain. Chem. Eng.*, 8, 9, 3494–3511, 2020.
29. Almutairi, M.M.A., Gong, C., Xu, Y.G., Chang, Y., Shi, H., Factors controlling permeability of the blood-brain barrier. *Cell. Mol. Life Sci.*, 73, 1, 57–77, 2016.
30. Oriekhova, O. and Stoll, S., Heteroaggregation of nanoplastic particles in the presence of inorganic colloids and natural organic matter. *Environ. Sci. Nano.*, 5, 3, 792–799, 2018.
31. Brun, E., Barreau, F., Veronesi, G., Fayard, B., Sorieul, S., Chanéac, C., Carapito, C., Rabilloud, T., Mabondzo, A., Herlin-Boime, N., Carrière, M., Titanium dioxide NP impact and translocation through *ex vivo*, *in vivo* and *in vitro* gut epithelia. *Part. Fibre Toxicol.*, 11, 1, 2014.
32. Li, Y., Tao, L., Wang, Q., Wang, F., Li, G., Song, M., Potential health impact of MPs: A review of environmental distribution, human exposure, and toxic effects. *Environ. Health*, 1, 4, 249–257, 2023.
33. Kumar, R., Manna, C., Padha, S., Verma, A., Sharma, P., Dhar, A., Ghosh, A., Bhattacharya, P., Micro(nano)plastics pollution and human health: How plastics can induce carcinogenesis to humans? *Chemosphere*, 298, 2022.
34. Wu, P., Lin, S., Cao, G., Wu, J., Jin, H., Wang, C., Wong, M.H., Yang, Z., Cai, Z., Absorption, distribution, metabolism, excretion and toxicity of MPs in the human body and health implications. *J. Hazard. Mater.*, 437, 2022.
35. Praveena, S.M. and Laohaprapanon, S., Quality assessment for methodological aspects of MPs analysis in bottled water – A critical review. *Food Control*, 130, 2021.
36. Plochocka, P., Excitons in a twisted world. *Nat. Nanotechnol.*, 15, 9, 727–729, 2020.

37. Canesi, L., Ciacci, C., Bergami, E., Monopoli, M.P., Dawson, K.A., Papa, S., Canonico, B., Corsi, I., Evidence for immunomodulation and apoptotic processes induced by cationic polystyrene NPs in the hemocytes of the marine bivalve Mytilus. *Mar. Environ. Res.*, 111, 34–40, 2015.
38. Wegner, A., Besseling, E., Foekema, E.M., Kamermans, P., Koelmans, A.A., Effects of nanopolystyrene on the feeding behavior of the blue mussel (Mytilus edulis L.). *Environ. Toxicol. Chem.*, 31, 11, 2490–2497, 2012.
39. Bergami, E., Bocci, E., Vannuccini, M.L., Monopoli, M., Salvati, A., Dawson, K.A., Corsi, I., Nano-sized polystyrene affects feeding, behavior and physiology of brine shrimp Artemia franciscana larvae. *Ecotoxicol. Environ. Saf.*, 123, 18–25, 2016.
40. Rainieri, S., Conlledo, N., Larsen, B.K., Granby, K., Barranco, A., Combined effects of MPs and chemical contaminants on the organ toxicity of zebrafish (Danio rerio). *Environ. Res.*, 162, 135–143, 2018.
41. Agarwal, N., Solanki, V., Singh, N., Shah, M., *Nanotechnology in Environmental Remediation: Prespectives and Prospects*, 2025, 10.2174/97898153229411250101.
42. Liu, E.K., He, W.Q., Yan, C.R., "White revolution" to "white pollution" – Agricultural plastic film mulch in China. *Environ. Res. Lett.*, 9, 9, 2014.
43. Ng, E.L., Huerta Lwanga, E., Eldridge, S.M., Johnston, P., Hu, H.W., Geissen, V., Chen, D., An overview of MP and nanoplastic pollution in agroecosystems. *Sci. Total Environ.*, 627, 1377–1388, 2018.
44. Steinmetz, Z., Wollmann, C., Schaefer, M., Buchmann, C., David, J., Tröger, J., Muñoz, K., Frör, O., Schaumann, G.E., Plastic mulching in agriculture. Trading short-term agronomic benefits for long-term soil degradation? *Sci. Total Environ.*, 550, 690–705, 2016.
45. ter Halle, A., Ladirat, L., Martignac, M., Mingotaud, A.F., Boyron, O., Perez, E., To what extent are MPs from the open ocean weathered? *Environ. Pollut.*, 227, 167–174, 2017.
46. Bandara, S., Du, H., Carson, L., Bradford, D., Kommalapati, R., Agricultural and biomedical applications of chitosan-based NMs. *Nms*, 10, 10, 1–31, 2020.
47. Nasery, M.M., Abadi, B., Poormoghadam, D., Zarrabi, A., Keyhanvar, P., Khanbabaei, H., Ashrafizadeh, M., Mohammadinejad, R., Tavakol, S., Sethi, G., Curcumin delivery mediated by bio-based NPs: A review. *Molecules*, 25, 3, 2020.
48. Abdul Khalil, H.P.S., Davoudpour, Y., Islam, M.N., Mustapha, A., Sudesh, K., Dungani, R., Jawaid, M., Production and modification of nanofibrillated cellulose using various mechanical processes: A review. *Carbohydr. Polym.*, 99, 649–665, 2014.
49. Rahman, A., Kumar, S., Nawaz, T., Biosynthesis of NMs using algae, in: *Microalgae Cultivation for Biofuels Production*, pp. 265–279, Elsevier, 2019.
50. Lee, J.W., Park, J.H., Robinson, J.R., Bioadhesive-based dosage forms: The next generation. *Mini Rev.*, 2000.

51. Jayakumar, R., Prabaharan, M., Nair, S.V., Tamura, H., Novel chitin and chitosan nanofibers in biomedical applications. *Biotechnol. Adv.*, 28, 1, 142–150, 2010.
52. Al Nadhari, S., Al-Enazi, N.M., Alshehrei, F., Ameen, F., A review on biogenic synthesis of metal NPs using marine algae and its applications. *Environ. Res.*, 194, 2021.
53. Agarwal, N., Srivastava, A., Solanki, V., Singh, N., Yadav, R., *Emerging trends in the use of microbial isolates for the treatment of dye containing wastewater*, 2024, 10.1016/B978-0-443-13561-3.00010-7.
54. Alves, A.A., Silva, W.E., Belian, M.F., Lins, L.S.G., Galembeck, A., Bacterial cellulose membranes for environmental water remediation and industrial wastewater treatment. *Int. J. Environ. Sci. Technol.*, 17, 9, 3997–4008, 2020.
55. Sulaeva, I., Henniges, U., Rosenau, T., Potthast, A., Bacterial cellulose as a material for wound treatment: Properties and modifications – A review. *Biotechnol. Adv.*, 33, 8, 1547–1571, 2015.
56. Prozorov, T., Bazylinski, D.A., Mallapragada, S.K., Prozorov, R., Novel magnetic NMs inspired by magnetotactic bacteria: Topical review. *Mater. Sci. Eng. R Rep.*, 74, 5, 133–172, 2013.
57. Nagireddi, S., Patel, P.K., Golder, A.K., Uppaluri, R.V.S., Evaluation of a triethylenetetramine-chitosan derivative for the removal of palladium ion from an intricate adsorbate system. *J. Mol. Struct.*, 1323, 2025.
58. Nagireddi, S., Patel, P.K., Golder, A.K., Uppaluri, R.V.S., Recovery of $PdEDTA^{2-}$ complex from electroless plating solutions using nitrogen-functionalized chitosan derivative. *Polym. Bull.*, 2025.
59. Oza, H. and Pathak, T., A review on micro and nano plastics removal from aqueous environment through bioNMs. *Interactions*, 245, 1, 2024.
60. Jalvo, B., Aguilar-Sanchez, A., Ruiz-Caldas, M.X., Mathew, A.P., Water filtration membranes based on non-woven cellulose fabrics: Effect of nanopolysaccharide coatings on selective particle rejection, antifouling, and antibacterial properties. *Nms*, 11, 7, 2021.
61. Bahi, A., Shao, J., Mohseni, M., Ko, F.K., Membranes based on electrospun lignin-zeolite composite nanofibers. *Sep. Purif. Technol.*, 187, 207–213, 2017.
62. Peydayesh, M., Suta, T., Usuelli, M., Handschin, S., Canelli, G., Bagnani, M., Mezzenga, R., Sustainable removal of MPs and natural organic matter from water by coagulation-flocculation with protein amyloid fibrils. *Environ. Sci. Technol.*, 55, 13, 8848–8858, 2021.
63. Khan, W.S., Asmatulu, R., Davuluri, S., Dandin, V.K., Improving the economic values of the recycled plastics using nanotechnology associated studies. *J. Mater. Sci. Technol.*, 30, 9, 854–859, 2014.

64. Mehra, K. and Sarkar, N., NMs for the management of plastic waste: A way combating white pollution. *Nano. Environ. Eng.*, 2024.
65. López de Dicastillo, C., Velásquez, E., Rojas, A., Guarda, A., Galotto, M.J., The use of nanoadditives within recycled polymers for food packaging: Properties, recyclability, and safety. *Compr. Rev. Food Sci. Food Saf.*, 19, 4, 1760–1776, 2020.
66. Zdiri, K., Elamri, A., Hamdaoui, M., Harzallah, O., Khenoussi, N., Brendlé, J., Reinforcement of recycled PP polymers by NPs incorporation. *Green Chem. Lett. Rev.*, 11, 3, 296–311, 2018.
67. Bolade, O.P., Williams, A.B., Benson, N.U., Green synthesis of iron-based NMs for environmental remediation: A review. *Environ. Nanotechnol. Monit. Manag.*, 13, 2020.
68. Safieddin Ardebili, S.M., Taghipoor, A., Solmaz, H., Mostafaei, M., The effect of nano-biochar on the performance and emissions of a diesel engine fueled with fusel oil-diesel fuel. *Fuel*, 268, 2020.
69. Maria Ribeiro Guimarães, B., Vanoli Scatolino, M., Alice Martins, M., Rocha Ferreira, S., Marin Mendes, L., Tarcísio Lima, J., Guimarães Junior, M., Henrique Denzin Tonoli, G., Luis, S.V., Bio-based films/nanopapers from lignocellulosic wastes for production of added-value micro/NMs. *Ind. Crops Prod.*, 2024.
70. Li, H., Yoshida, S., Mitani, N., Egusa, M., Takagi, M., Izawa, H., Matsumoto, T., Kaminaka, H., Ifuku, S., Disease resistance and growth promotion activities of chitin/cellulose nanofiber from spent mushroom substrate to plant. *Carbohydr. Polym.*, 284, 2022.
71. Patel, P.K., Pandey, L.M., Uppaluri, R.V.S., Cyclic adsorption and desorption characteristics of citric acid–chitosan variant resins for Pb, Fe, and Zn removal from simulated mining and agricultural wastewater system. *J. Polym. Environ.*, 32, 5750–5770, 2024.
72. Patel, P.K., Pandey, L.M., Uppaluri, R.V.S., Highly effective removal of multi-heavy metals from simulated industrial effluent through an adsorption process employing carboxymethyl-chitosan composites. *Environ. Res.*, 240, 2024.
73. Catalán, J. and Norppa, H., Safety aspects of bio-based nanomaterials. *Bioengineering*, 4, 4, 2017.
74. Padmanabhan, J. and Kyriakides, T.R., Nanomaterials, inflammation, and tissue engineering. *Wiley Interdiscip. Rev. Nanomed. Nanobiotechnol.*, 7, 3, 355–370, 2015.
75. Ong, K.J., Shatkin, J.A., Nelson, K., Ede, J.D., Retsina, T., Establishing the safety of novel bio-based cellulose nanomaterials for commercialization. *Nanoimpact*, 6, 19–29, 2017.
76. Nath, D. and Das, B., "Green nanomaterial" – How green they are as biotherapeutic tool. *J. Nanomed. Biotherapeutic Discov.*, 4, 2, 2014.

77. Patel, P.K., Pandey, L.M., Uppaluri, R.V.S., Conceptual techno-economic analysis of chitosan derivative resin and commercial sorbents for multi-heavy metal removal. *Sep. Purif. Technol.*, 131732, 2025.
78. Putro, J.N., Santoso, S.P., Ismadji, S., Ju, Y.H., Investigation of heavy metal adsorption in binary system by nanocrystalline cellulose–bentonite nanocomposite: Improvement on extended Langmuir isotherm model. *Microporous Mesoporous Mater.*, 246, 166–177, 2017.
79. Saraswat, P., Singh, S., Prasad, M., Misra, R., Rajput, V.D., Ranjan, R., Applications of bio-based nanomaterials in environment and agriculture: A review on recent progresses. *Hybrid Adv.*, 4, 100097, 2023.
80. Jozala, A.F., de Lencastre-Novaes, L.C., Lopes, A.M., de Carvalho Santos-Ebinuma, V., Mazzola, P.G., Pessoa-Jr, A., Grotto, D., Gerenutti, M., Chaud, M.V., Bacterial nanocellulose production and application: a 10-year overview. *Appl. Microbiol. Biotechnol.*, 100, 5, 2063–2072, 2016.

7

Metallic Oxides as a Catalytic Tool to Depolymerize the Plastic Wastes

Anindra Sharma[1] and Abhay Nanda Srivastva[2*]

[1]P.G. Department of Chemistry, Lalit Narayan Mithila University, Darbhanga, Bihar, India
[2]Laboratory of Bioactive Material Synthesis, University Department of Chemistry, B.R.A. Bihar University, Muzaffarpur, Bihar, India

Abstract

Plastic waste, being one of the most threatening forms of waste that contaminates our surroundings, is a very serious issue and requires the best possible approach to eliminate this hazard. The incessant deposition of plastics in environment is creating serious issues in living beings including human. A lethal quantity of microplastics has been detected in various human organs like the brain, kidney, liver, and lungs, causing organ malfunctions, disturbed metabolic activities, cancers, or organ failure, sometimes ending in death. Controlling the damage caused by plastic waste can only be achieved by decomposing it into safer molecules with potential industrial applications. Metallic oxides serve as highly effective catalysts for various chemical processes involved in plastic depolymerization. The catalytic potential of metallic oxides is being studied by researchers worldwide for plastic decomposition, which can help mitigate environmental issues and the hazards associated with plastic waste accumulation. The promising results obtained by scientists and environmental chemists have persuaded the authors to compile the key findings on metal oxide-based catalytic depolymerization of commonly used polymers such as polyolefins, polystyrene, polyethylene terephthalate, and polyvinyl chloride, in order to promote future research in this area.

***Keywords*:** Plastic hazard, nanometal particles, plastic polymer, environment sustainability, plastic degradation, metallic catalysts

**Corresponding author*: abhay.chem.brabu@gmail.com

Neha Agarwal, Brijesh Pare, Vijendra Singh Solanki and Sankara Rao Miditana (eds.) Plastic Remediation: Transformative Pathways for a Rejuvenated Sustainable Environment, (171–186)

7.1 Introduction

The widespread use of synthetic plastics, derived primarily from petrochemical feedstocks, has revolutionized material science and consumer industries due to their excellent mechanical properties, chemical resistance, durability, and cost-efficiency [1]. As of 2022, global plastic production exceeded 390 million metric tons annually, a figure that is expected to grow significantly over the coming decades due to increasing demands across packaging, textiles, automotive, and electronic sectors [2]. However, these benefits are accompanied by a substantial environmental burden, primarily because conventional plastics are inherently resistant to microbial degradation and persist in ecosystems for hundreds of years. Only a small fraction, less than 10% of produced plastics, is effectively recycled, while the remainder accumulates in landfills, aquatic environments, and even within biological systems in the form of microplastics and nanoplastics [3]. Recent studies have detected plastic microparticles in human blood, lungs, placenta, and even the brain, raising significant public health concerns [4]. Traditional recycling strategies such as mechanical reprocessing and incineration suffer from limitations including polymer property degradation, energy intensiveness, release of greenhouse gases, and generation of toxic byproducts [5]. Consequently, the need for alternative, sustainable plastic waste management strategies is critical for environmental and human health protection. Chemical recycling has emerged as a technologically advanced and environmentally promising route, involving the cleavage of polymer chains into monomers or valuable chemical intermediates that can be reused in industrial processes [6, 7]. Among various chemical recycling approaches, catalytic depolymerization using metal oxide-based systems has shown considerable potential. Metal oxides are particularly attractive due to their high thermal and chemical stability, tunable Lewis and Brønsted acid–base characteristics, redox versatility, and surface modifiability. These features allow for the selective cleavage of robust C–C, C–O, and C–Cl bonds under relatively mild reaction conditions [8]. Recent advancements demonstrate the applicability of metal oxides in diverse depolymerization schemes such as pyrolysis, hydrocracking, hydrogenolysis, and solvolysis. This chapter presents a comprehensive review of the recent advances in the catalytic depolymerization of plastic waste using metal oxide-based systems, with emphasis on mechanistic insights, catalyst performance, and product distributions across major plastic classes. The discussion includes strategies for the conversion of polyolefins, polystyrene, PET, and PVC using metal oxide catalysts under varied reaction conditions. This synthesis of current findings is expected to inform future research directions and contribute to the design of more efficient, scalable, and environmentally sustainable plastic waste recycling technologies.

7.2 Polyolefins (Polyethylene and Polypropylene)

Polyolefins, including polyethylene (PE) and polypropylene (PP), are the most widely produced plastics globally. Their saturated hydrocarbon backbones make them highly resistant to degradation, posing a major challenge for recycling. Conventional pyrolysis processes require temperatures exceeding 500°C and often yield broad, low-value product distributions [9, 10]. To overcome these limitations, researchers have explored the use of metal oxide catalysts to improve selectivity and reduce energy requirements.

7.2.1 Catalytic Pyrolysis with Metal Oxides

Catalytic pyrolysis is a thermal degradation technique wherein plastic waste is decomposed in the absence of oxygen using solid acid catalysts, typically operated within a temperature range of 300–500°C [11]. Reaction durations usually range from 20 to 60 minutes, depending on the nature of the polymer and the specific catalyst employed. The introduction of catalysts significantly improves the efficiency of chain scission processes, resulting in enhanced selectivity and a more favorable distribution of liquid hydrocarbon products compared to conventional thermal pyrolysis. These catalytic systems function by offering acidic and/or redox-active sites that facilitate the cleavage of C–C and C–H bonds more effectively. Furthermore, the incorporation of metal oxides into the catalyst matrix can suppress the formation of undesirable byproducts, thereby improving process sustainability, reducing environmental impact, and enhancing the economic feasibility of converting plastic waste into valuable fuel-range hydrocarbons. Ahmad *et al.* reported the catalytic pyrolysis of polyolefins polypropylene (PP) and high-density polyethylene (HDPE)—using acid-washed bentonite clay (AWBC) impregnated with transition metal oxides such as ZnO, Co_2O_3, Fe_2O_3, NiO, and MnO_2 [12]. The pyrolysis was conducted under a nitrogen atmosphere at 300°C for PP and 350°C for HDPE in a fixed-bed batch reactor. The incorporation of metal oxides into the clay matrix significantly improved the catalytic efficiency, resulting in enhanced liquid hydrocarbon yields and selectivity compared to both thermal pyrolysis and unmodified clay. Among the catalysts tested, Co/AWBC and

Zn/AWBC exhibited the highest activity, producing up to 92.76 wt% and 91.00 wt% liquid products from PP and HDPE, respectively. GC–MS and FTIR analyses revealed that the pyrolysis oils were predominantly composed of paraffins, olefins, and naphthenes, with negligible amounts of aromatics or oxygenated compounds. The improved performance was attributed to the increased acidity and favorable surface characteristics conferred by the metal oxide impregnation. These results highlight the effectiveness of metal oxide-supported bentonite clays as low-cost and efficient catalysts for converting plastic waste into olefin-rich liquid fuels. Similarly, Mehdi *et al.* investigated the catalytic degradation of waste HDPE using Fuller's earth clay modified with transition metal oxides to enhance its catalytic behavior [13]. Conventional clay-based catalysts often suffer from low activity and tend to yield heavier hydrocarbon fractions. In this study, blends of clay with ZnO, Fe_2O_3, and NiO were employed to overcome these limitations. Pyrolysis was performed at 300°C and 22 psi for 30 minutes in a custom-designed cylindrical reactor. Among the tested formulations, the Fe_2O_3–clay composite exhibited superior catalytic activity, producing a high proportion of light hydrocarbons, including propene (55.10 wt%), butyl isocyanide (30.10 wt%), and 1-pentene (15.30 wt%), as confirmed by GC–MS analysis. The enhanced performance was ascribed to the moderate acidity and higher density of active sites provided by the Fe-based catalyst. This approach offers a promising and economically viable route for the conversion of HDPE waste into low-molecular-weight hydrocarbons, with potential applications in sustainable fuel and petrochemical sectors. Lin *et al.* reported the catalytic fast pyrolysis of wood–polypropylene composites (WPPs) using metal oxides such as ZnO, CaO, MgO, and Fe_2O_3 [14]. Pyrolysis experiments performed at 600°C using Py-GC/MS demonstrated that these oxides substantially influenced product distribution, particularly enhancing the generation of olefins and deoxygenated compounds. ZnO was especially effective in promoting the formation of alkenes and monomeric phenols, while CaO significantly increased ketone content and facilitated deoxygenation due to its strong basicity. MgO promoted cracking reactions, yielding smaller aliphatic hydrocarbons, and Fe_2O_3 favored the formation of aromatic compounds. Notably, a synergistic interaction between the biomass and polyolefin fractions was observed, which was further amplified in the presence of ZnO, leading to a higher yield of light olefins and phenols through hydrogen transfer and radical-mediated mechanisms. Together, these studies underscore the versatility of metal oxide-mediated catalytic pyrolysis as a cost-effective and sustainable strategy for converting plastic and composite wastes into olefin-rich liquid fuels and chemical feedstocks.

7.2.2 Hydrocracking Using Bifunctional Metal Oxides

Hydrocracking is a catalytic process that operates under comparatively milder conditions than pyrolysis, typically at temperatures ranging from 300°C to 450°C and under hydrogen pressures of 20–150 bar. The presence of hydrogen not only facilitates shorter reaction times but also contributes to improved control over product selectivity. The process employs bifunctional catalysts composed of a hydrogenation metal such as Pt, Ni, or Ru dispersed on an acidic support, often metal oxides like Al_2O_3 or ZrO_2. The acidic sites catalyze the cracking of polymeric chains, while the metallic component promotes hydrogenolysis and subsequent hydrogenation of unsaturated intermediates. This dual functionality enables the efficient conversion of long-chain hydrocarbons into shorter, saturated liquid products, primarily alkanes, which are well-suited for use as fuels or fuel oils. In some cases, hydrogen treatment is applied post-pyrolysis through hydrotreating to enhance product quality by removing oxygenates, aromatics, or unsaturated compounds. Overall, hydrocracking represents a highly selective and efficient route for transforming plastic waste into high-quality fuel-range hydrocarbons. Venkatesh *et al.* reported the hydrocracking of long-chain alkanes and polyolefins such as high-density polyethylene (HDPE) and polypropylene (PP) using bifunctional catalysts composed of platinum or nickel supported on anion-modified zirconium oxides (AZOs), particularly sulfated ($Pt/ZrO_2/SO_4$) and tungstated ($Pt/ZrO_2/WO_3$) systems [15]. These catalysts exhibited high activity for hydrocracking under hydrogen pressures of 500–1200 psig and moderate temperatures (300–375°C), yielding a high proportion of branched alkanes in the gasoline range (C_5–C_{12}). The Pt-based catalysts promoted both C–C bond cleavage and isomerization, resulting in isoalkane-rich products with minimal formation of olefins or aromatics. Interestingly, the study suggested that hydrocracking of heavier alkanes does not proceed through the classical bifunctional mechanism involving olefinic intermediates; instead, a unimolecular pathway is likely operative. Compared to sulfated systems, WO_3-modified catalysts demonstrated better thermal stability and retained their anionic character even under severe reducing conditions. Both HDPE and PP were converted with over 98% efficiency, and product selectivity could be tuned by adjusting metal type, loading, and reaction parameters. These findings underscore the potential of metal-promoted AZOs as robust and efficient catalysts for converting polyolefin waste into high-value liquid fuels with controlled branching and boiling point profiles.

7.2.3 Hydrogenolysis with Supported Metal Oxides

Hydrogenolysis is a promising chemical recycling strategy for the depolymerization of plastic waste, particularly polyolefins like PE and PP. This process involves the catalytic cleavage of C–C bonds in the presence of hydrogen, yielding valuable hydrocarbon products such as gasoline, diesel, and light lubricants. Supported metal catalysts, including Ru/C, Ru/CeO_2, and Ru/Al_2O_3, have shown high efficiency under relatively mild conditions [16]. Compared to mechanical recycling, which is energy-intensive and produces low-quality materials, hydrogenolysis offers a more sustainable and effective route for polymer upcycling. Celik *et al.* reported the selective hydrocracking of polyethylene (PE) into high-quality liquid hydrocarbons using a bifunctional catalyst composed of platinum nanoparticles supported on strontium titanate (Pt/$SrTiO_3$) [17]. The catalyst was synthesized *via* atomic layer deposition (ALD), affording highly dispersed Pt particles on well-defined $SrTiO_3$ nanocuboids. Under solvent-free conditions at 300°C and 170 psi hydrogen pressure, the Pt/$SrTiO_3$ catalyst effectively converted various grades of PE including commercial single-use plastic bags into lubricant-like products with narrow molecular weight distributions (Mn = 590–990 Da, Đ = 1.1–1.3). The bifunctional catalytic system exhibited enhanced performance due to the synergy between the redox-active Pt sites, responsible for C–C bond cleavage *via* hydrogenolysis, and the structurally stable $SrTiO_3$ support, which suppressed sintering and promoted uniform nanoparticle dispersion. Notably, the control over Pt nanoparticle size and surface coordination allowed fine-tuning of product selectivity, with smaller Pt NPs leading to over-hydrogenolysis and gas formation, while optimal particle sizes (2–3 nm) favored liquid product yields. The study highlights the importance of catalyst structure, metal-support interactions, and adsorption behavior in achieving efficient upcycling of polyolefin waste into value-added liquid fuels under hydrocracking conditions. Perras and co-workers demonstrated the hydrogenolysis of high-density polyethylene (HDPE) using a mesoporous silica-supported platinum catalyst (mSiO_2/Pt/SiO_2) at 250°C [18]. Solid-state NMR revealed that the polymer chains are adsorbed onto the mesoporous silica and move through its pores, where platinum nanoparticles catalyze C–C bond cleavage. This mechanism efficiently converts long-chain polymers into smaller alkanes (C_6–C_{12}). The method was applied not only to commercial polyethylene but also to post-consumer plastics like shopping bags and isotactic polypropylene (PP). In follow-up work, the same group studied the effect of platinum nanoparticle size and showed that smaller particles enhance the selective breakdown of molten polyethylene into valuable products [19].

Leshkov and co-workers developed a solvent-free hydrogenolysis method using Ru/C as a catalyst to convert low-density polyethylene (LDPE) into hydrocarbons at 200°C [20, 21]. Initial tests on octadecane confirmed that Ru/C outperformed other supports like Al_2O_3, CeO_2, SiO_2, and TiO_2. Applying this to commercial and post-consumer LDPE (Mw ~4,000 Da), they achieved up to 45% yield of liquid hydrocarbons under 20 bar H_2. Increasing the temperature to 250°C led to methane formation, while milder conditions produced longer alkanes (C_7–C_{45}), demonstrating the catalyst's potential for plastic waste valorization. They further extended this system to PP and mixed polyolefins, achieving over 68% yield of isoalkanes (C_5–C_{32}) from high-molecular-weight PP (Mw ~340,000 Da). In another report, Ru/CeO_2 catalysts enabled the conversion of LDPE into fuels (C_5–C_{21}) and waxes (C_{22}–C_{45}) at 475 K and 2 MPa H_2, delivering up to 92% total yield [22]. This strategy was also effective for HDPE and PP, yielding 83%–90% of useful hydrocarbon products. Liu *et al.* employed a Ru/TiO_2 catalyst to degrade waste PP into lubricant-range hydrocarbons at 250°C under moderate hydrogen pressure [23]. Remarkably, this system required low catalyst loading and achieved up to 80% yield. Advanced techniques such as NMR, ATR-IR, GC-MS, and isotopic labeling confirmed the product structures and degradation mechanism.

Abu-Omar and colleagues introduced a different approach for polyolefin depolymerization, targeting the production of α,ω-divinyl-functionalized oligomers [24]. Their method involved a three-step cascade: bromination under visible light (400–410 nm), dehydrobromination using *t*-BuOK, followed by olefin metathesis under ethylene pressure with the Grubbs catalyst. This sequence successfully converted polyethylene samples (Mn = 1.5–6.6 kg/mol) into functionalized oligomers. The strategy presents an efficient route for transforming plastic waste into high-value chemical feed stocks and specialty olefins.

7.3 Polystyrene (PS)

Polystyrene is another challenging plastic due to its aromatic content and high thermal stability [25]. The catalytic depolymerization of polystyrene using metal oxides has emerged as an efficient method for converting plastic waste into valuable monomers, particularly styrene. In a study by Anwar *et al.*, various metal oxides and carbonates were tested for their ability to catalyze the thermal degradation of waste polystyrene at 300–350°C [26]. Among the catalysts evaluated, CaO was found to be the most effective and economical, yielding up to 77% styrene with high purity (~91.5%). Other oxides such as MgO, MnO_2, ZnO, Al_2O_3, and SiO_2 also showed moderate activity. While metal carbonates like Na_2CO_3 and $CaCO_3$ produced relatively pure styrene, their yields were lower. The study demonstrated that basic metal oxides facilitate depolymerization through a carbanion mechanism, while acidic media tend to promote secondary reactions. The use of paraffin oil as a heat transfer medium did not significantly affect the yields. Overall, CaO stood out as a low-cost and highly effective catalyst for the pyrolytic recovery of styrene from polystyrene waste. Similarly, Tiwary and Guria reported that metal oxide-mediated cleavage of waste polystyrene in benzene under hydrogen at 240°C enhances degradation efficiency [27]. Alkali metal oxides like CaO exhibited higher catalytic activity than transition metal oxides, with degradation rates inversely related to metal electronegativity. Metal oxide semiconductors have shown promising potential in the degradation of polystyrene, particularly in its nano-sized form, which poses emerging environmental and health concerns. Polystyrene nanoplastics (PS-NPs), typically around 350 nm in size, are persistent pollutants due to their stability and ability to adsorb toxic substances. Hernández-López *et al.* demonstrated the application of immobilized copper oxide (Cu_2O/CuO) semiconductors, prepared *via* anodization in NH_4F and NaOH electrolytes, for the photocatalytic degradation of PS-NPs under visible light [28]. These copper oxides, with bandgaps between 1.6 and 2.0 eV, facilitated polymer chain scission, leading to a 23% reduction in nanoplastic concentration, significantly outperforming photolysis alone. Analytical techniques including TOC, ATR-FTIR, turbidity measurements, and gas chromatography confirmed degradation, with evidence of carbonyl-containing intermediates and up to 15% mineralization. This approach underscores the effectiveness of copper-based metal oxides in utilizing solar energy to break down PS nanoplastics, offering a sustainable pathway for mitigating micro- and nanoplastic pollution.

7.4 Polyethylene Terephthalate (PET)

PET is commonly used in beverage bottles and textiles. Its ester linkages make it suitable for depolymerization *via* glycolysis, hydrolysis, and hydrogenolysis. Solvolysis, particularly glycolysis, is the most widely studied method for PET depolymerization. Depending on the solvent and conditions, PET can be converted into various intermediates such as terephthalic acid (TPA), dimethyl terephthalate (DMT), and bis(hydroxyethyl) terephthalate (BHET), which serve as precursors for re-synthesizing PET or producing value-added chemicals (Figure 7.1) [29, 30]. While hydrolysis

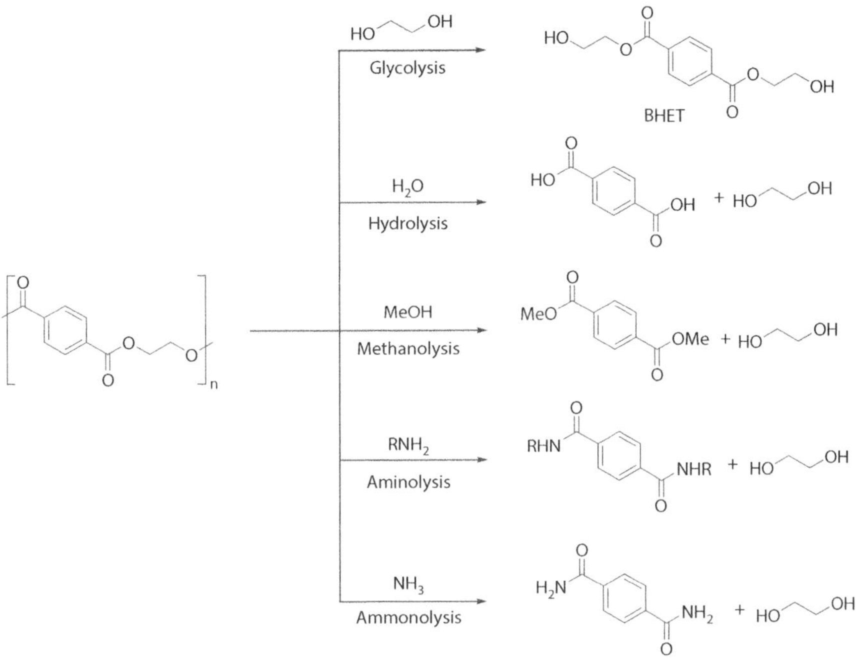

Figure 7.1 Solvolysis depolymerization routes for PET.

and methanolysis yield TPA and DMT, respectively, BHET from glycolysis is versatile and finds applications in textile softeners and polyester resins. Products from aminolysis and ammonolysis are used in plasticizers, adhesives, dyes, drugs, and corrosion inhibitors. Catalyst design is crucial for the industrial implementation of PET glycolysis, particularly in addressing challenges related to catalyst recovery and reusability. Zinc ferrite magnetic nanoparticles (Zn-MNPs), engineered by doping Zn^{2+} into a Fe_3O_4 spinal structure, have shown remarkable catalytic performance in PET glycolysis [31]. The Zn incorporation enhances the number of surface-active sites and catalytic selectivity. These Zn-MNPs enabled 100% PET depolymerization and 79.8% monomer yield under mild conditions (2 h reaction time) with low ethylene glycol consumption and demonstrated efficient recyclability over five cycles.

Nabid *et al.* reported a highly efficient PET glycolysis process, achieving complete BHET yield (100%) using a bifunctional catalyst composed of γ-Fe_2O_3 supported on N-doped graphene at 195°C [32]. In a related study, Fe_2O_3 nanoparticles functionalized with organic ligands on mesoporous silica (SiO_2) demonstrated superior catalytic performance at 190°C compared to bare Fe_2O_3, Fe^{3+} ions, and $FeCl_3$ salts with equivalent iron content, though with a slightly reduced BHET yield [33]. Furthermore, Son *et al.* utilized exfoliated ultrathin MnO_2 nanosheets (e-MON) as catalysts, attaining 100% BHET yield within 30 minutes at 200°C. This performance significantly exceeded that of bulk MnO_2, which yielded 77.6%, and the catalyst maintained high efficiency over five reuse cycles [34]. Recent advances have highlighted the effectiveness of metal oxide catalysts in the glycolytic depolymerization of polyethylene terephthalate (PET). Among these, zinc oxide (ZnO) supported on γ-alumina (γ-Al_2O_3) has been extensively investigated [35]. It was reported that the catalytic depolymerization of waste PET in supercritical ethanol (SCE) using ZnO/γ-Al_2O_3 achieved high efficiency, with a maximum diethyl terephthalate (DET) yield of 92.2% at 270°C within 60 minutes, employing 7.5 wt% ZnO on acidic γ-Al_2O_3 at 5 wt% catalyst dosage. Notably, the reaction resulted in complete depolymerization with no residual PET as confirmed by NMR analysis, and the catalyst maintained its activity over five reuse cycles, indicating good stability and reusability. In parallel, alternative depolymerization approaches involving hydrogenolysis have also reported from metal oxide catalysis. Marks and co-workers reported a carbon-supported single-site molybdenum dioxide catalyst (C/MoO_2) capable of cleaving PET's ester linkages under solvent-free conditions [36]. The catalyst exhibited high activity for both virgin and post-consumer PET, achieving complete conversion to terephthalic acid (TA) at 260°C and 0.1 MPa hydrogen pressure. This efficiency is attributed to the uniform dispersion of catalytically active Mo sites on the carbon support, which selectively activate the ester bonds. Expanding the scope of metal oxide-based strategies beyond conventional glycolysis and hydrogenolysis, Wang and co-workers developed a Ru/Nb_2O_5 catalytic system capable of H_2-free reductive depolymerization of PET by utilizing the inherent hydrogen present in the ethylene glycol (EG) segments of the polymer itself [37]. The reaction pathway involves a sequence of hydrolysis, *in situ* EG reforming, hydrogenolysis, and decarboxylation steps. Notably, the NbOx support plays a crucial role by promoting hydrogenolysis while simultaneously suppressing unwanted decarboxylation, owing to its strong affinity for oxygen-containing intermediates. The presence of zero-valent Ru species facilitates the activation and cleavage of C–O bonds, leading to efficient conversion. Under relatively mild conditions and without external hydrogen, this system achieved 92.4% yield of aromatic hydrocarbons, demonstrating a novel, sustainable route for PET upcycling. Therefore, metal oxides offer a versatile and effective platform for PET depolymerization. Their ability to facilitate key transformations, including ester bond cleavage and *in situ* hydrogen transfer, under mild or solvent-free conditions underscores their strong potential for advancing scalable and sustainable PET recycling.

7.5 Polyvinyl Chloride (PVC)

Polyvinyl chloride is a widely used and highly versatile polymer, ranking as the third most-produced plastic globally. With an annual production of approximately 61 million tons, its demand is expected to grow at a rate of 3.2% per year. Pyrolysis is the primary method used for the depolymerization of PVC, but the release of HCl during the process poses significant challenges, including catalyst degradation and equipment corrosion. To address this, a two-step approach is commonly employed: first, PVC undergoes dehydrochlorination to form polyenes, followed by pyrolysis to produce gases, oils, and char with reduced chlorine content. Catalysts are typically introduced to enhance one or both of these steps. Ji *et al.* explored the role of transition metal oxides in enhancing the pyrolysis behavior of polyvinyl chloride (PVC) [38]. Two groups of metal oxides were investigated: Group IVB–VIB (TiO_2, V_2O_5, and MoO_3) and Group VIIB–IIB (MnO_2, CuO, Fe_2O_3, and ZnO). Thermogravimetric analysis (TG), pyrolysis-GC-MS, SEM, and XRD techniques revealed that these oxides significantly influence the thermal decomposition profile of PVC. Their presence lowered the onset temperature of dehydrochlorination, promoted earlier and more complete pyrolysis, and enhanced HCl capture, thereby reducing equipment corrosion risks and toxic byproduct formation. Importantly, the addition of these oxides suppressed the generation of polycyclic aromatic hydrocarbons (PAHs), increased aliphatic content in volatiles, and improved the residual carbon structure. Among the catalysts, V_2O_5, MoO_3, and CuO showed superior performance in reducing harmful emissions and stabilizing the carbon matrix. Overall, the findings provide a strong foundation for metal oxide-mediated, environmentally safer, and more efficient PVC recycling through catalytic pyrolysis. He *et al.* explored the low-temperature (310°C) dehydrochlorination of PVC using various metal oxides (ZnO, CuO, CaO, MgO, Fe_2O_3, and Al_2O_3), focusing on chlorine capture and product characteristics [39]. ZnO and CuO notably reduced decomposition temperatures, indicating direct interaction with PVC, with ZnO achieving 97.6% chlorine capture and minimal HCl release (0.5%). In contrast, MgO showed poor thermal stability of chloride salts, and CuO produced secondary HCl *via* side reactions. Fe_2O_3 and Al_2O_3 were less effective in chlorine retention. MOx-modified PVC showed enhanced unsaturation, with ZnO, CaO, and Al_2O_3 increasing volatile hydrocarbon release.

These findings highlight ZnO as a promising candidate for efficient PVC dechlorination and for converting PVC waste into more useful and less harmful products. Recent studies demonstrated that Zn-based mono-, bi-, and tri-metal oxides (ZnO, $ZnFe_2O_4$, $NiZnFe_4O_4$) significantly enhance the dechlorination and pyrolysis of PVC by capturing HCl and converting it to metal chlorides [40]. Among them, ZnO showed the highest HCl capture efficiency (91.13%) and suppressed toxic emissions (e.g., PAHs), favoring the formation of valuable aliphatic hydrocarbons like octane, making it a promising candidate for sustainable PVC recycling.

7.6 Conclusion and Future Perspectives

The application of metal oxide-based catalysts in the depolymerization of plastic wastes represents a promising frontier in sustainable waste management and circular chemical processing. The collected literature demonstrates that metal oxides such as ZnO, Fe_2O_3, Co_2O_3, TiO_2, and CuO, when used either in pure form or supported on porous matrices, exhibit excellent catalytic activity across diverse polymer substrates including polyolefins, polystyrene, polyethylene terephthalate, and polyvinyl chloride. These catalysts enable efficient C–C, C–O, and C–Cl bond cleavage under relatively mild reaction conditions through processes such as pyrolysis, hydrocracking, hydrogenolysis, and glycolysis. Their acid–base tunability and redox capabilities have been instrumental in achieving high conversion efficiencies, selective product distributions, and reduced formation of undesirable byproducts. The use of metal oxide-supported systems has also shown promise in suppressing toxic emissions such as polycyclic aromatic hydrocarbons during the degradation of chlorine-containing polymers like PVC. Furthermore, bifunctional and nanostructured metal oxide catalysts have demonstrated enhanced surface activity and recyclability, particularly in the upcycling of polyolefins to high-value fuels and lubricants, and in the conversion of PET to industrially useful monomers such as BHET and TA. Despite these advances, several critical challenges remain to be addressed. The translation of laboratory-scale processes to industrial applications requires significant optimization in terms of catalyst lifetime, tolerance to mixed plastic feed stocks, and economic viability. Moreover, the efficient recovery and reuse of catalysts, particularly in continuous or semi-continuous operations, is essential to ensure process sustainability. Understanding the mechanistic pathways of depolymerization and identifying the roles of active sites through advanced characterization and modeling techniques are crucial for the rational design of next-generation catalytic systems. Emphasis should also be placed on integrating these catalytic depolymerization processes with existing mechanical and chemical recycling infrastructures to create comprehensive and adaptable plastic waste valorization technologies. With continued innovation in catalyst development and process engineering, metal oxide-based systems hold strong potential to advance the global efforts in mitigating plastic pollution through scalable and environmentally benign chemical recycling approaches.

AI disclosure statement: The author confirms that no artificial intelligence (AI)-generated content or AI-assisted writing tools were used in the preparation of this chapter (including summaries and the abstract). All text, analysis, and conclusions are the result of the author's original work.

References

1. Andrady, A.L. and Neal, M.A., Applications and societal benefits of plastics. *Philos. Trans. R. Soc. Lond. B Biol. Sci.*, 364, 1977, 2009, https://doi.org/10.1098/rstb.2008.0304.
2. Market, Size Value of Plastics Worldwide from 2021 to 2030, 2022, https://www.statista.com/statistics/1060583/global-market-value-of-plastic/.
3. Pilapitiya, P.G.C.N.T. and Ratnayake, A.S., The world of plastic waste: A review. *Clean. Mater.*, 11, 100220, 2024, https://doi.org/10.1016/j.clema.2024.100220.
4. Chen, S. and Hu, Y.H., Advancements and future directions in waste plastics recycling: From mechanical methods to innovative chemical processes. *Chem. Eng. J.*, 493, 152727, 2024, https://doi.org/10.1016/j.cej.2024.152727.
5. Meneses, R.A.M., Cabrera-Papamija, G., Machuca-Martínez, F., Rodríguez, L.A., Diosa, J.E., Mosquera-Vargas, E., Plastic recycling and their use as raw material for the synthesis of carbonaceous materials. *Heliyon*, 8, e09028, 2022, https://doi.org/10.1016/j.heliyon.2022.e09028.
6. Chin, M.T. and Diao, T., Industrial and Laboratory Technologies for the Chemical Recycling of Plastic Waste. *ACS Catal.*, 14, 12437, 2024, https://doi.org/10.1021/acscatal.4c03194.
7. Bohre, A., Jadhao, P.R., Tripathi, K., Pant, K.K., Likozar, B., Saha, B., Chemical recycling processes of waste polyethylene terephthalate using solid catalysts. *Chemsuschem*, 16, e202300142, 2023, https://doi.org/10.1002/cssc.202300142.
8. Chen, S. and Hu, Y.H., Chemical recycling of plastic wastes with alkaline earth metal oxides: A review. *Sci. Total Environ.*, 905, 167251, 2023, https://doi.org/10.1016/j.scitotenv.2023.167251.
9. Zhang, F., Wang, F., Wei, X., Yang, Y., Xu, S., Deng, D., Wang, Y.-Z., From trash to treasure: Chemical recycling and upcycling of commodity plastic waste

to fuels, highvalued chemicals and advanced materials. *J. Energy Chem.*, 69, 369, 2022, https://doi.org/10.1016/j.jechem.2021.12.052.
10. Vollmer, I., Jenks, M.J.F., Roelands, M.C.P., White, R.J., van Harmelen, T., de Wild, P., van der Laan, G.P., Meirer, F., Keurentjes, J.T.F., Weckhuysen, B.M., Beyond mechanical recycling: Giving new life to Plastic waste. *Angew. Chem. Int. Ed. Engl.*, 59, 15402, 2020, https://doi.org/10.1002/anie.%20201915651.
11. Miandad, R., Barakat, M.A., Aburiazaiza, A.S., Rehan, M., Nizami, A.S., Catalytic pyrolysis of plastic waste: A review. *Process Saf. Environ. Prot.*, 102, 822, 2016, https://doi.org/10.1016/j.psep.2016.06.022.
12. Ahmad, I., Khan, M.I., Khan, H., Ishaq, M., Tariq, R., Gul, K., Ahmad, W., Influence of Metal-Oxide-Supported Bentonites on the Pyrolysis Behavior of Polypropylene and High-Density Polyethylene. *J. Appl. Polym. Sci.*, 132, 41221, 2015, https://doi.org/10.1002/app.41221.
13. Mehdi, M.S., Mehran, M.T., Naqvi, S.R., Zaman, S.U., Khoja, A.H., Bahadar, A., Catalytic pyrolysis of polyethylene waste with Fuller's earth clay and metal oxides under mild conditions. *Mater. Today: Proc.*, 57, 1322, 2022, https://doi.org/10.1016/j.matpr.2021.12.552.
14. Lin, X., Zhang, Z., Zhang, Z., Sun, J., Wang, Q., Pittman, C.U., Catalytic fast pyrolysis of a wood-plastic composite with metal oxides as catalysts. *Waste Manage. (Oxford)*, 79, 38, 2018, http://dx.doi.org/10.1016/j.wasman.2018.07.021.
15. Venkatesh, K.R., Hu, J., Wang, W., Holder, G.D., Tierney, J.W., Wender, I., Hydrocracking and hydroisomerization of long-chain alkanes and polyolefins over metal-promoted anion-modified Zirconium oxides. *Energy Fuels*, 10, 1163, 1996, https://doi.org/10.1021/ef960049j.
16. de Souza, A.S., Ferreira, P.G., de Jesus, I.S., de Oliveira, R.P.R.F., de Carvalho, A.S., Futuro, D.O., Ferreira, V.F., Recent progress in polyolefin plastic: Polyethylene and polypropylene transformation and depolymerization techniques. *Molecules*, 30, 87, 2025, https://doi.org/10.3390/molecules30010087.
17. Celik, G., Kennedy, R.M., Hackler, R.A., Ferrandon, M., Tennakoon, A., Patnaik, S., Lapointe, A.M., Ammal, S.C., Heyden, A., Perras, F.A., Marek Pruski, M., Scott, S.L., Poeppelmeier, K.R., Sadow, A.D., Delferro, M., Upcycling single-use polyethylene into high-quality liquid products. *ACS Cent. Sci.*, 5, 1795, 2019, https://doi.org/10.1021/acscentsci.9b00722.
18. Tennakoon, A., Wu, X., Paterson, A.L., Patnaik, S., Pei, Y., LaPointe, A.M., Ammal, S.C., Hackler, R.A., Heyden, A., Slowing, I.I., Coates, G.W., Delferro, M., Peters, B., Huang, W., Sadow, A.D., Perras, F.A., *Catal*ytic upcycling of high-density polyethylene *via* a processive mechanism. *Nat. Catal.*, 3, 893, 2020, https://doi.org/10.1038/s41929-020-00519-4.
19. Wu, X., Tennakoon, A., Yappert, R., Esveld, M., Ferrandon, M.S., Hackler, R.A., Lapointe, A.M., Heyden, A., Delferro, M., Peters, B., Sadow, A.D., Huang, W., Size-Controlled nanoparticles embedded in a mesoporous architecture leading to efficient and selective hydrogenolysis of polyolefins. *J. Am. Chem. Soc.*, 144, 5323, 2022, https://doi.org/10.1021/jacs.1c11694.

20. Rorrer, J.E., Beckham, G.T., Roman-Leshkov, Y., Conversion of polyolefin waste to liquid alkanes with Ru-based catalysts under mild conditions. *JACS Au*, 1, 8, 2021, https://doi.org/10.1021/jacsau.0c00041
21. Rorrer, J.E., Troyano-Valls, C., Beckham, G.T., Roman-Leshkov, Y., Hydrogenolysis of polypropylene and mixed polyolefin plastic waste over Ru/C to produce liquid alkanes. *ACS Sustain. Chem. Eng.*, 9, 35, 11661, 2021, https://doi.org/10.1021/acssuschemeng.1c03786.
22. Nakaji, Y., Tamura, M., Miyaoka, S., Kumagai, S., Tanji, M., Nakagawa, Y., Yoshioka, T., Tomishige, K., Low-temperature catalytic upgrading of waste polyolefinic plastics into liquid fuels and waxes. *Appl. Catal. B Environ.*, 285, 119805, 2021, https://doi.org/10.1016/j.apcatb.2020.119805.
23. Kots, P.A., Liu, S., Vance, B.C., Wang, C., Sheehan, J.D., Vlachos, D.G., Polypropylene plastic waste conversion to lubricants over Ru/TiO2 Catalysts. *ACS Catal.*, 11, 8104, 2021, https://doi.org/10.1021/acscatal.1c00874.
24. Zeng, M., Lee, Y.H., Strong, G., Lapointe, A.M., Kocen, A.L., Qu, Z., Coates, G.W., Scott, S.L., Abu-Omar, M.M., Chemical upcycling of polyethylene to value-added α,ω-divinyl-functionalized oligomers. *ACS Sustain. Chem. Eng.*, 9, 13926, 2021, https://doi.org/10.1021/acssuschemeng.1c05272.
25. Yeung, C.W.S., Teo, J.Y.Q., Loh, X.J., Lim, J.Y.C., Polyolefins and polystyrene as chemical resources for a sustainable future: challenges, advances, and prospects. *ACS Mater. Lett.*, 3, 1660, 2021, https://doi.org/10.1021/acsmaterialslett.1c00490.
26. Anwar, J., Munawar, M.A., Waheed-uz-Zaman, Dar, A., Tahira, U., Catalytic Depolymerisation of Polystyrene. *Prog. Rubber Plast. Recycl. Technol.*, 24, 1, 47, 2008, https://doi.org/10.1177/147776060802400104.
27. Tiwary, P. and Guria, C., Effect of Metal Oxide Catalysts on Degradation of Waste Polystyrene in Hydrogen at Elevated Temperature and Pressure in Benzene Solution. *J. Polym. Environ.*, 18, 298, 2010, https://doi.org/10.1007/s10924-010-0235-7.
28. Acuna-Bedoya, J.D., Luevano-Hipolito, E., Cedillo-Gonzalez, E.I., Domínguez-Jaimes, L.P., Hurtado, A.M., Hernandez-Lopez, J.M., Boosting visible-light photocatalytic degradation of polystyrene nanoplastics with immobilized CuxO obtained by anodization. *J. Environ. Chem. Eng.*, 9, 5, 106208, 2021, https://doi.org/10.1016/J.JECE.2021.106208.
29. Khopade, K.V., Chikkali, S.H., Barsu, N., Metal-Catalyzed Plastic Depolymerization. *Cell Rep. Phys. Sci.*, 4, 5, 101341, 2023, https://doi.org/10.1016/j.xcrp.2023.101341.
30. Spychaj, T., Chemical Recycling of PET: Methods and Products, in: *Handbook of Thermoplastic Polyesters*, Wiley-VCH Verlag GmbH & Co, Weinheim, 2002, https://doi.org/10.1002/3527601961.ch27.
31. Li, Y., Li, K., Li, M., Ge, M., Zinc-doped ferrite nanoparticles as magnetic recyclable catalysts for scale-up glycolysis of poly(ethylene terephthalate) wastes. *Adv. Powder Technol.*, 33, 103444, 2022, https://doi.org/10.1016/j.apt.2022.103444.

32. Nabid, M.R., Bide, Y., Fereidouni, N., Etemadi, B., Maghemite/nitrogen-doped graphene hybrid material as a reusable bifunctional catalyst for glycolysis of polyethylene terephthalate. *Polym. Degrad. Stab.*, 144, 434, 2017, https://doi.org/10.1016/j.polymdegradstab.2017.08.033.
33. Casey, É., Breen, R., Gómez, J.S., Kentgens, A.P., Pareras, G., Rimola, A., Holmes, J.D., Collins, G., Ligand-Aided Glycolysis of PET Using Functionalized Silica-Supported Fe2O3 Nanoparticles. *ACS Sustain. Chem. Eng.*, 11, 15544, 2023, https://doi.org/10.1021/acssuschemeng.3c03585.
34. Son, S.G., Jin, S.B., Kim, S.J., Park, H.J., Shin, J., Ryu, T., Jeong, J.M., Choi, B.G., Exfoliated manganese oxide nanosheets as highly active catalysts for glycolysis of polyethylene terephthalate. *Flatchem*, 36, 100430, 2022, https://doi.org/10.1016/j.flatc.2022.100430.
35. Yang, Y., Chen, F., Shen, T., Pariatamby, A., Wen, X., Yan, M., Kanchanatip, E., Catalytic depolymerization of waste polyethylene terephthalate plastic in supercritical ethanol by ZnO/γ-Al2O3 catalyst. *Process Saf. Environ. Prot.*, 173, 881, 2023, https://doi.org/10.1016/j.psep.2023.04.001.
36. Kratish, Y., Li, J., Liu, S., Gao, Y., Marks, T.J., Polyethylene terephthalate deconstruction catalyzed by a carbon-supported single-site Molybdenum-dioxo complex. *Angew. Chem. Int. Ed.*, 59, 19857, 2020, https://doi.org/10.1002/anie.202007423.
37. Lu, S., Jing, Y., Feng, B., Guo, Y., Liu, X., Wang, Y., H2-free Plastic Conversion: Converting PET back to BTX by Unlocking Hidden Hydrogen. *ChemSusChem*, 14, 4242, 2021, https://doi.org/10.1002/cssc.202100196.
38. Ji, M., Chen, L., Que, J., Zheng, L., Chen, Z., Wu, Z., Effects of transition metal oxides on pyrolysis properties of PVC. *Process Saf. Environ. Prot.*, 140, 211, 2020, https://doi.org/10.1016/j.psep.2020.04.010.
39. Meng, T.-T., Zhang, H., Lü, F., Shao, L.-., He, P.-J., Comparing the effects of different metal oxides on low temperature decomposition of PVC. *J. Anal. Appl. Pyrolysis*, 159, 105312, 2021, https://doi.org/10.1016/j.jaap.2021.105312.
40. Ali, L., Kuttiyathil, M.S., Mousa, H.A., Teoh, W.Y., Altarawneh, M., Conversion of PVC wastes to fuel-based chemicals *via* thermal recycling with Zn mono-, Bi-, and Tri-metal oxides. *J. Hazard. Mater. Adv.*, 18, 100717, 2025, https://doi.org/10.1016/j.hazadv.2025.100717.

8

Advanced Catalytic Methods for Plastic Depolymerization with Metallic Moieties

Abhay Nanda Srivastva[1]*, Kriti Milan[1], Nisha Saxena[2], Anindra Sharma[3] and Maneesh Kumar Gupta[4]

[1]*Laboratory of Bioactive Material Synthesis, University Department of Chemistry, B.R.A. Bihar University, Muzaffarpur, Bihar, India*
[2]*P.G. Department of Chemistry, M.R.M. College (A Constituent Unit of Lalit Narayan Mithila University), Darbhanga, Bihar, India*
[3]*P.G. Department of Chemistry, Lalit Narayan Mithila University, Darbhanga, Bihar, India*
[4]*Department of Chemistry, Hotilal Ramnath College (A Constituent Unit of J. P. University, Chapra), Chapra, Bihar, India*

Abstract
According to data from the UN Environment, approximately 400 million tons of plastic waste is produced globally each year, and this figure is projected to rise to about 1100 million tons by 2050. Plastic pollution is a serious global concern as it continues to pollute our environment. It affects human health, causing lethal diseases like cancer, cardiovascular diseases, respiratory disorders, and metabolic disorders. Preventing these hazardous effects is the need of the hour. Despite their wide range of applications, the non-biodegradability of plastics and the rising volume of plastic waste pose a significant threat to the global ecosystem. Depolymerization offers an efficient route to convert plastic polymers into simple monomers for recycling. Catalytic depolymerization is an emerging technique that breaks down long-chain polymers into simpler monomers, which can be reused as starting synthons for the preparation of materials with versatile applications. Numerous catalysts have been developed and gained recognition for their ability to depolymerize plastics, thereby reducing their harmful environmental impacts.

**Corresponding author*: abhay.chem.brabu@gmail.com

Neha Agarwal, Brijesh Pare, Vijendra Singh Solanki and Sankara Rao Miditana (eds.) Plastic Remediation: Transformative Pathways for a Rejuvenated Sustainable Environment, (187–210)

Metal-based compounds, particularly metal complexes and metal–organic frameworks have shown high efficiency in catalytic depolymerization, significantly mitigating the effects of plastic pollution. This chapter explores advanced catalytic strategies for plastic depolymerization, focusing on metal-based compounds such as metal complexes and metal–organic frameworks, which are promising catalysts for converting plastics into valuable, eco-friendly small molecules.

***Keywords*:** Plastic upcycling, transition metal, metal complexes, environment sustainability, plastic pollution management

8.1 Introduction

The escalating use of plastics in various fields is continuously increasing, leading to a corresponding amplification of plastic waste worldwide [1]. Millions of tons of plastic waste not only contribute to soil infertility but also adversely affect crops grown in polluted soil [2, 3]. Their harmful effects can lead to severe consequences, sometimes even resulting in death [4]. A simple approach to mitigating plastic pollution is encapsulated in the 5 R's—Refuse, Reduce, Reuse, Repurpose, and Recycle. Plastics are ubiquitously used polymeric materials, valued for their numerous properties such as cost-effectiveness, high durability, light weight, and the easy availability of raw materials for manufacturing [5]. The extensive use of plastics has become a major cause of plastic pollution, adversely affecting marine life, while microplastics have also emerged as a serious threat to human health [6, 7]. A review of studies on the consumption and properties of plastics and plastic-based products makes it clear that an alarming situation has arisen. This compels the global scientific community, particularly those in polymer industries and environmental chemistry, to develop concrete solutions for reducing the adverse impacts of plastic waste without compromising the benefits to industry and consumers. Plastics are formed through the polymerization of long chains of monomers, while depolymerization involves the reverse process [8]. Catalytic depolymerization simplifies these long-chain polymers into smaller monomers through the use of a catalyst [9].

In chemistry, a catalyst makes a reaction more feasible by lowering the activation energy, thereby reducing the time required and making the process more time-efficient [10]. Metals are effective catalysts in many chemical reactions due to their high selectivity and the fact that they remain chemically unchanged during the reaction [11]. Metal catalysts have been widely used to synthesize a plethora of compounds efficiently.

Consequently, depolymerizing plastics using metal complexes and metal–organic frameworks is a highly effective and emerging approach that is attracting significant attention from material scientists worldwide. Metal complexes, possessing versatile metallic moieties coordinated with a variety of lone-pair–donating ligands, are among the most effective tools for the catalytic depolymerization of plastics due to their unique structural properties. Similarly, metal organic frame works (MOFs) also provide a suitable platform for the depolymerization of plastics due to their unique mode of binding metal and organic moieties as well as structural layering of materials. The present chapter provides a synopsis of recent advancements in metal-based catalytic depolymerization of plastics, along with the emerging possibilities in this field. The discussion is supported by insights drawn from several contemporary research articles and reviews.

8.2 Metal-Based Catalytic Depolymerization

Metal plays an important role in the depolymerization of plastics [12]. Metal-based compounds such as metal complexes, metal oxides, and metal organic framework escalate the depolymerization to a greater extent [13–15]. In depolymerization of plastic polymers, metal moiety-based catalysts are expediting the reaction rate [16].

8.2.1 Metal Complex-Based Depolymerization

Metal complexes are the class of compounds having versatile applications [17–19]. Applications of metal complexes to catalyze the depolymerization of plastic have increased and shown promising results [20]. Their specificity, recyclability, and the ability to react under mild conditions provide a sustainable pathway for the depolymerization of plastics [21]. Many classes of polymers are depolymerized and upcycled to new useful products [22]. Poly-cyclopentene carbonate (PCPC) is a polymer having applications in several areas due to its high modulus and glass transition temperature [23]. Deng *et al.* reported selective depolymerization based on inner transition Zn(II) bimetallic complexes to depolymerize the PCPC into its monomers, cyclopentene oxide and carbon dioxide. The catalyst provides a high depolymerization yield. Thermogravimetric analysis and Arrhenius plots suggest a high energy barrier for the depolymerization of PCPC which was overcome using only 0.2 mol % of metal complex at 160°C temperature. The depolymerization mechanism is believed to proceed in two steps: initial chain scission followed by chain backbiting (Scheme 8.1). Among the

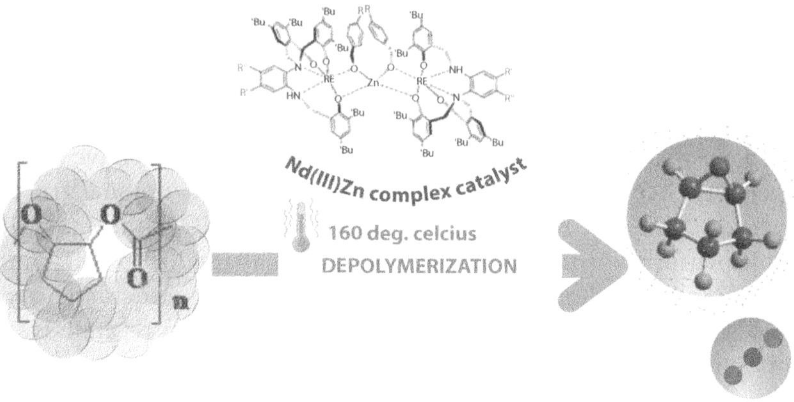

Scheme 8.1 Depolymerization of PCPC by Nd(III)/Zn(II) catalyst.

tested catalysts, the Nd(III)/Zn(II) complex exhibited the highest catalytic potency, achieving a 93% yield of depolymerized monomers. These monomers can be repolymerized into PCPC, thereby establishing a sustainable monomer–polymer–monomer–polymer cycle [24].

Bimetallic metal complexes show significant catalytic properties for the depolymerization of plastic [25]. Yu *et al.* used a Cr(III) bimetallic complex to depolymerize PCPC into cyclohexene oxide with more than 99% selectivity. *In situ* studies were performed to explain the depolymerization of PCPC *via* Cr(III) complexes. The thermal and robust nature of the catalyst complex provides a suitable mechanism which converts the polymer into monomer by heating at 120°C for 12 h. NMR and gel permeation chromatographic techniques were used to track the depolymerization process where it was observed that instead of random chain scission, an unzipping process was followed. DFT studies also revealed that the nucleophilic attack leading to epoxide and carbonate formation is catalyzed by a bimetallic process, resulting in a lower energy barrier that is easier to overcome, thereby making the process more cost-effective [26]. Depolymerization of numbers of aliphatic as well as aromatic polymers was reported using Mn(II) complex as catalysts. Borane and silane were utilized as reducing agent providing specific catalytic properties for the depolymerization. Four Mn(II) complexes viz. $Mn(OTf)_2$, $Mn_2(CO)_{10}$, $MnBr(CO)_5$, and $MeCpMn(CO)_3$, and SiMn were observed as a cheaper and sustainable depolymerizing catalysts as they were able to perform 14 cycles giving excellent yield. ^{1}H NMR spectroscopy was used to determine the yield, which suggested high range of yield 91–95% obtained in the depolymerization of polycaprolactone (PCL) by the $Mn(OTf)_2$ and $Mn(CO)_2$ catalysts along with reducing agents [27]. Monsigny *et al.* adopted a hydrosilylation technique for the depolymerization of polymers such as polyethylene terephthalate (PET) which is generally present in the plastic wastes and also of polylactic acid (PLA) which is a constituent of 3D printer filaments. Brookhart's Ir(III) pincer complex-based catalysis was performed for the depolymerization of both the polymers. Two different conditions were provided with the Brookhart's catalyst. In the first condition, the polymers of plastic wastes were depolymerized in the presence of aforesaid catalyst along with Et_3SiH (Scheme 8.2).

In other conditions, a methyl-substituted derivative of disiloxane (TMDS) which is a cost-effective source of hydrides was utilized along with Ir(III) complex-based catalyst for the depolymerization of PLA and PCL (Scheme 8.3).

The reductive depolymerization of PLA results in 100% conversion and gives hydrocarbons as product which can be used as biofuels. The Lewis base type cationic pincer complex of Ir(III) provides a suitable path for the depolymerization of plastic wastes (Scheme 8.4) [28].

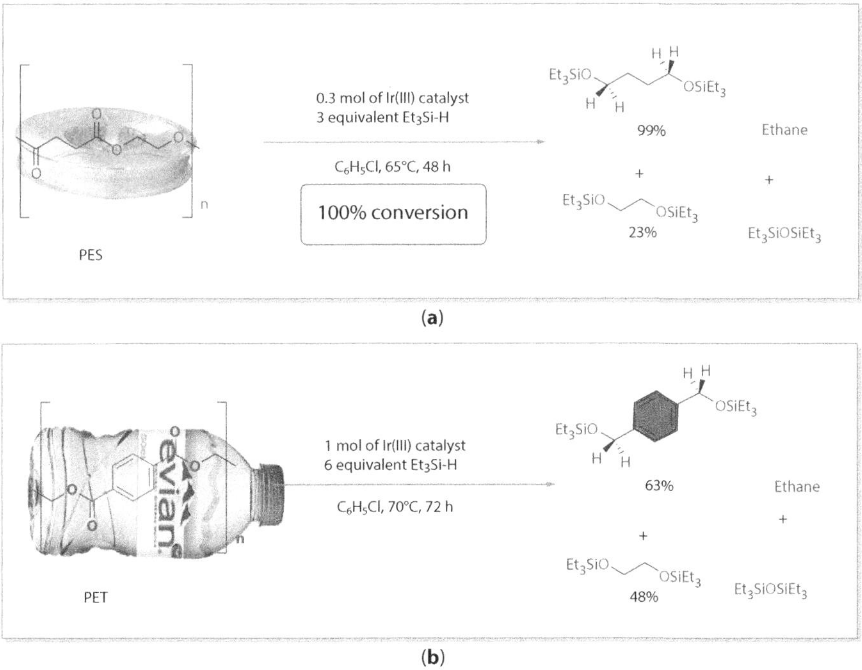

Scheme 8.2 Depolymerization of (a) PES from wiring and (b) PET from a plastic bottle using Brookhart's catalyst.

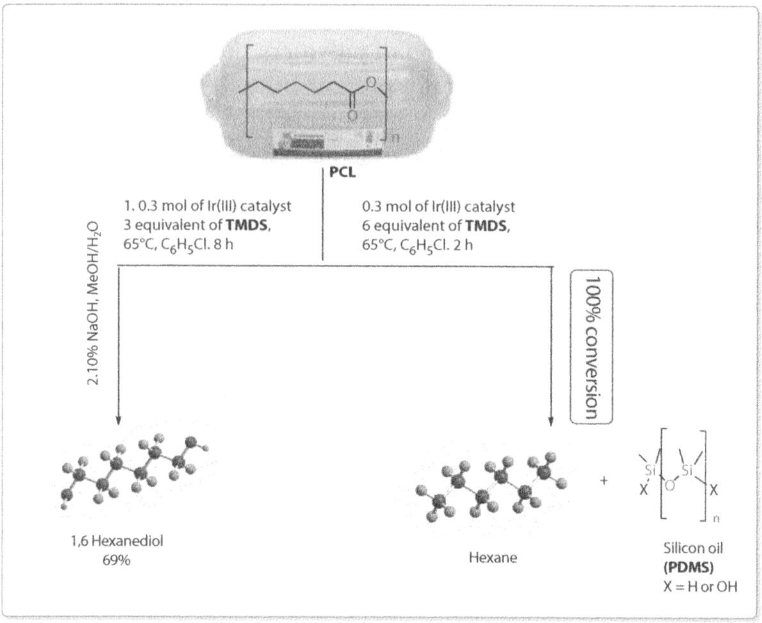

Scheme 8.3 Reductive depolymerization of PCL by two routes catalyzed by Ir(III) complexes.

Polyurethanes are widely found in our environment as they are being used as foam in car and furniture seats, as dampers in shoes, in medical devices, and also in packaging [29]. Zubar *et al.* performed depolymerization of polyurethane catalyzed by a series of Mn-pincer complexes. The hydrogenation route was adopted for the depolymerization which results in the formation of 89% polyetherols and 76% diaminotoluene.

Hydrogenation of self-made polyurethane is shown in Scheme 8.5. After this, the group also performed depolymerization of polyurethane in real-life samples. The depolymerization of 2g PU in standard foam yields 76% diaminotoluene and 89% Lupranol (Scheme 8.6).

A black packaging material made of polyurethane was depolymerized under same conditions showing 66% conversion (Scheme 8.7). Similar results were obtained when depolymerization of PU samples from white packaging material, end-of-life mattresses, and rigid PU foam were done. The proposed mechanism of hydrogenation-assisted depolymerization suggests polyol was formed by the cleavage of C-O bond in the first step followed by hydrogenation of formate into formaldehyde and amine.

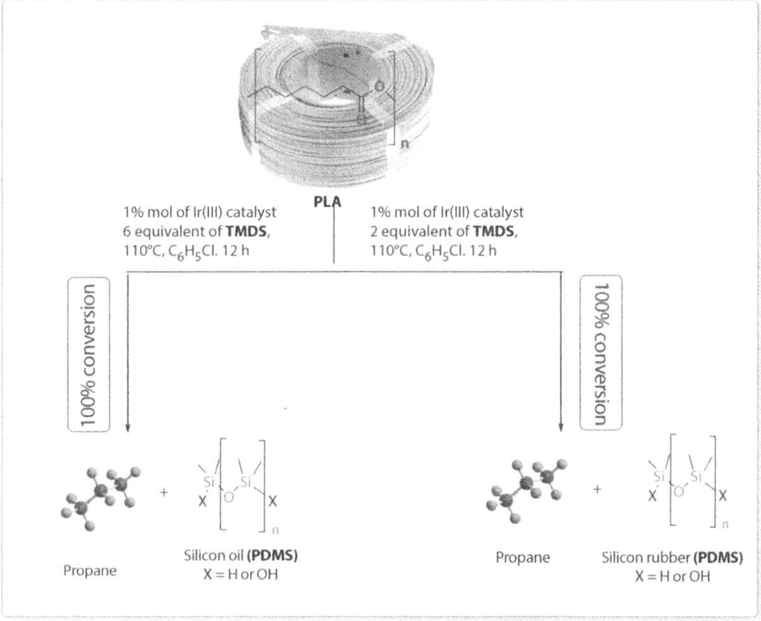

Scheme 8.4 Reductive depolymerization of PLA into propane by two different routes catalyzed by Ir(III) complexes.

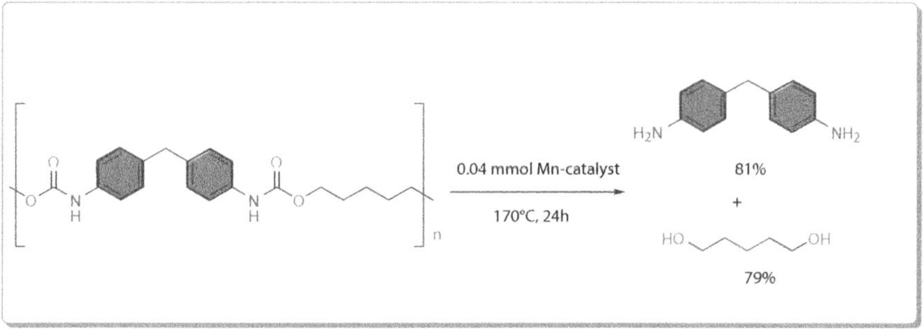

Scheme 8.5 Depolymerization of PU in the presence of a Mn-complex-based catalyst.

Elevation in temperature also results in effective depolymerization of polyurethanes making this approach more attractive and significant [30].

Nylon-6 is utilized in various areas due to its robustness and durability and is one of the classes of polymer which are non-biodegradable [31]. Thus, waste generated from Nylon-6 is a matter of concern for our environment. To mitigate the hazardous effects caused by Nylon-6 wastes,

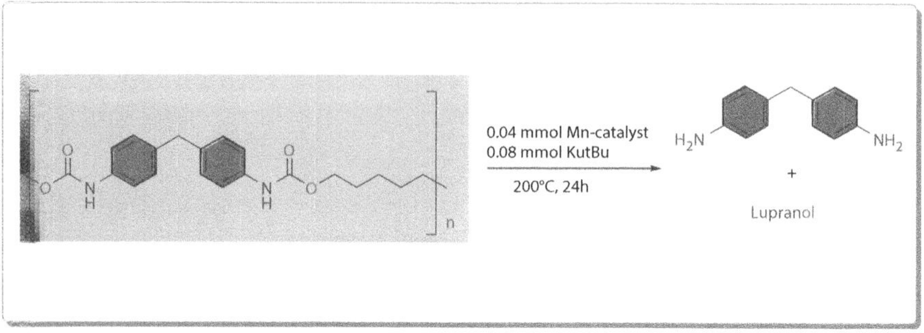

Scheme 8.6 Depolymerization of PU from foam in the presence of a Mn-complex-based catalyst.

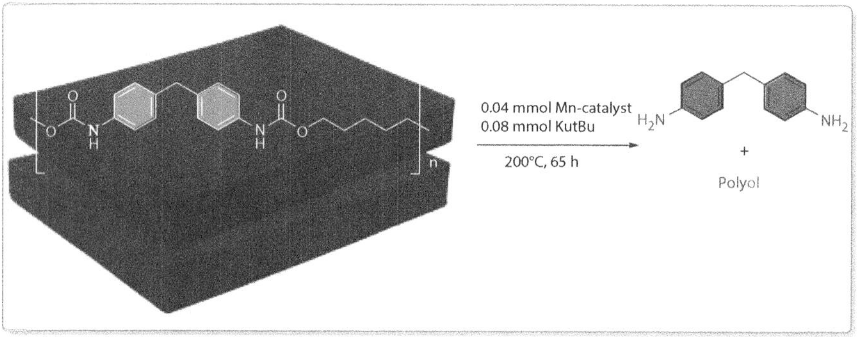

Scheme 8.7 Depolymerization of PU from black packaging material in the presence of a Mn-complex-based catalyst.

Ye *et al.* reported a solvent-free approach for the depolymerization of Nylon-6. More than 99% yield of monomer caprolactam was achieved using lanthanide and yttrium metallocene and ansa-metallocene catalyst. Four metal-based catalysts were used to depolymerize Nylon-6 in different plastic wastes giving different yields (Figure 8.1).

Recyclability of catalyst was also tested which shows that the catalyst can work 6 cycles without affecting the depolymerization properties. The proposed mechanism of depolymerization suggests that unzipping, followed by a π-ligand effect and subsequent interchain hopping of La^{3+} within seconds, is responsible for the depolymerization of Nylon-6 [32]. Smite *et al.* further reported the depolymerization of poly(cyclohexene carbonate) (PCHC) using a heterodinuclear metal complex as the catalyst (Scheme 8.8).

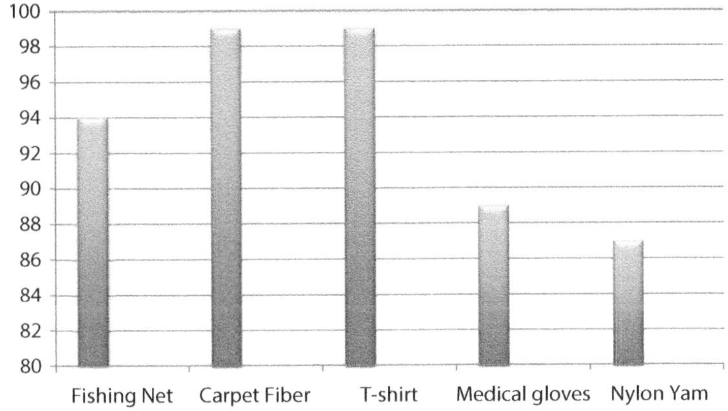

Figure 8.1 % Yield obtained of monomer after depolymerization of Nylon-6 from various plastic waste sources.

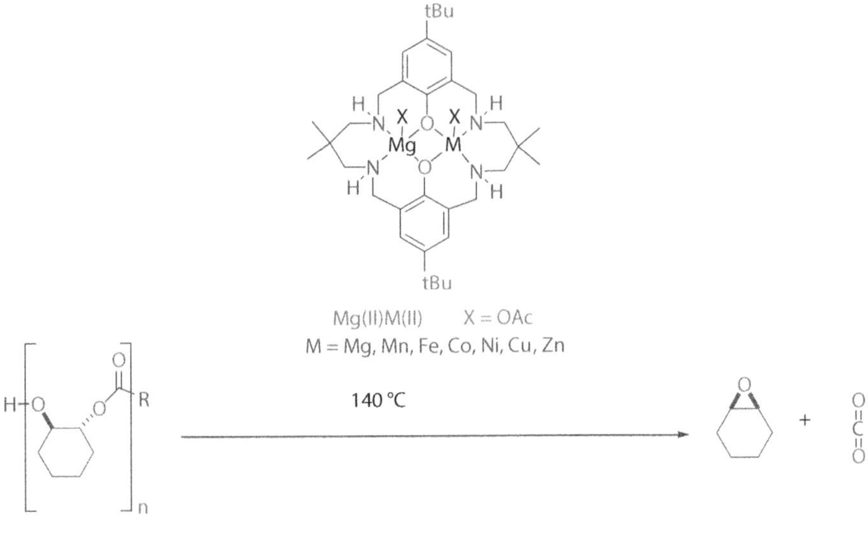

Scheme 8.8 Depolymerization of PCHC in the presence of a Mg(II)M(II) complex-based catalyst.

A series of Mg(II)M(II) where M = Mn, Fe, Ni, Cu, and Zn type complexes were used as catalyst. All the complexes depolymerize (PCHC) successfully in less than 20 min and achieve more than 90% mass loss. The polymer reacts with catalyst and forms a metal alkoxide intermediate which is referred to as the true catalyst which results in the monomers (Scheme 8.9).

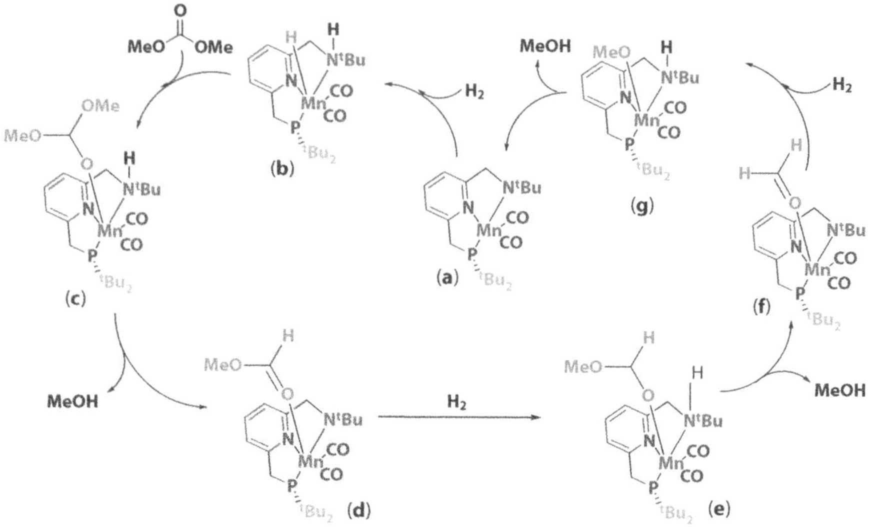

Scheme 8.9 Depolymerization of PPC in presence of Mn complex based catalyst.

The cycle of depolymerization continues as the intermediate again reacts with chain. Nucleophilic attack is rate limiting in the reaction. The Mg(II) M(II) complex catalysts are plausible for depolymerization of plastics due to high selectivity active and are capable to work at low temperature [33]. Two catalysts, both pincer-type ruthenium complexes (Figure 8.2), were employed under optimized reaction conditions using a solvent mixture of anisole and THF at 160°C with a catalyst loading of 1 mol %. Under these conditions, near-complete conversion of both poly(propylene carbonate) (PPC) and polyethylene carbonate (PEC) to their corresponding diols was achieved, demonstrating the broad substrate scope and high efficiency of the catalytic system [34].

The Ir-pincer complex (Figure 8.3), in combination with either triethylsilane (Et_3SiH) or 1,1,3,3-tetramethyldisiloxane (TMDS), was employed as an efficient catalyst for the depolymerization of polycarbonates (Scheme 8.2). When applied to poly(propylene carbonate) (PPC), the system yielded disilylether glycol (PG–Si) and $MeOSiEt_3$ in high yield. Similarly, depolymerization of bisphenol A-based polycarbonate (PC–BPA) produced a BPA–Si intermediate in 88% yield within 6 hours. Subsequent alkaline workup of the TMDS-based reaction further afforded 83% pure BPA. These results highlight the effectiveness of this catalytic protocol for aromatic polycarbonate recycling under relatively mild conditions [28].

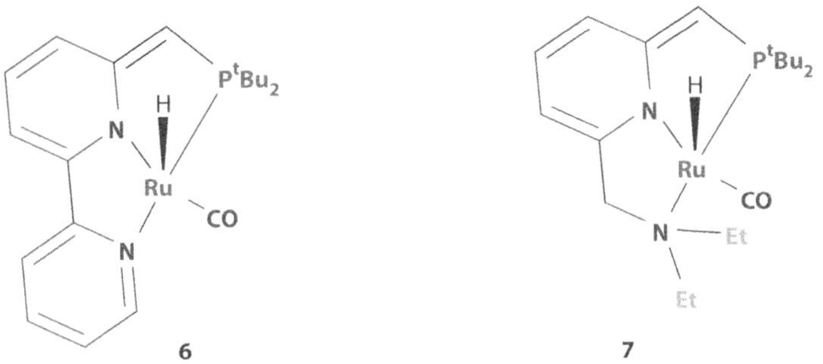

Figure 8.2 Structure of Ru-pincer complexes.

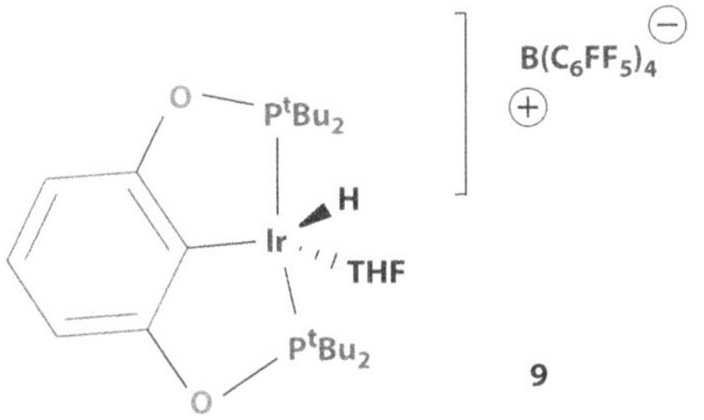

Figure 8.3 Structure of Ir-pincer complex.

Using 2 mol % of the manganese PNP pre-catalyst under 50 bar of hydrogen pressure, poly(propylene carbonate) (PPC) was effectively depolymerized, yielding 1,2-propanediol (68%) and methanol (59%), along with 30% unconverted propylene carbonate (Scheme 8.9). The proposed mechanism begins with the dehydrobromination of the pre-catalyst, generating the active amido complex (a). Subsequent addition of H_2 forms the hydride-containing species 26 (b), which facilitates hydride transfer to the carbonate, producing intermediate 27 (c). Through sequential cycles of hydride and proton transfer, the catalyst is regenerated, and the alcohol products are released, completing the catalytic cycle [35].

The Ru complex (Figure 8.4 (a)), used in conjunction with the co-catalyst $HNTf_2$, demonstrated exceptional efficiency in the depolymerization of bisphenol A-based polycarbonate (PC–BPA), yielding bisphenol A (BPA) and methanol in quantitative amounts (Figure 8.4). Notably, this catalytic system maintained high performance even when applied to post-consumer polycarbonate waste, such as compact discs (CDs), underscoring its robustness and practical relevance for real-world plastic recycling applications [36].

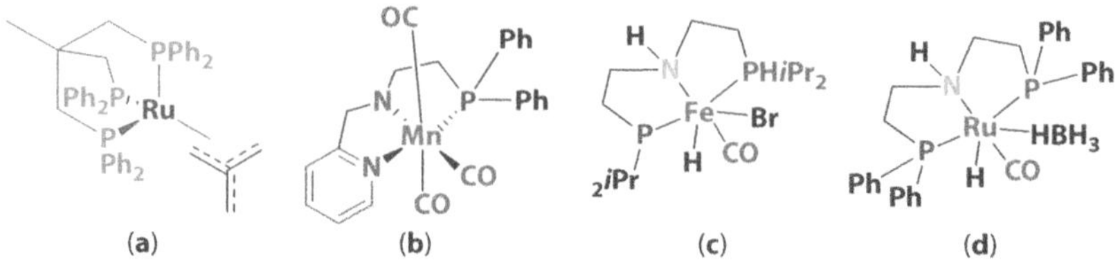

Figure 8.4 Structure of metal complex-based catalysis applied to depolymerize plastic polymers.

An air- and moisture-tolerant Mn–PNN complex (b) was employed at 1 mol % loading with 2.5 mol % KOtBu under 50 bar of hydrogen pressure, demonstrating high catalytic efficiency in the depolymerization of polycarbonates. This system yielded 1,2-propanediol in 91% and methanol in 84% yield. Mechanistically, the reaction proceeds *via* a stepwise hydrogenation pathway in which the carbonate intermediate is first converted to a formate species, followed by further hydrogenation to ethylene glycol and formaldehyde, and ultimately to methanol [37].

The Fe–PNP pincer complex (c) was utilized at a 1 mol % loading with 5 mol % KOtBu in iso-propanol as the solvent at 140°C, facilitating the depolymerization of polycarbonates. Under these conditions, the reaction yielded 1,2-propanediol in 65% and methanol in 43%. The proposed catalytic mechanism involves initial alcohol dehydrogenation of iso-propanol to generate a metal hydride species, which subsequently transfers the hydride to the polycarbonate substrate. This process culminates in the release of 1,2-propanediol and methanol, along with the formation of acetone as a byproduct derived from iso-propanol [38].

The MACHO-type Ru pincer complex (d) demonstrated outstanding catalytic performance in the depolymerization of bisphenol A-based polycarbonate (PC–BPA), achieving bisphenol A (BPA) yields of up to 99% under mild conditions (5 bar H_2, 80°C) with ligand loadings ranging from 0.5 to 5 mol %. The system exhibited an excellent turnover number (TON) of approximately 142 h^{-1}. Notably, this catalyst operates efficiently without the need for a base and significantly outperforms the classical Milstein-type pincer complexes in terms of both reaction mildness and overall efficiency [39, 40].

8.2.2 Metal Organic Framework-Based Depolymerization

The exponential growth in plastic production and its environmental impact have made plastic waste management a major global concern.

Among various emerging solutions, the chemical recycling of plastics *via* depolymerization holds immense promise for converting waste polymers back into valuable monomers. Metal-organic frameworks (MOFs), with their high surface area, tunable porosity, and functional catalytic sites, have recently emerged as promising materials in this context [73]. MOF-based depolymerization represents a significant leap toward sustainable plastic recycling, particularly in challenging polymers like polyethylene (PE), polypropylene (PP), polyethylene terephthalate (PET), and polycarbonates [41–45]. This portion highlights recent advancements in MOF-assisted depolymerization processes, focusing on pyrolytic, photochemical, and catalytic approaches while linking experimental outcomes to structural MOF properties. Recent breakthroughs have shown that Zr-based MOFs, particularly the UiO-66 family, can catalyze the thermal depolymerization of polyolefins like PE and PP. Heng *et al.* demonstrated that defective UiO-66, enriched with Lewis-acidic zirconium nodes, facilitates C–C bond cleavage during pyrolysis at 400°C. Their study revealed that such catalytic activity significantly enhances gas and liquid yields while reducing solid residue formation compared to uncatalyzed pyrolysis (Scheme 8.10).

The catalytic performance is attributed to the MOF's structural robustness at high temperatures, combined with its ability to stabilize transition states through Zr–O coordination. The presence of missing-linker defects further improves active-site accessibility, enhancing reaction kinetics. These features make UiO-66 a viable catalyst for polyolefin recycling, offering an alternative to traditional, energy-intensive cracking methods [46]. Furthermore, an efficient example of MOF is the defective sulfonated

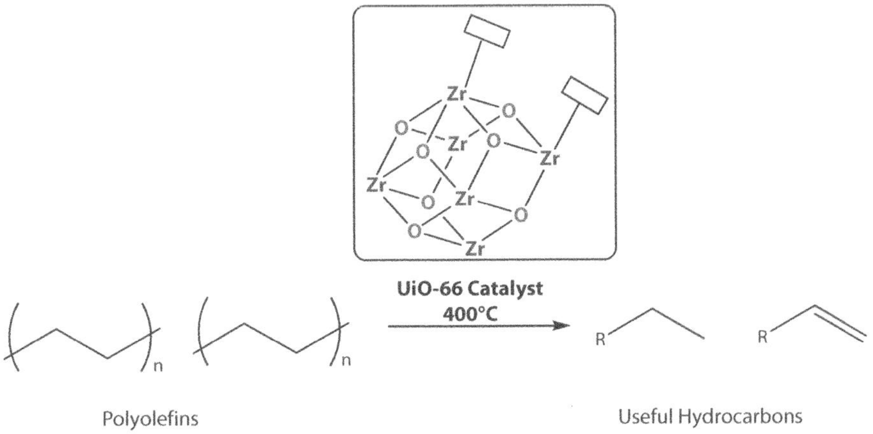

Scheme 8.10 Depolymerization of polyolefins catalyzed by Zr-based MOF.

UiO-66$_{def-}$SO$_3$H-supported Ru catalyst reported by Ou *et al.* that was utilized for the direct conversion of polymeric waste polyethylene terephthalate (PET) to dimethyl cyclohexane-1,4-dicarboxylate (DMCD) resulting in a yield of 97.7% at 170°C after 6 hours (Scheme 8.11) [47].

Another promising strategy is photochemical depolymerization using MOFs as photoreactive scaffolds. In-Hyeok Park *et al.* reported the formation and cleavage of cyclobutane-based polymers within a Zn-based MOF *via* reversible [2+2] cycloaddition and thermal depolymerization. The unique feature of this system lies in the precise spatial confinement provided by the MOF's pores, which ensures orderly polymerization and enables selective depolymerization upon heating to 250°C. This method exemplifies a controlled, stimuli-responsive depolymerization process, showcasing the potential of MOFs not just as catalysts but as dynamic reaction vessels [48]. Similarly, work by Chuanyin Tang *et al.* explored light-driven depolymerization of photosensitive polymers using porphyrinic MOFs as photochemical antennas, achieving depolymerization under ambient conditions (Scheme 8.12). The capability of photocatalytic degradation of organic pollutants over porphyrin-based MOFs can be modified by the choice and assembly of porphyrin ligands along with metal nodes. PCN-224 (Zr/Ti), incorporating bimetallic nodes, was synthesized through a simple cation exchange process to partially substitute Zr with Ti in the Zr-porphyrin-based MOF (PCN-224) [49].

Suo *et al.* demonstrated MOF-catalyzed PET glycolysis using ZIF-8 to convert PET into BHET and its dimer (Scheme 8.13). Under the influence of ZIF-8, PET pellets can be entirely degraded in 1.5 hours at 197°C, with BHET monomer and its dimers being established as the primary product and by-products, respectively [50].

Wang *et al.* also reported CoNi-based metal–organic framework PET glycolysis as Co0.8Ni-MOF that is applicable as anode for lithium-ion batteries [51]. Furthermore, Trzebiatowska *et al.* explored the use of MOFs and post-synthetically modified MOFs in the catalytic depolymerization of

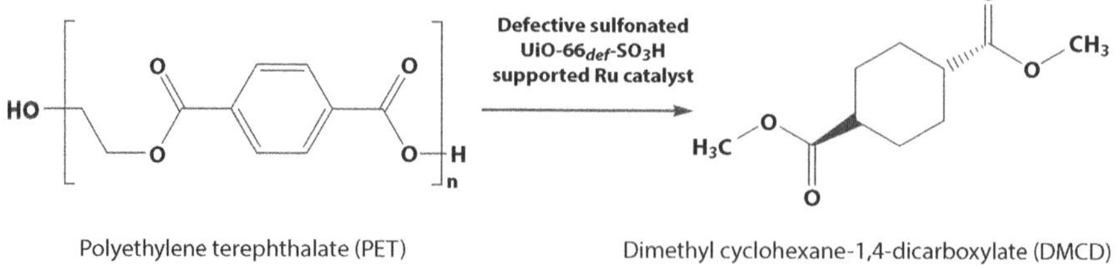

Scheme 8.11 Depolymerization of PET catalyzed by U-based MOF supported by Ru.

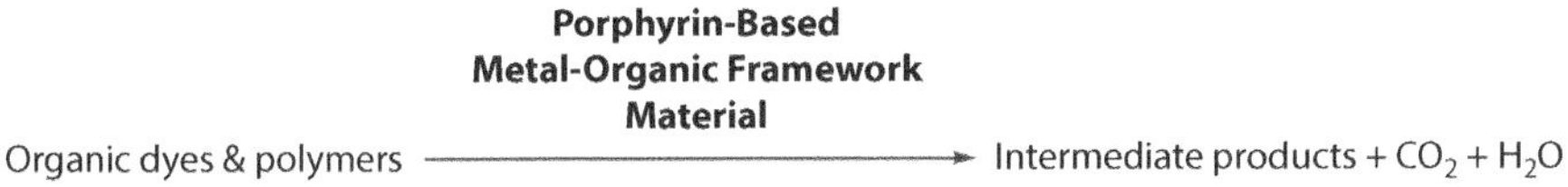

Scheme 8.12 Depolymerization of organic pollutant in the presence of porphyrin-based MOF.

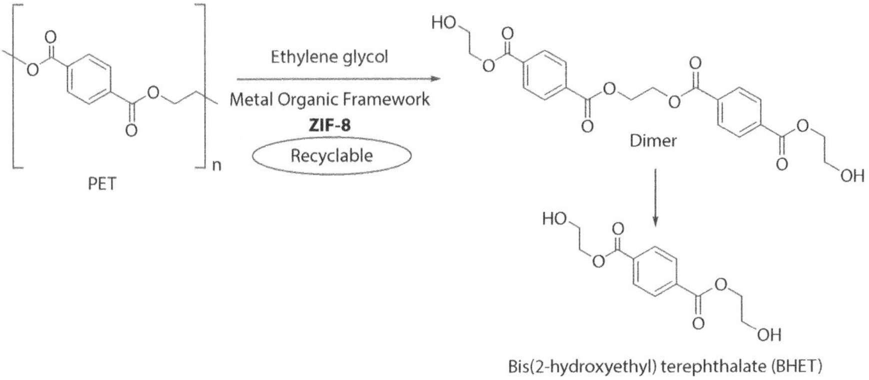

Scheme 8.13 Depolymerization of PET polymer.

polycarbonates using methanolysis. Their study identified various frameworks that could effectively promote the breakdown of carbonate linkages, enabling monomer recovery under milder conditions. The study revealed that catalysts based on MOFs (UiO-66-NH2, MAF-6, ZIF-8) exhibited catalytic activity for the selective cleavage of ester bonds in the polymer backbone, resulting in 49%, 89%, and 94% conversion of PC, respectively. An increase in reaction temperature consistently improved PC conversion and BPA yield for all catalysts, with notable enhancements observed at 130°C (Scheme 8.14). The superior performance of ZIF-8 and MAF-6 can be attributed to their basic active sites and advantageous morphologies, whereas the lower yields of UiO-66-NH2 are likely due to limited mass transfer and fewer exposed active sites. Furthermore, ZIF-8 showed high efficiency in larger-scale reactions and with post-consumer waste, highlighting the practical potential of MOF-based catalysts for PC recycling [52].

Yufang Wu *et al.* reported MOF which is Zirconium-based represented as (Zr-MOF) UiO-66, effectively catalyzed and de-polymerized polyethylene terephthalate (PET) into terephthalic acid (TA) under hydrogenolysis conditions (Scheme 8.15). During the de-polymerization and degradation

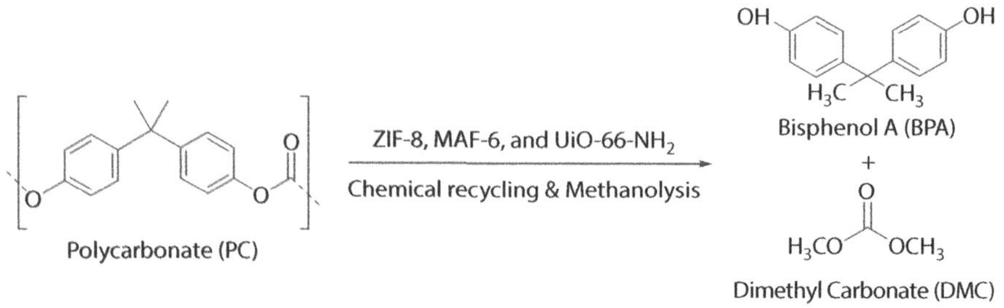

Scheme 8.14 Depolymerization of PC in presence of three different MOFs.

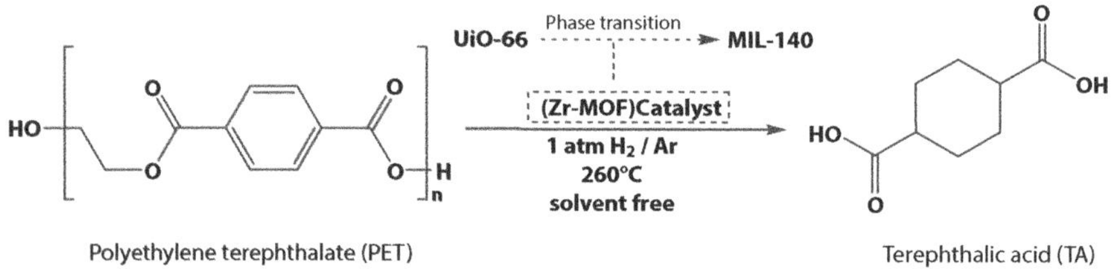

Scheme 8.15 PET depolymerization by Zr-MOF catalyst.

process, the UiO-66 framework undergoes phase transition into the MIL-140 framework that also effectively catalyzes and degrades the polymer in good yield [53].

Enzyme-MOF composites can be created by attaching enzymes to the outer surface of the MOF, which allows for enzyme heterogenization while keeping them accessible for reactions with plastic substrates. For example, *Candida rugosa* lipase (CrL) anchored on UiO-66-NH$_2$ was able to decompose PET into BHET. Various MOFs were examined as potential supports (such as UiO-66, UiO-66-COOH, UiO-66-NO2, MIL-125, MIL-100(Fe), and ZnBDP) using BHET as a model substrate, but UiO-66-NH$_2$ was selected due to its ability to facilitate high surface enzyme attachment, outstanding chemical and structural stability that preserved crystallinity post-CrL immobilization, and effectiveness in BHET hydrolysis (Scheme 8.16) [54]. Importantly, PET degradation was exclusively noted for the CrL-anchored MOF, and not when CrL or UiO-66-NH$_2$ was used independently, which was linked to the complex's preferential adhesion to the plastic, enhancing interfacial contact for hydrolysis [54, 55].

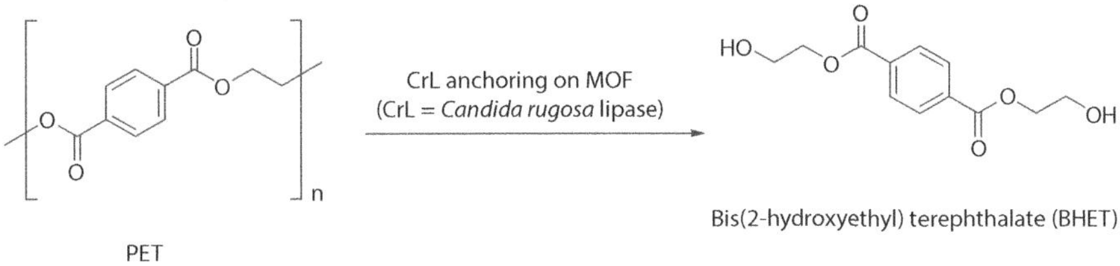

Scheme 8.16 Catalytic depolymerization of PET in the presence of CrL anchoring MoF.

8.3 Conclusions and Future Prospects

The present recapitulation suggests that metallic moieties in the form of metal complex and metal organic framework are a promising catalytic agent to depolymerize the plastic into valuable eco-friendly small industrial molecules and thus can reduce the environmental disruption caused by plastic wastes. Two different moieties show excellent catalytic properties due to their numerous characteristics favoring the surface creation for plastic depolymerization reactions. These metallic compounds thus have the potency to eliminate the threat of environment and promote sustainability. The responsible factors of governing the catalytic potential of these metallic moieties are the oxidation state of metal ions, arrangements of ligand moieties around the metallic ions, variety of ligands coordinated to metallic moieties, and/or types of cluster formation. The mechanistic approaches of plastic depolymerization also move around these factors. Though, many effective metallic catalysts have been discovered, but still the rigorous research in the designing of highly potent metal complexes and MoFs are in need of time to develop the best-fit catalyst for specific types of polymeric plastics decomposition. Such kind of specific catalyst will be easier to be opted by polymer industries for plastic waste management in a cost-effective and non-lethargic manner. The integration of structural design, catalysis, and green chemistry principles in metal-based catalysts will likely pave the way for more efficient, selective, and environmentally friendly depolymerization technologies. Future research should focus on catalyst recyclability, process scalability, and integration with existing waste management infrastructure to promote industrial adoption and environmental sustainability.

Acknowledgements

All the authors would like to convey their thanks to authorities of their respective institutions for providing the necessary facilities to carry out the present work.

AI disclosure statement: The author confirms that no artificial intelligence (AI)-generated content or AI-assisted writing tools were used in the preparation of this chapter (including summaries and the abstract). All text, analysis, and conclusions are the result of the author's original work.

References

1. Pilapitiya, P.G.C.N.T. and Ratnayake, A.S., The world of plastic waste: A review. *Cleaner Mater.*, 11, 100220, 2024, https://doi.org/10.1016/j.clema.2024.100220.
2. Sa'adu, I. and Farsang, A., Plastic contamination in agricultural soils: a review. *Environ. Sci. Eur.*, 35, 13, 2023, https://doi.org/10.1186/s12302-023-00720-9.
3. Hartmann, G.F., Ricachenevsky, F.K., Silveira, N.M., Barbosa, A.P., Phytotoxic effects of plastic pollution in crops: what is the size of the problem?. *Environ. Pollut.(b)*, 292, 118420, 2022, https://doi.org/10.1016/j.envpol.2021.118420.
4. Sheriff, S.S., Yusuf, A.A., Akiyode, O.O., Hallie, E.F., Odoma, S., Yambasu, R.Al., Thompson-Williams, K., Asumana, C., Gono, S.Z., Kamara, Md.A., A comprehensive review on exposure to toxins and health risks from plastic waste: Challenges, mitigation measures, and policy interventions. *Waste Manag. Bull.*, 3, 100204, 2025, https://doi.org/10.1016/j.wmb.2025.100204.
5. Caux, B., Jores, C.D.S., Naccoul, R.A., Horie, S., West, C., Advantages of online supercritical fluid extraction and chromatography hyphenated to mass spectrometry to analyse plastic additives in laboratory gloves. *J. Chromatogr. A*, 1735, 465323, 2024, https://doi.org/10.1016/j.chroma.2024.465323.
6. Tursi, A., Baratta, E.M.T., Chatzisymeon, E., Chidichimo, F., Biase, M.D., Filpo, G.D., Microplastics in aquatic systems, a comprehensive review: origination, accumulation, impact, and removal technologies. *RSC Adv.*, 12, 28318, 2022, https://doi.org/10.1039/D2RA04713F.
7. Winiarska, E., Jutel, M., Z.-Wiacek, M., The potential impact of nano- and microplastics on human health: Understanding human health risks. 251, Part 2, 118535, 2024, https://doi.org/10.1016/j.envres.2024.118535.
8. Ribeiro, A.P.C., Martins, M.O., *et al.*, Introduction to Polymerization and Depolymerization, in: *Depolymerization: Concepts, Progress, and Challenges Volume 1: Core Concepts and Fundamentals*, ACS Symposium Series, R.K., Gupta (Ed.), vol. 1498, 1, pp. 1–23, Washington, DC, USA, 2025, https://doi.org/10.1021/bk-2025-1498.ch001.

9. Jablonski, P., Irgum, K., Mikkola, J.-P., Wärnå, J., Khokarale, S.G., Brønsted Acid Ionic Liquid Catalyzed Depolymerization of Poly-(3-hydroxybutyrate) to 3-Hydroxybutyric Acid: Highly Selective and Sustainable Transformation in Methyl Isobutyl Ketone and Water-Containing Phase-Separable Reaction Media. *ACS Sustainable Chem. Eng.*, 12, 13946, 2024, https://doi.org/10.1021/acssuschemeng.4c04723.
10. Jiang, W., Li, Q., Cao, M., Lou, H., Li, Z., Qiu, X., High-Efficiency Dual-Site Biomimetic Catalyst for Lignin Depolymerization. *ACS Catal.*, 15, 2595, 2025, https://doi.org/10.1021/acscatal.4c07113.
11. Yang, R., Xu, G., Lv, C., Dong, B., Zhou, L., Wang, Q., Zn(HMDS)2 as a Versatile Transesterification Catalyst for Polyesters Synthesis and Degradation toward a Circular Materials Economy Approach. *ACS Sustainable Chem. Eng.*, 8, 18347, 2020, https://doi.org/10.1021/acssuschemeng.0c07595.
12. Parker, J., Jimenez, E.S., Ware, A.E.H., Umenweke, G.C., Heintz, O., Caboche, G., Crocker, M., Reductive Depolymerization of Lignin to Aromatic Compounds over Promoted Nickel Catalysts in Sub- and Supercritical Methanol. *Chemnanomat.*, 11, e202400222, 2025, https://doi.org/10.1002/cnma.202400222.
13. Chang, T., Wang, Y., Li, Z., Hu, Y., Shi, N., Zou, X., Li, Xu, Y., Li, N., Guo, K., Double metal cyanide complex as a heterogeneous catalyst for depolymerization of polyesters. *Catal. Today*, 452, 115241, 2025, https://doi.org/10.1016/j.cattod.2025.115241.
14. Le, S., Zhang, Z., Chen, C., Wang, J., Chen, L., Liu, Q., Wang, S., Zhu, C., Regulated charge transfer by S-scheme heterojunction of MOF(Fe)-methyl blue for efficient photocatalytic lignite depolymerization. *Chem. Eng. J.*, 503, 158182, 2025, https://doi.org/10.1016/j.cej.2024.158182.
15. Soni, S., Kumari, S., Sharma, B.S.K., Sharma, A.K., Tetravalent metals modulated Zn-based layered double hydroxides and their mixed metal oxides for catalytic depolymerization of carbonyl-coordinating plastic waste. *Catal. Today*, 446, 115136, 2025, https://doi.org/10.1016/j.cattod.2024.115136.
16. Saegher, T.D., Atanasova, B., Vermeir, P., Geem, K.M.V., Verberckmoes, J.D.A., Lauwaert, J., Enhancing the performance of heterogeneous palladium based catalysts in the mild reductive depolymerization of soda lignin through addition of a non-noble metal and tuning of the preparation strategy. *RSC Sustain.*, 2, 1551, 2024, https://doi.org/10.1039/d4su00054d.
17. Scalcon, V., Bonsignore, R., Aupič, J., Thomas, S.R., Folda, A., Heidecker, A.A., Pöthig, A., Magistrato, A., Casini, A., Rigobello, M.P., *J. Med. Chem.*, 66, 14, 9823–9841, 2023, https://doi.org/10.1021/acs.jmedchem.3c00617.
18. Srivastava, A.N., Singh, N.P., Shriwastaw, C.K., Synthesis and characterization of bioactive binuclear transition metal complexes of Schiff base ligand derived from 4-amino-pyrimidine-2-one, diacetyl and glycine. *J. Serb. Chem. Soc.*, 79, 421, 2014, https://doi.org/10.2298/JSC130123148S.
19. Srivastva, A.N., Milan, K., Singh, N.P., Singh, D.V., Kumar, M., Chaudhary, V., Dabas, S., Electrochemical Biosensors Designed with Coordination

Compounds: Recent Advancements and Future Perspectives. *ECS Sens. Plus*, 4, 011601, 2025, https://doi.org/10.1149/2754-2726/ada394.
20. Dolai, S., Behera, C.K., Patra, S.K., Depolymerization by transition metal complexes: strategic approaches to convert polymeric waste into feedstocks. *Dalton Trans.*, 54, 3977, 2025, https://doi.org/10.1039/D4DT02555E.
21. Liu, Q., Yang, R., Dong, B., Sun, H., Xu, G., Wang, Q., Ultrafast and selective chemical recycling of PLA to methyl lactate by using MHMDS as simple catalysts. *Polym. Degrad. Stab.*, 222, 110706, 2024, https://doi.org/10.1016/j.polymdegradstab.2024.110706.
22. Parkatzidis, K., Wang, H.S., Anastasaki, A., Photocatalytic Upcycling and Depolymerization of Vinyl Polymers. *Angew. Chem.*, 63, e202402436, 2024, https://doi.org/10.1002/anie.202402436.
23. Yang, G.-W., Wang, Y., Qi, H., Zhang, Y.-Y., Zhu, X.-F., Lu, C., Yang, L., Wu, G.-P., Highly Selective Preparation and Depolymerization of Chemically Recyclable Poly(cyclopentene carbonate) Enabled by Organoboron Catalysts. *Angew. Chem.*, 61, e202210243, 2022, https://doi.org/10.1002/anie.202210243.
24. Deng, Z., Yuan, D., Yao, Y., Efficient Polymerization and Selective Depolymerization of Poly(cyclopentene carbonate) Mediated Solely by Heterometallic Rare-Earth(III)/Zinc(II) Complexes. *Macromolecules*, 57, 10041, 2024, https://doi.org/10.1021/acs.macromol.4c02086.
25. Santulli, F., D'Auria, I., Boggioni, L., Losio, S., Proverbio, M., Costabile, C., Mazzeo, M., Bimetallic Aluminum Complexes Bearing Binaphthyl-Based Iminophenolate Ligands as Catalysts for the Synthesis of Polyesters. *Organometallics*, 39, 1213, 2020, https://doi.org/10.1021/acs.organomet.0c00016.
26. Yu, Y., Ren, B.-H., Liu, Y., Lu, X.-B., High-performance depolymerization of lignin by bimetallic Cu-Ni@C catalysts prepared with MOF as a carrier under mild conditions. *Appl. Catal. a Gen.*, 656, 119120, 2023, https://doi.org/10.1016/j.apcata.2023.119120.
27. Lourenço, D.L., Oliveira, D.F., Fernandes, A.C., Efficient Depolymerization of Polyester and Polycarbonate Plastic Waste Catalyzed by Commercially Available Homogeneous and Heterogeneous Manganese Catalysts. *Adv. Sustain. Syst.*, 8, 2300381, 2024, https://doi.org/10.1002/adsu.202300381.
28. Monsigny, L., Berthet, J.-C., Cantat, T., Depolymerization of Waste Plastics to Monomers and Chemicals Using a Hydrosilylation Strategy Facilitated by Brookhart's Iridium(III) Catalyst. *ACS Sustain. Chem. Eng.*, 6, 10481, 2018, https://doi.org/10.g1021/acssuschemeng.8b01842.
29. de Souza, F.M., Kahol, P.K., *et al.*, Introduction to Polyurethane Chemistry, in: *Polyurethane Chemistry: Renewable Polyols and Isocyanates*, ACS Symposium Series, P.K. Kahol and R.K. Gupta, (Eds.), vol. 1380, 1, pp. 1–24, 2021, https://doi.org/10.1021/bk-2021-1380.ch001.
30. Zubar, V., Haedler, A.T., Schutte, M., Hashmi, A.S.K., Schaub, T., Hydrogenative Depolymerization of Polyurethanes Catalyzed by a Manganese

Pincer Complex. *Chemsuschem*, 15, e202101606, 2022, https://doi.org/10.1002/cssc.202101606.

31. Massey, L.K., Nylon 6, in: *The Effects of UV Light and Weather on Plastics and Elastomers (Second Edition), Plastics Design Library*, L.K. Massey, (Ed.), pp. 117–125, William Andrew Publishing, Norwish, New York, 2007, https://doi.org/10.1016/B978-081551525-8.50024-5.
32. Ye, L., Liu, X., Beckett, K.B., Rothbaum, J.O., Lincoln, C., Broadbelt, L.J., Kratish, Y., Marks, T.J., Catalyst metal-ligand design for rapid, selective, and solventless depolymerization of Nylon-6 plastics. *Chem*, 10, 172, 2024, https://doi.org/10.1016/j.chempr.2023.10.022.
33. Smith, M.L., McGuire, T.M., Buchard, A., Williams, C.K., Evaluating Heterodinuclear Mg(II)M(II) (M = Mn, Fe, Ni, Cu, and Zn) Catalysts for the Chemical Recycling of Poly(cyclohexene carbonate). *ACS Catal.*, 13, 15770, 2023, https://doi.org/10.1021/acscatal.3c04208.
34. Krall, E.M., Klein, T.W., Andersen, R.J., Nett, A.J., Glasgow, R.W., Reader, D.S., Dauphinais, B.C., McIlrath, S.P., Fischer, A.A., Carney, M.J., *et al.*, Controlled hydrogenative depolymerization of polyesters and polycarbonates catalyzed by ruthenium(II) PNN pincer complexes. *Chem. Commun.*, 50, 4884, 2014, https://doi.org/10.1039/C4CC00541D.
35. Kumar, A., Janes, T., Espinosa-Jalapa, N.A., Milstein, D., Manganese Catalyzed Hydrogenation of Organic Carbonates to Methanol and Alcohols. *Angew. Chem. Int. Ed. Engl.*, 57, 12076, 2018, https://doi.org/10.1002/anie.201806289.
36. Stefan, W., Jasmine, I., Jürgen, K., Molecular catalyst systems as key enablers for tailored polyesters and polycarbonate recycling concepts. *Sci. Adv.*, 4, eaat9669, 2022, https://doi.org/10.1126/sciadv.aat9669.
37. Zubar, V., Lebedev, Y., Azofra, L.M., Cavallo, L., El-Sepelgy, O., Rueping, M., Hydrogenation of CO2-derived carbonates and polycarbonates to methanol and diols by metal–ligand cooperative manganese catalysis. *Angew. Chem. Int. Ed. Engl.*, 57, 13439–13443, 2018, https://doi.org/10.1002/anie.201805630.
38. Kim, S.H. and Hong, S.H., Transfer hydrogenation of organic formates and cyclic carbonates: an alternative route to methanol from carbon dioxide. *ACS Catal.*, 4, 3630, 2014, https://doi.org/10.1021/cs501133m.
39. Alberti, C., Eckelt, S., Enthaler, S., Ruthenium-catalyzed hydrogenativedepolymerization of end-of-life poly(bisphenolAcarbonate). *Chemistryselect*, 4, 12268, 2019, https://doi.org/10.1021/cs501133m.
40. Kindler, T.-O., Alberti, C., Sundermeier, J., Enthaler, S., Hydrogenativedepolymerization of end-of-life poly (bisphenol A carbonate) catalyzed by a ruthenium-MACHO-complex. *Chemistryopen*, 8, 1410, 2019, https://doi.org/10.1002/open.201900319.
41. Nguyen, B.-N.T., Tan, T.T.Y., Otake, K.-I., Kitagawa, S., Lim, J.Y.C., MOF Catalysts for Plastic Depolymerization. *Angew. Chem.*, 64, e202504017, 2025, https://doi.org/10.1002/anie.202504017.

42. Chauhan, M., Thadhani, C., Rana, B., Gupta, P., Ghosh, B., Manna, K., Hydrogenolysis of Polyethylene by Metal–Organic Framework Confined Single-Site Ruthenium Catalysts. *Chem. Mater.*, 36, 10670, 2024, https://doi.org/10.1021/acs.chemmater.4c02186.
43. Qin, J., Xie, J., Zheng, Y., Yang, J., Yamashita, H., Shang, J., Hu, Y., Construction of Co-MOF-derivative modified TiO2 photocatalyst with uniform Co site: A synergistic route for CO2 reduction and waste polypropylene plastic treatment. *J. Environ. Chem. Eng.*, 13, 116767, 2025, https://doi.org/10.1016/j.jece.2025.116767.
44. Guo, Z., Wu, J., Wang, J., Chemical degradation and recycling of polyethylene terephthalate (PET): a review. *RSC Sustain.*, 3, 2111, 2025, https://doi.org/10.1039/D4SU00658E.
45. Padhi, G., Pansare, V.R., Bajpai, P., Krishna, G.R., Vanka, K., Barsu, N., Depolymerization of Waste Polycarbonates to Value-Added Products. *Chemsuschem*, 18, e202400756, 2025, https://doi.org/10.1002/cssc.202400756.
46. Heng, J.Z.X., Tan, T.T.Y., Li, X., Loh, W.W., Chen, Y., Xing, Z., Lim, Z., Ong, J.L.I.Y., Lin, K.S., Nishiyama, Y., Yoshida, T., Zhang, L., Otake, K.-I, Kitagawa, S., Loh, X.J., Ye, E., Lim, J.Y.C., Pyrolytic Depolymerization of Polyolefins Catalysed by Zirconium-based UiO-66 Metal–Organic Frameworks. *Angew. Chem.*, 136, e202408718, 2024, https://doi.org/10.1002/ange.202408718.
47. Ou, W., Ye, Y., Zhang, Y., Zhao, H., Du, W., Hou, Z., Catalytic upgrading of waste PET to dimethyl cyclohexane-1,4-dicarboxylate over defective sulfonated UiO-66def-SO3H supported Ru catalyst. *Chin. J. Catal.*, 71, 363, 2025, https://doi.org/10.1016/S1872–2067(24)60242–1, https://doi.org/10.1016/S1872-2067(24)60242-1.
48. Park, I.-H., Sasaki, K., Quah, H.S., Lee, E., Ohba, M., Lee, S.S., Vittal, J.J., Metal Organo-Polymeric Framework *via* [2 + 2] Cycloaddition Reaction: Influence of Hydrogen Bonding on Depolymerization. *Cryst. Growth Des.*, 19, 1996, 2019, https://doi.org/10.1021/acs.cgd.9b00018.
49. Tang, C., Li, X., Hu, Y., Du, X., Wang, S., Chen, B., Wang, S., Porphyrin-Based Metal-Organic Framework Materials: Design, Construction, and Application in the Field of Photocatalysis., *Molecules*, 29, 467, 2024, https://doi.org/10.3390/molecules29020467.
50. Suo, Q., Zi, J., Bai, Z., Qi, S., The Glycolysis of Poly(ethylene terephthalate) Promoted by Metal Organic Framework (MOF) Catalysts. *Catal. Lett.*, 147, 240, 2017, https://doi.org/10.1007/s10562-016-1897-0.
51. Wang, Y., Wang, H., Li, S., Sun, S., Waste PET Plastic-Derived CoNi-Based Metal–Organic Framework as an Anode for Lithium-Ion Batteries. *ACS Omega*, 7, 39, 35180, 2022, https://doi.org/10.1021/acsomega.2c04264.
52. Trzebiatowska, P.J., Maramorosz, Z., Baluk, M.A., Gazda, M., Eceiza, A., Zaleska-Medynska, A., The catalytic activity of metal–organic frameworks (MOFs) and post-synthetic modified MOF towards depolymerisation of polycarbonate. *Appl. Surf. Sci.*, 673, 160894, 2024, https://doi.org/10.1016/j.apsusc.2024.160894.

53. Wu, Y., Wang, X., Kirlikovali, K.O., Gong, X., Atilgan, A., Ma, K., Schweitzer, N.M., Gianneschi, N.C., Li, Z., Zhang, X., Farha, O.K., Catalytic Degradation of Polyethylene Terephthalate Using a Phase-Transitional Zirconium-Based Metal–Organic Framework. *Angew. Chem., Int. Ed.*, 61, e202117528, 2022, https://doi.org/10.1002/anie.202117528.
54. Rincon, I., Hidgola, T., Armani, G., Rojas, S., Horcajada, P., Enzyme_Metal-Organic Framework Composites as Novel Approach for Microplastic Degradation. *Chemsuschem*, 17, e202301350, 2024, https://doi.org/10.1002/cssc.202301350.
55. Sha, F., Wu, A., Su, S., Farha, O.K., Recyclable Enzymatic Hydrolysis with Metal–Organic Framework Stabilized Humicola insolens Cutinase (HiC) for Potential PET Upcycling. *Chem. Bio. Eng.*, 1, 798, 2024, https://doi.org/10.1021/cbe.4c00101.

9

Converting Plastic Waste to Energy through Pyrolysis and Gasification

Reetu Raj[1*], Neeraj Kumbhakarna[1] and Jeewan Vachan Tirkey[2]

[1]Department of Mechanical Engineering, Indian Institute of Technology Bombay, Mumbai, Maharashtra, India
[2]Department of Mechanical Engineering, IIT-BHU (Varanasi), Varanasi, Uttar Pradesh, India

Abstract
The increasing accumulation of plastic waste (PW) poses a significant environmental challenge worldwide. Pyrolysis and gasification, including co-gasification with other feedstocks such as biomass, are promising thermochemical routes to convert PW into valuable energy products, like liquid fuels, synthetic gas, and char. These methods help minimize PW while producing energy-rich outputs suitable for power generation and chemical production. A key focus is utilizing syngas from gasification as fuel for internal combustion engines, providing a sustainable solution for decentralized energy generation. The current chapter provides an extensive review of the fundamental principles, technological advancements, and associated challenges with pyrolysis, gasification, and co-gasification of plastic waste. Key parameters influencing plastic blend percentages and output performances (oil yield, char yield, gas compositions, cold gas efficiency, carbon conversion, tar yield) are discussed, along with the environmental and economic benefits of integrating these technologies into circular economy frameworks. Case studies and real-world applications are also highlighted, demonstrating the capability of these technologies in reducing plastic pollution and fostering sustainable energy production, including their viability for IC engine applications.

Keywords: Plastic waste, pyrolysis, co-gasification, gasification efficiency, plastic blend

**Corresponding author*: reeturaj.rs.mec20@itbhu.ac.in

Neha Agarwal, Brijesh Pare, Vijendra Singh Solanki and Sankara Rao Miditana (eds.) Plastic Remediation: Transformative Pathways for a Rejuvenated Sustainable Environment, (211–244)

9.1 Introduction

Plastic waste (PW) has emerged as a major environmental issue, with global production exceeding 400 million tons each year [1, 2]. Several developed countries have introduced regulatory frameworks for plastic waste management, like Circular Economy Action Plan of European Union and National Sword policy of China, which limits PW imports [3, 4]. Advanced nations like Germany, Japan, and the Netherlands have invested in waste-to-energy (WTE) technologies, including pyrolysis and gasification, to manage plastic waste sustainably [5]. Moreover, research institutions worldwide are exploring hybrid processes to enhance conversion efficiency and minimize environmental impact. Despite extensive recycling initiatives, below 10% of PW is effectively recycled, while the remaining is either incinerated, landfilled, or mismanaged, leading to severe ecological impacts [6]. In contrast, emerging economies like India, China, and Brazil face considerable challenges in plastic waste management, primarily due to insufficient recycling infrastructure and high plastic consumption rates. India, for example, produces around 62 million tons of PW each year, with only a small proportion being recycled [7]. Government initiatives such as the Swachh Bharat Abhiyan and Plastic Waste Management Rules aim to improve waste management practices [7]. Several pilot projects have demonstrated the feasibility of thermochemical conversion processes, like pyrolysis and gasification for transforming PW into invaluable energy carriers, such as syngas, liquid fuels, and char. However, large-scale adoption is hindered by high capital costs and regulatory uncertainties. Addressing these challenges requires an integrated approach that considers technological advancements, policy frameworks, and economic feasibility.

Pyrolysis is a thermal degradation technique conducted without oxygen, in which plastic polymers are broken down into smaller hydrocarbon chains [8]. Depending on the operating temperature (ranges from 300 to 600°C) and heating rate, pyrolysis can yield liquid oil, gas products, and solid char. The main reactions involved include depolymerization, chain scission, and aromatization [9]. There are different types of pyrolysis like (i) slow pyrolysis: it is outlined by low heating rates and longer residence durations and primarily favors the char production; (ii) fast pyrolysis: it involves rapid heating and short residence durations, resulting in a high output of liquid oil; and (iii) catalytic pyrolysis: it employs catalysts to enhance product selectivity and quality, reducing the formation of undesirable compounds whereas gasification transforms PW into syngas by means of partial oxidation at elevated temperatures (700–1500°C). This technique involves a combination of drying, pyrolysis, combustion, and reduction reactions [10]. The produced syngas, abundant in carbon monoxide (CO) and hydrogen (H_2), can be utilized for power generation, hydrogen production, or as a feed-material in chemical synthesis. There are various gasifying agents that can be employed, including air, steam, and CO_2. The gasifiers can be classified as (a) fixed-bed gasifiers: it is simple in design but limited scalability; (b) fluidized-bed gasifiers: this offers uniform temperature distribution and better feedstock flexibility; and (c) plasma gasifiers: this utilizes plasma arcs for high-temperature operation, ensuring complete feedstock conversion. The producer gas from gasification depends on the feedstock and gasification agent. It typically contains CO, H_2, CH_4, N_2, and trace impurities. Syngas serves as a versatile energy carrier, suitable for applications like power generation, Fischer–Tropsch (FT) synthesis, methanol production, and as a precursor for hydrogen [11]. Figure 9.1 presents a schematic representation of PW valorization pathways, illustrating how pyrolysis and gasification routes transform waste into valuable products like syngas, which can then be utilized for power generation, Fischer–Tropsch synthesis, and methanol production. Moreover, Table 9.1 provides a comparative overview of pyrolysis and gasification in the context of plastic waste conversion, highlighting their distinct characteristics and applications. However, despite significant advancements, several challenges remain in the field of pyrolysis and gasification [12, 13]. Limited understanding of the co-gasification behavior of PW with biomass/coal hampers the optimization of feedstock blending strategies [14]. Additionally, inconsistent syngas quality due to feedstock variability affects the reliability of the process. The high operational costs and energy consumption associated with pyrolysis and gasification technologies further pose economic barriers to their large-scale implementation. Moreover, the development of advanced catalysts and reactor configurations is crucial for enhancing process efficiency and product yield [15]. These gaps underscore the necessity for targeted research initiatives aimed at enhancing the efficiency and commercial feasibility of PW conversion technologies.

This chapter aims to address these research gaps by investigating the impact of PW co-gasification with biomass and coal on syngas composition and quality. A key objective is to evaluate the effect of critical process parameters, like waste plastic blending ratios, on the performance of pyrolysis, and gasification process. Additionally, this study examines important performance indicators, including oil yield, carbon conversion efficiency, gas compositions, cold gas efficiency, and tar reduction strategies, to provide insights into process optimization. Furthermore, the chapter explores techno-economic aspects of plastic waste-to-energy (WTE) technologies

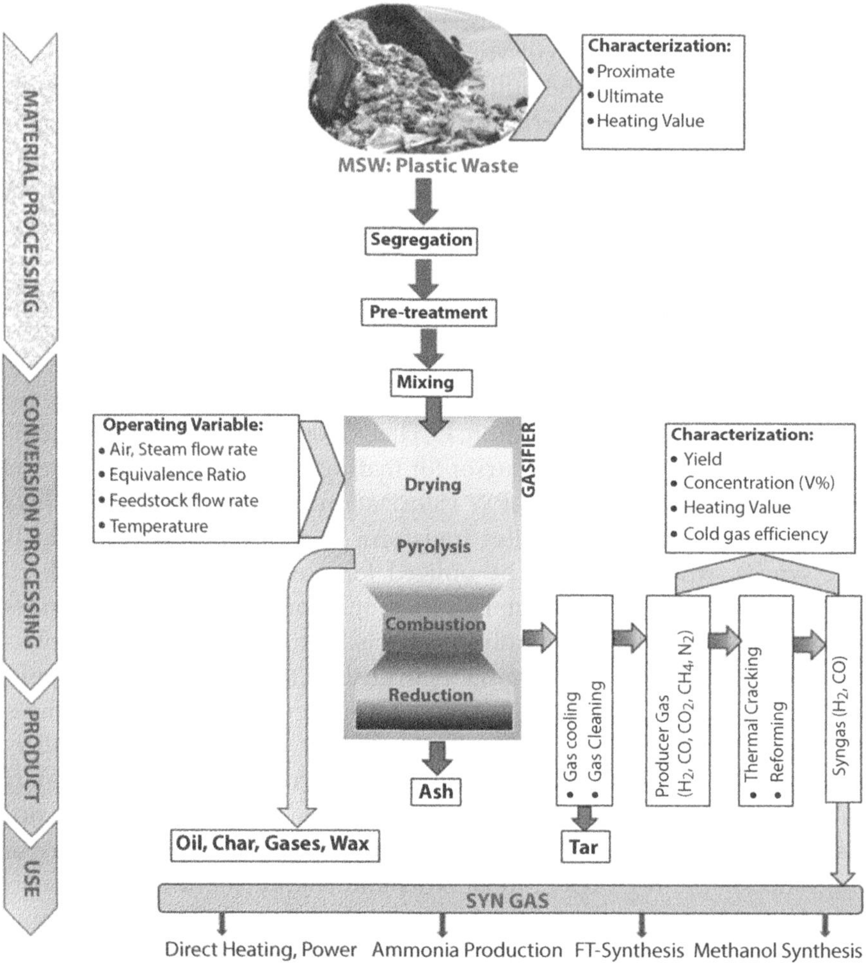

Figure 9.1 Schematic diagram of plastic waste conversion through pyrolysis and gasification.

and assesses potential challenges and opportunities in industrial applications for plastic waste management.

The novelty of this work stems from its detailed assessment of plastic blending proportions and their impact on pyrolysis and gasification performance. By providing a detailed examination of syngas composition, calorific value, and energy recovery efficiency, this study offers valuable insights for optimizing co-gasification strategies. Moreover, the integration of recent advancements in reactor design, catalyst application, and hybrid processing techniques enhances the practical relevance of the findings.

Table 9.1 Comparison of pyrolysis and gasification for plastic waste conversion.

Parameter	Pyrolysis	Gasification
Process conditions	Absence of oxygen, 300–800°C	Partial oxidation, 700-1500°C
Main products	Liquid oil, syngas, char	Syngas (H_2, CO, CH_4, CO_2), tar, char
Gasifying agent	Not required	Air, oxygen, steam, CO_2
Syngas composition	Low H_2, high hydrocarbons	High H_2, CO-rich
Energy efficiency	Moderate	Higher than pyrolysis
Tar formation	High	Lower than pyrolysis
Carbon conversion	Lower compared to gasification	Higher carbon conversion
Application	Liquid fuel production, chemical feedstock	Power generation, hydrogen production, chemicals
Environmental impact	Lower emissions, but high tar content	Reduced emissions, cleaner gas production

A comparative assessment of international and national waste-to-energy strategies further contributes to the policy and industrial discourse on sustainable waste management. By bridging the existing research gaps and offering practical insights into process optimization, this chapter contributes to the ongoing efforts to develop sustainable and economically viable solutions for plastic waste management. This chapter seeks to offer an in-depth overview of pyrolysis and gasification technologies, focusing on their mechanisms, feedstock requirements, operating conditions, and potential applications.

9.2 Thermochemical Processing of Plastic Waste

This section offers a detailed overview of pyrolysis and co-gasification, highlighting their mechanisms, key inputs, reactor technologies, and output products for sustainable plastic WTE transformation.

9.2.1 Pyrolysis

Pyrolysis is a thermochemical route that breaks down organic waste, such as biomass, PW in the absence of oxygen, usually occurring at temperature 300°C to 800°C [16]. It transforms plastic polymers into useful products like liquid oil (used as a fuel), syngas (a fuel gas), and solid char (a carbon-rich residue) [8]. This process offers a viable way to manage PW and recover energy. The standard stages of pyrolysis include (a) feedstock preparation: sorting and shredding plastic waste; (b) thermal treatment: heating under an inert atmosphere to prevent combustion; and (c) product recovery: collecting and separating gas, oil, and char. Moreover, the chemical mechanisms of pyrolysis primarily include the following:

- **Depolymerization**: The plastic polymers are broken down into their original monomers [17].
- **Cracking**: Larger hydrocarbon molecules are split into smaller, more volatile compounds [18].
- **Isomerization**: Structural rearrangement of molecules without altering their molecular formula [19].
- **Aromatization**: Formation of stable aromatic compounds, often found in fuel oils [20].

Plastic types like polypropylene (PP), polystyrene (PS), and polyethylene (PE) are particularly fit for pyrolysis due to their high carbon and hydrogen content. However, plastics like PET and PVC require extra care because they release acidic gases and toxic substances during processing [21, 22]. Moreover, Table 9.2 provides the comparative list of different plastics with their respective pyrolysis products, gasification strategies, and emissions. The most common pyrolysis reactor types used are as follows:

- **Fixed-bed reactors**: simple design, compatible for pilot-mode deployment [23].
- **Fluidized-bed reactors**: allow uniform temperature distribution and high heat transfer [24].
- **Rotary kiln reactors**: enable continuous operation and handling of various feedstock types [25].
- **Microwave-assisted pyrolysis**: offers rapid heating and energy efficiency through electromagnetic radiation [26].

Pyrolysis is further classified by heating rate and vapor residence time into slow (300–700°C, favoring char), fast (400–500°C, maximizing bio-oil),

Table 9.2 Comparison of different plastic types for pyrolysis and gasification.

Plastic type	Pyrolysis product	Gasification synergy	Emission concern
HDPE/LDPE	Aliphatic-rich gas/oil	High syngas yield	Low
PP	Mixed hydrocarbons	Good blending agent	Low
PS	Aromatic-rich oil	Enhances tar formation	Moderate
PET/PVC	Oxygenated/ chlorinated	Requires pretreatment	High (HCl, CO)

flash (800–1000°C, enhancing volatile yields), and intermediate (500–600°C) pyrolysis [27, 28]. Product yields are temperature-dependent—higher temperatures typically enhance oil yields (up to 47 wt% at 500–600°C) while reducing char content. Rapid vapor condensation forms bio-oil, whereas extended vapor residence promotes secondary reactions leading to more char and tar. The resultant bio-oil contains a mix of acids, alcohols, phenolics, and aqueous compounds and typically requires upgrading to improve quality and stability. Upgrading strategies involve direct techniques such as zeolite cracking, hydrodeoxygenation, and emulsification, along with indirect methods like reforming into syngas followed by FT synthesis to produce clean fuels, including alcohols and green diesel [29, 30]. Plastic pyrolysis yields a spectrum of products, including light and heavy oils, naphtha, gaseous hydrocarbons (methane, ethylene, and butane), and solid carbon-rich residues, thereby offering a viable route for plastic waste valorization and sustainable energy generation [8, 31].

9.2.2 Co-Gasification

Co-gasification is an advanced thermochemical route that combines plastic waste with biomass or coal to produce syngas, a versatile energy carrier composed mainly of H_2, CO, CO_2, and CH_4 [32]. This process is conducted at temperatures between 700°C to 1500°C with a regulated input of air or steam. Co-gasification leverages the complementary properties of plastics namely their high calorific value and hydrogen content and the reactivity and catalytic nature of biomass or coal. The resulting synergy improves gas quality, enhances thermal efficiency, and helps manage tar formation issues typically associated with plastic-only gasification [33, 34].

The first step in co-gasification is feedstock preparation. Plastic waste like LDPE, HDPE, PS, and PP are sorted, shredded, and dried to reduce moisture content. These plastics are chosen due to their high hydrocarbon content, which significantly boosts the energy output of the process. However, plastics like PET and PVC contain oxygen and halogenated compounds (e.g., chlorine), requiring pretreatment steps like dechlorination or the use of sorbents to minimize the emission of harmful substances like HCl during gasification [35]. In parallel, biomass (e.g., wood chips, agricultural residues) or low-rank coal is also prepared, usually by grinding and drying. The prepared plastic and biomass/coal are then blended in appropriate ratios to ensure a homogeneous mixture for consistent feed into the gasifier. After feedstock preparation, it is fed into a gasification reactor, like fluidized-bed, fixed-bed, or entrained-flow type depending on scale, efficiency, and feedstock properties. Moreover, the main reactor designs available for co-gasification include the following:

- **Entrained flow gasifiers**: high temperature, short residence time, uniform syngas output [36].
- **Fluidized bed gasifiers**: excellent mixing and heat transfer, suitable for varied feedstocks [37].
- **Downdraft and updraft gasifiers**: lower cost, fit for small- to medium-scale uses [38].

In the reactor, several sequential regions facilitate the transformation of feedstock into syngas [10]. Initially, in the drying region, moisture is evaporated. Within the pyrolysis region, the feedstock undergoes thermal breakdown, generating volatile hydrocarbons and solid char. These volatiles then move into the oxidation region, where partial combustion with limited oxygen generates the heat required to drive the endothermic reactions. Finally, in the reduction region, char reacts with oxidizing agents like CO_2 and H_2O to generate syngas *via* the Boudouard reaction ($C + CO_2 \rightarrow 2CO$), the water–gas reaction ($C + H_2O \rightarrow CO + H_2$), and methanation ($C + 2H_2 \rightarrow CH_4$). The water–gas shift reaction ($CO + H_2O \leftrightarrow CO_2 + H_2$) further adjusts the H_2 content in the syngas. The raw syngas obtained contains impurities, including tar, particulates, HCl, and sulfur compounds, which must be removed before energy utilization. This is achieved through a gas cleaning process involving cyclone separators for particulates, tar cracking using catalysts like dolomite or nickel-based materials, and acid gas scrubbing to remove harmful emissions. For feeds containing PET or

PVC, additional sorbents like calcium-based materials may be added to capture chlorine-containing species [39]. Effective gas cleaning is essential to ensure that the syngas fulfils the required quality standards for subsequent deployments.

The clean syngas produced from co-gasification of PW has multiple energy applications. One such application in IC engine power production has been shown in Figure 9.2. Syngas can be utilized for power generation in turbines, gas engines, or in IGCC systems, which provide high efficiency and reduced emissions. The syngas can also serve as a fuel in boilers and burners for direct heat applications. Another major use of syngas is in synthetic fuel

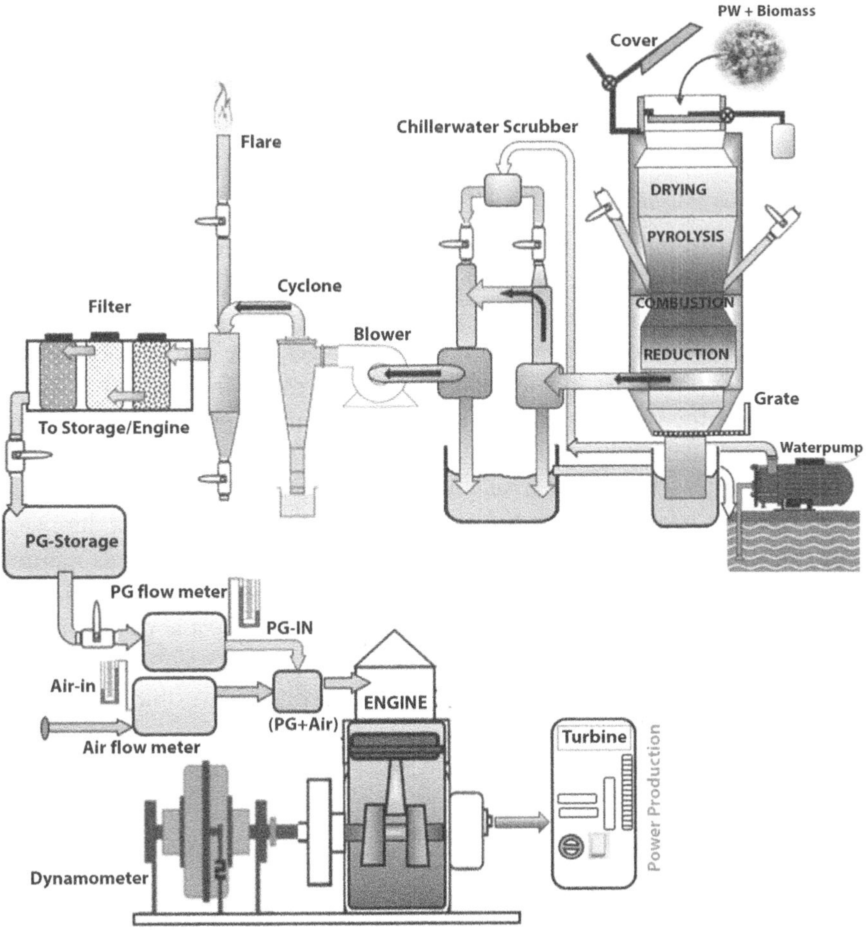

Figure 9.2 Schematic of plastic waste co-gasification to syngas for IC engine energy application.

Table 9.3 Product and application potential of co-gasification of plastic waste.

Input blend	Syngas composition	End use	Remarks
HDPE + biomass	H_2 (25–30%), CO (20–30%)	Power, hydrogen, FT fuels	High gas yield, low ash
PP + coal	H_2 (20–28%), CO (25–35%)	Electricity, methanol	Stable process, enhanced calorific value
PS + biomass	H_2 (15–22%), CO (30–40%)	Process heat, diesel	Higher tar yield—needs better scrubbing
PET/PVC + biomass	H_2 (10–20%), CO (20–30%)	Syngas + chemical feedstock	Requires halogen treatment

production through the Fischer–Tropsch process, which converts syngas into diesel, methanol, and other valuable chemicals. Additionally, hydrogen-rich syngas can undergo further processing through the water–gas shift reaction and membrane separation methods to generate high-purity hydrogen, fit for applications in fuel cells and the chemical industry [40]. Table 9.3 illustrates the various products and their potential applications resulting from the co-gasification of PW. A key benefit of co-gasification lies in the synergistic interactions that arise when plastics are combined with biomass. While plastics provide a high heating value and hydrogen content, biomass contributes oxygenated compounds and mineral ash, which act as natural catalysts to reduce tar formation and enhance gas conversion [41]. This balance leads to improved thermal efficiency and a more stable gas composition. Furthermore, co-gasification allows for the simultaneous disposal of two major types of waste plastics and biomass while recovering energy in a cleaner and more sustainable form.

9.3 Results and Discussions

9.3.1 Pyrolysis Performance Parameters

Figure 9.3 illustrates the influence of varying plastic content specifically polyvinyl chloride (PVC) and polystyrene (PS) on key pyrolysis output

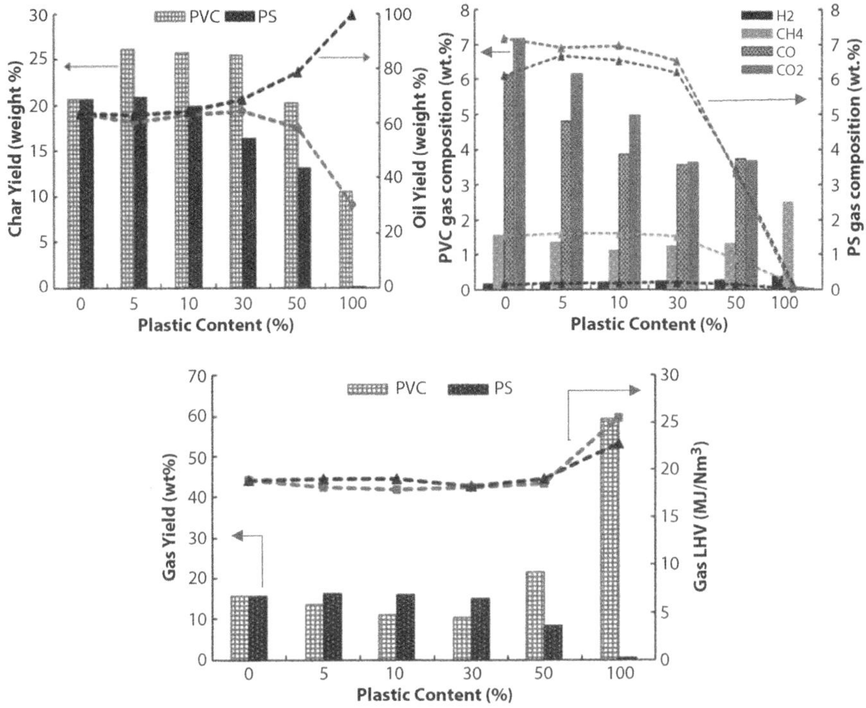

Figure 9.3 Impact of plastic content on pyrolysis output parameter.

parameters such as char yield, oil yield, gas composition, gas yield, and the lower heating value (LHV) of the produced gas. With increasing plastic content, particularly beyond 30%, a decline in char yield is observed, more prominently in the case of PVC [42]. This reduction is due to plastics having higher volatility and lower fixed carbon content than biomass, which contains lignocellulosic compounds that promote higher char production. PVC, in particular, undergoes dehydrochlorination, releasing HCl and disrupting carbon structure formation, thereby reducing char yield [43, 44]. Oil yield trends differ between the plastics; PS slightly enhances oil production at moderate concentrations due to its depolymerization into monomeric products like styrene [45, 46], while PVC leads to a significant drop in oil yield as its decomposition favors the formation of gases and tar-like residues [47]. Gas composition analysis reveals that with increasing plastic content, particularly PS, there is an increase in hydrocarbon gases like CH_4 and CO, while CO_2 levels decline due to the lower oxygen content in plastics [48]. Hydrogen production remains minimal due to limited water–gas shift reactions in plastic pyrolysis [49, 50]. The gas yield initially decreases slightly but increases sharply at 100% plastic content, with a concurrent rise in gas LHV. This is due to the enhanced formation of energy-rich hydrocarbon gases [50]. Consequently, the incorporation of plastics into biomass pyrolysis alters the product distribution and improves the calorific value of the produced gas [51, 52]. Similar observations have been documented in the literature, highlighting the synergistic effects of biomass-plastic co-pyrolysis in optimizing fuel properties and energy recovery [52–55].

9.3.2 Co-Gasification Performance Parameters

Figure 9.4 presents the variation in major producer gas components (H_2, CO, CO_2, and CH_4) as a function of increasing plastic content during the co-gasification of PW and biomass based on data compiled from multiple studies. A consistent trend observed across several authors, including Lopez *et al.* [56], Baloch *et al.* [57], and Pinto *et al.* [58], is the increase in H_2 yield with rising plastic content, especially beyond 50%. This enhancement in hydrogen concentration is because of the thermal cracking of long-chain hydrocarbons in plastics and secondary reactions like steam reforming and water gas shift, which are supported by the moisture or steam in the biomass [56, 57, 59]. However, some studies, such as Wilk *et al.* [60], observed a more moderate increase due to differences in plastic type, gasification conditions, and reactor configurations.

In contrast, CO concentrations generally decline with higher plastic content. Biomass gasification typically generates more CO due to the breakdown of cellulose and hemicellulose. As plastics lack oxygen in their structure, increasing their share reduces the contribution of oxygenated volatiles, resulting in lower CO formation (Xu *et al.* [61], Yu *et al.* [62]). Similarly, a decreasing trend in CO_2 content is evident across most studies with higher plastic proportions. Since CO_2 primarily results from oxidation reactions and the decarboxylation of oxygen-rich biomass, its concentration diminishes as plastic, which contains minimal oxygen, replaces the biomass [63]. Some studies also suggest that higher plastic ratios suppress char gasification reactions, further reducing CO_2 output (Pinto *et al.* [58]).

On the other hand, CH_4 levels increase significantly with plastic addition, as shown by Wilk *et al.* [60, 64], Alvarez *et al.* [63], and Yu *et al.* [62]. This results from the thermal cracking and pyrolysis of polymers such as polyethylene and polystyrene, which encourage the formation of CH_4 and other light hydrocarbons [65, 66]. Additionally, the suppression of water gas shift and reforming reactions under lower steam or oxygen availability in plastic-rich blends allows more CH_4 to remain in the gas phase.

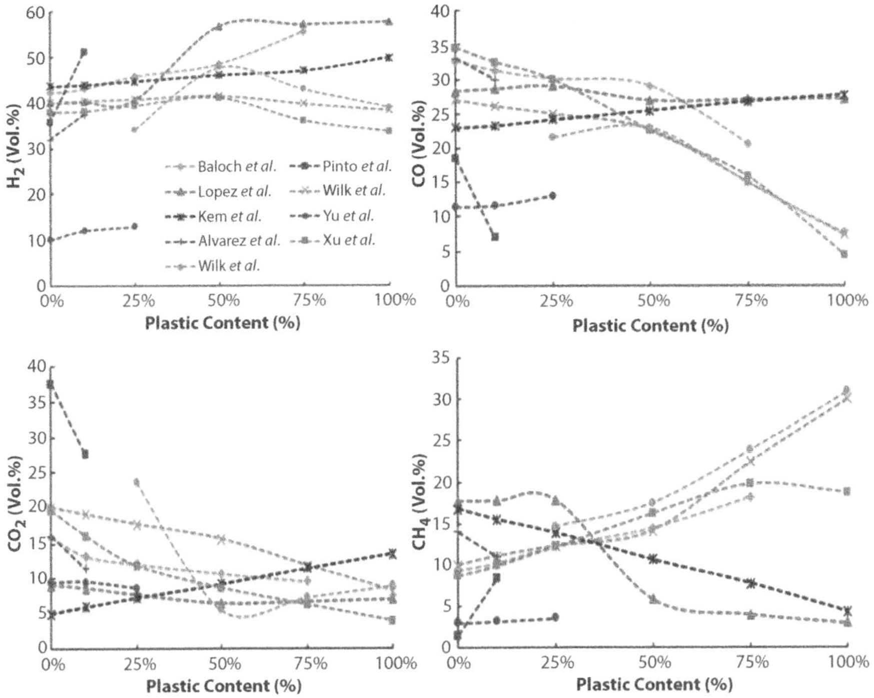

Figure 9.4 Impact of plastic content on co-gasification gas compositions.

These findings align well with previous research, confirming that co-gasification with plastic waste alters the gas composition toward a more hydrocarbon-rich, energy-dense syngas, offering promising prospects for fuel upgrading and utilization [67–69].

Figure 9.5 illustrates the impact of increasing plastic content on other important key output parameters during co-gasification of PW and biomass including tar content, LHV, cold gas efficiency (CGE), carbon conversion efficiency, and gas yield. A prominent observation is the overall reduction in tar content with increasing plastic ratio, as noted by Lopez *et al.* [56] and Wilk *et al.* [64]. Plastics, particularly polyolefins, tend to thermally crack into lighter hydrocarbons under gasification conditions, reducing the formation of complex, high-molecular-weight tars commonly produced from lignocellulosic biomass [68, 70]. However, studies like those of Zaccariello *et al.* [71] and Cho *et al.* [72] observed a rise in tar content at very high plastic loads, potentially due to incomplete cracking of volatile intermediates when plastic dominates the feedstock.

Simultaneously, the LHV of the syngas increases with higher plastic content, as observed in studies by Kern *et al.* [67], Baloch *et al.* [57], and Zaccariello *et al.* [71]. This enhancement stems from the higher volatile content and lack of oxygen in plastic, which shifts the product gas toward

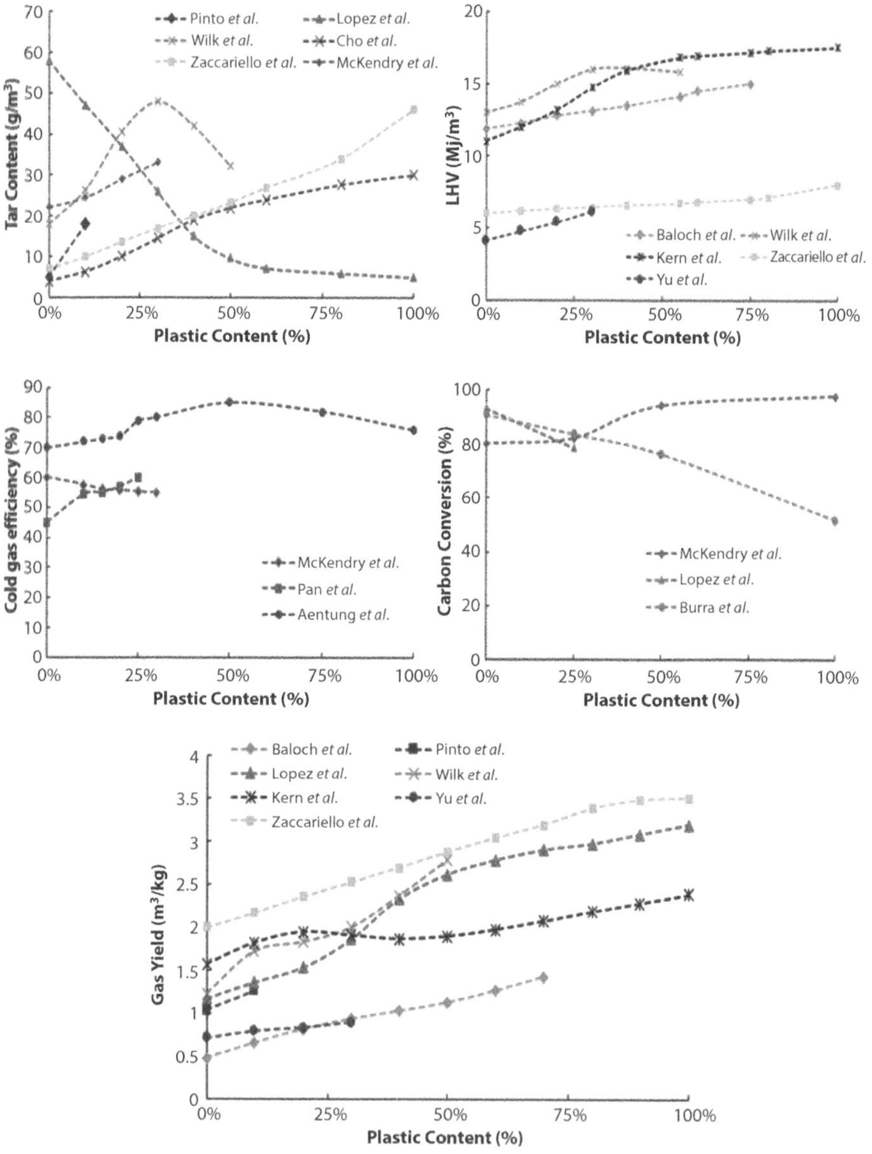

Figure 9.5 Impact of plastic content on co-gasification output parameter.

hydrocarbon-rich and energy-dense species such as CH_4 and higher alkanes [67, 73]. Moreover, CGE which measures the energy efficiency of converting feedstock into gas, tends to improve modestly or plateau with plastic addition, as shown by Pan *et al.* [74] and Aentung *et al.* [75], indicating favorable thermochemical synergy in co-processing.

The carbon conversion efficiency, however, exhibits mixed trends. While McKendry *et al.* [76] report near-complete conversion with rising plastic content, Burra *et al.* [77] show a decreasing pattern. This discrepancy could be linked to differences in reactor temperature, residence time, and gasifying agents, where high plastic content may suppress char gasification reactions due to lower fixed carbon availability in plastics [70, 77].

Furthermore, gas yield increases significantly with plastic content, as consistently demonstrated by recent studies [61, 70]. This is because plastics have higher volatile content and minimal ash compared to biomass, promoting greater generation of gaseous products per unit mass [71]. Overall, these trends support the view that controlled addition of plastic waste in biomass gasification can enhance gas quality and energy recovery, a finding that aligns with earlier research on co-gasification synergies [58, 69, 78].

9.3.3 Techno-Economic Analysis of Plastic Waste-to-Energy

The techno-economic viability of plastic waste-to-energy (WTE) systems *via* pyrolysis and gasification plays a critical role in their commercial deployment and environmental sustainability. Several factors, including capital investment, operational costs, process efficiency, product market value, and regulatory policies, dictate the success of these technologies [79–81]. A systematic evaluation of these aspects is essential to understand the competitiveness of pyrolysis and gasification relative to primitive waste management methods like landfilling or incineration. Table 9.4 shows a comparative feasibility of pyrolysis and co-gasification of plastic WTE system.

9.3.3.1 Capital and Operational Costs

The initial capital cost for plastic pyrolysis and gasification plants varies significantly depending on the scale, technology type, reactor design, and feedstock preparation needs. Pyrolysis plants are generally less capital-intensive compared to gasification units because of simpler reactor designs and lower temperature operations. A typical small-scale pyrolysis unit

Table 9.4 Comparative feasibility of plastic waste to energy system.

Parameter	Pyrolysis	Co-gasification
CAPEX	Low to medium	Medium to high
Feedstock flexibility	Moderate (requires sorting)	High (biomass/plastic mix)
Product type	Liquid fuels, gas, char	Syngas
Energy efficiency	Moderate	High (with steam reforming)
Emission control	Moderate (condensation needed)	Requires advanced gas cleaning
Scalability	Medium (modular designs)	High (centralized facilities)
Profitability [82]	Higher	Lower
	USD 80.7/ton PP NPV: USD 149 million IRR: 22.25%	USD 51.9/ton PP NPV: USD 96 million IRR: 16.91%

(processing ~10 tons/day) may require \$2–5 million in capital investment, while a similar-scale gasification plant might need \$5–8 million [79, 82]. Cost components include feedstock sorting equipment, pretreatment facilities (dryers, shredders), reactors, product recovery systems (condensers, separators), pollution control units, and auxiliary infrastructure such as utilities and automation systems. Modular and skid-mounted systems are being increasingly adopted to reduce installation time and cost, enhancing project viability.

Operating costs mainly consist of labor, energy input (electricity, auxiliary fuel), catalyst/reagent consumption (for upgrading and tar removal), maintenance, and waste disposal. Energy self-sufficiency is a key determinant: for instance, using a portion of produced syngas or pyro-gas for heating can drastically lower operational energy expenses [83, 84]. Typical operational expenses range between \$70 and 120 per ton of processed plastic waste [85]. However, plants treating mixed plastic streams often incur additional costs for sorting and decontamination, particularly when removing halogenated plastics like PVC that otherwise cause environmental compliance issues. Catalyst replacement frequency and cost are significant in catalytic pyrolysis and gasification processes, impacting the long-term economic performance.

9.3.3.2 *Energy and Product Yields*

One of the key techno-economic advantages of pyrolysis and gasification lies in the high-energy yield of plastic-derived products. The liquid oil yield from pyrolysis can reach 60–80 wt% for polyolefin-rich plastics (HDPE, LDPE, and PP), with calorific values comparable to conventional diesel (~42–45 MJ/kg) [86, 87]. Similarly, gasification processes can yield syngas with an energy content ranging from 5 to 18 MJ/Nm^3, based on feed-material composition and operating conditions [88]. The market value of the products whether sold as fuels, chemical feedstocks, or hydrogen significantly influences the economic outlook. High-value outputs such as hydrogen and Fischer–Tropsch liquids fetch premium prices but often require additional upgrading steps, increasing overall project cost and complexity.

9.3.3.3 *Economic Indicators*

The economic viability of plastic WTE plants is commonly assessed using parameters like net present value (NPV), internal rate of return (IRR), payback period, and levelized cost of energy (LCOE). For medium-scale pyrolysis plants, a typical payback period is estimated to be 4-7 years, while for gasification plants, it can extend to 6–10 years depending on energy prices and policy incentives [53, 82]. Sensitivity analyses using NPV and IRR suggest that policy incentives and high oil prices improve the investment attractiveness of these technologies. Financial performance improves with economies of scale; larger plants (>100 tons/day) often achieve better margins by optimizing operating costs and negotiating favorable contracts for electricity, fuels, or chemical products. Carbon credit revenues under emission trading schemes or clean development mechanism frameworks can further improve economic returns, especially when plastics otherwise destined for landfilling are diverted toward energy recovery.

However, scalability remains a crucial challenge in both pyrolysis and gasification of plastics. Small-scale units are easier to install and operate but suffer from higher per-unit processing costs and limited product sales volume. Large-scale centralized plants benefit from better economics but require robust feedstock supply chains and sophisticated waste logistics [89]. Distributed modular systems, where multiple smaller pyrolysis or gasification units are deployed near waste-generation sites and the outputs are aggregated for refining or utilization, are gaining traction as a promising hybrid approach. Commercial ventures such as Plastic Energy (UK) and Enerkem (Canada) have demonstrated successful pilot-to-commercial scale transitions, proving the potential viability of plastic WTE plants when backed by strong policy support and corporate investment [66, 90].

9.3.4 Environmental and Policy Considerations for Plastic Pyrolysis and Gasification

The environmental and policy dimensions of plastic WTE technologies are critical in shaping their acceptance, funding, and scaling across different regions. Although pyrolysis and gasification offer significant advantages over conventional waste disposal methods like landfilling and incineration, they also introduce new challenges related to emissions, life-cycle impacts, and regulatory compliance. An integrated understanding of these issues is necessary for the sustainable deployment of WTE solutions [70, 91, 92]. Table 9.5 illustrates the interaction between environmental impacts and policy drivers influencing plastic waste-to-energy projects. It highlights how technological optimization, regulatory support, and public perception must align to achieve sustainable outcomes.

9.3.4.1 Environmental Emissions and Byproducts

Pyrolysis and gasification effectively reduce plastic waste volume while recovering valuable energy-rich products. However, their environmental performance depends largely on operational conditions and feedstock composition. Pyrolysis can generate pollutants like volatile organic compounds, polycyclic aromatic hydrocarbons, and fine particulate matter if vapors are not adequately captured and treated [96]. Similarly, gasification may produce trace amounts of dioxins, furans, and heavy metals, particularly when processing chlorine-containing plastics like PVC without appropriate gas cleaning systems [35, 97]. Modern gas cleaning technologies including cyclones, scrubbers, tar reformers, and catalytic filters are thus essential to minimize harmful emissions and meet environmental standards.

9.3.4.2 Carbon Footprint and Life Cycle Assessment

Life cycle assessment (LCA) studies show that plastic pyrolysis and gasification typically have a lower carbon footprint than landfilling or uncontrolled

Table 9.5 Summary of environmental issues, mitigation measures, and policy frameworks for plastic waste-to-energy technologies.

Aspect	Pyrolysis	Gasification	Mitigation strategy	Relevant policy initiatives
Major emissions	VOCs, PAHs, Fine particulates	CO, CO_2, CH_4, trace dioxins and furans	Advanced gas cleaning, catalytic tar reforming	EU industrial emissions directive; US clean air act
Greenhouse gas footprint	Moderate (depends on oil upgrading energy)	Low to moderate (can be near-zero with CCS)	Integration with CCS; renewable energy use	Paris agreement; Net-zero targets [93]
Waste diversion	High (diverts plastics from landfill/ incineration)	High (valorizes plastics and biomass)	Material recovery prioritization	EU waste framework directive
Circular economy contribution	Chemical feedstock for new plastics	Hydrogen and energy carriers	Promote chemical recycling certification	European plastics strategy [94]
Public perception issues	Concerns about oil toxicity, micro-pollutants	Fears about syngas composition, plant safety	Transparency in reporting; public engagement	Extended producer responsibility (EPR)
Regulatory challenges	Product classification (fuel vs waste)	Permit complexity for syngas utilization	Standardized quality guidelines	Basel convention amendments [95]

incineration [98, 99]. Gasification, in particular, can achieve near-zero net emissions when integrated with carbon capture and storage or if syngas is utilized for low-carbon H_2 production. Moreover, using pyrolysis oil or syngas as substitutes for fossil fuels can contribute to substantial greenhouse gas (GHG) emission savings estimated at 1.5–3.0 tons of CO_2-eq. per ton of PW processed [100, 101]. However, if energy recovery efficiency is low or product upgrading is highly energy-intensive, the environmental benefits may be offset. Continuous optimization of process integration, energy recovery, and waste heat utilization is therefore critical to maximizing net environmental gains.

9.3.4.3 *Policy Support and Regulatory Frameworks*

The advancement of WTE projects is highly dependent on supportive policy environments. Many governments have introduced Extended Producer Responsibility (EPR) schemes, landfill bans, carbon taxes, and renewable energy incentives that indirectly benefit plastic pyrolysis and gasification initiatives [102–104]. For instance, the Waste Framework Directive of European Union emphasizes waste hierarchy, encouraging recycling and energy recovery over landfilling. Countries like Japan and Germany have successfully implemented plastic-to-fuel and plastic-to-hydrogen projects under national waste-to-energy programs. Conversely, regulatory ambiguity or lack of clear classification for pyrolysis oil and syngas (e.g., waste-derived fuel vs. chemical product) can delay permitting and commercialization. Clear standards for product quality, emissions, and plant operation are necessary to encourage investments and public trust.

9.3.4.4 *Circular Economy and Resource Recovery*

Plastic WTE technologies play a vital role in a circular economy by transforming waste materials into valuable resources. Pyrolysis-derived oils can serve as feedstocks for virgin plastic production *via* chemical recycling routes, enabling a closed-loop system [33, 105]. Gasification outputs, especially hydrogen, can power low-carbon industries, vehicles, and electricity grids, further promoting resource efficiency. Policy frameworks that recognize chemical recycling and WTE as contributors to circular economy targets will likely accelerate innovation and adoption of advanced technologies.

9.4 Challenges and Opportunities

Converting waste into fuels provides an effective approach to improve waste management and reduce reliance on landfilling. However, the pyrolysis and gasification of PW face several hurdles, including variability in feedstock composition, inconsistent quality of end products, challenges in effective sorting of plastics, and ambiguous regulatory frameworks. Table 9.6 shows a summary of different challenges and opportunities encountered during plastic WTE technologies. Collaborative efforts between suppliers and systematic sorting could help streamline recycling operations and reduce costs. According to Qureshi *et al.* [106], a major barrier is the complex legislative environment governing the pyrolysis of plastic materials.

Table 9.6 Different challenges and opportunities encountered during plastic WTE technologies.

Challenges	Details	Opportunities
Feedstock variability	Mixed plastics and contaminants affect product quality	Advanced sorting and AI-driven classification
Technological maturity	Tar formation, temperature control, limited pilot-scale plants	Reactor optimization and catalyst innovations
Regulatory frameworks	Pyrolysis not recognized as recycling in many regions	Policy reform and recognition in circular economy targets
Economic constraints	High CAPEX, uncertain markets	Market creation, subsidies, and scale-up strategies
Environmental & public concerns	Emissions and public resistance	High-temperature technologies and transparent impact assessments

When pyrolysis-derived oil is intended for fuel use, it must comply with greenhouse gas (GHG) reduction targets specified in the EU's revised Renewable Energy Directive. Furthermore, the EU Waste Framework Directive defines "recycling" strictly as processes that transform waste into new products or materials for original or alternative purposes, excluding energy recovery. As a result, pyrolysis is currently not classified as a recycling method if its output is used for energy generation. Qureshi *et al.* [106] suggest that this regulatory approach should be reconsidered to support the commercial viability of pyrolysis. Another limitation is the scarcity of case studies from full-scale commercial operations, indicating a need for comprehensive economic evaluations and research focused on the scalability of pyrolysis technologies.

Moreover, challenges like feedstock heterogeneity, process control, and product quality standardization hinder large-scale deployment. Table 9.7 shows a comparative overview of the TRLs, feedstock suitability, and maturity across various plastic WTE technologies. Gasification technologies

Table 9.7 Comparative overview of TRLs, feedstock suitability, and maturity across plastic WTE technologies.

Technology	TRL	Temperature range (°C)	Feedstock suitability	Maturity level	Key advantages	Key challenges
Pyrolysis	6–8	300–800	Plastics (PE, PP, PS)	Emerging commercial	Tailored for plastic waste; liquid fuel production	Product standardization; feedstock sorting
Slagging gasification	8–9	>1600–2000	Coal, MSW, adaptable to plastics	Near commercial	High efficiency; low emissions; vitrified slag	High operating cost; plastic-specific design gaps
Non-slagging gasification	7–8	800–1200	Biomass, low-ash waste	Demo phase	Flexible feedstocks; moderate efficiency	Needs optimization for plastics
Plasma gasification	6–7	>3000 (plasma torch)	Mixed waste, plastics	Early commercial	Handles hazardous and diverse waste; vitrified output	High CAPEX/OPEX; technical complexity
Supercritical water gasification	5–6	>374°C + high pressure	Wet biomass, organics	Pilot/demo phase	Suited for wet waste; PFAS destruction	Limited scale; plastics compatibility issues

are advancing, with slagging gasification nearing full commercial maturity (TRL 8–9), especially for high-ash feedstocks such as coal and MSW. FastOx Gasification by Sierra Energy, operating above 2000°C with oxygen and steam, demonstrates high efficiency and low emissions, with ongoing scaling from 10 to 100 tons/day capacity [107]. Japanese variants operate below 1600°C, showing lower efficiency due to additional fuel input, thus increasing carbon intensity [108, 109]. Non-slagging gasification (TRL 7–8) is well-suited for biomass and low-ash waste. Developed systems by GTI and commercial initiatives like Enerkem in Canada and Europe highlight its potential [110, 111], though adaptation for plastics is still in progress. Plasma gasification (TRL 6–7), while effective for diverse wastes, faces economic and operational constraints despite installations like Alter NRG's

Tees Valley plant [112, 113]. Supercritical water gasification (SCWG), currently at TRL 5–6, uses supercritical water to treat wet organic waste and is being piloted by firms like 374Water [114]. Compared to these technologies, pyrolysis optimized specifically for plastic waste offers targeted benefits in reducing plastic pollution. While slagging and non-slagging gasification are more technologically established, advancements in processing and product upgrading are making pyrolysis increasingly viable. Plasma and SCWG approaches, though flexible, require cost reductions and technical enhancements. All technologies aim to convert waste into energy-rich products while minimizing environmental impacts. Rigorous emissions monitoring and life-cycle assessments are essential for ensuring long-term sustainability and regulatory compliance.

9.5 Case Studies and Industrial Applications

Several pilot and commercial projects have demonstrated the feasibility of pyrolysis and gasification for plastic waste conversion. Examples include the following.

9.5.1 Pyrolysis-Based Industrial Applications

Pyrolysis has gained considerable momentum as an effective route for transforming PW into useful chemicals and fuels. One of the notable examples is Quantafuel ASA, a Norwegian company operating in Denmark, which converts mixed plastic waste into synthetic diesel and naphtha. Their Skive plant processes 20,000 tons of PW annually and is integrated with downstream refining processes for enhanced fuel quality [115]. Another prominent player is Plastic Energy, based in the UK and Spain, which utilizes a patented thermal anaerobic conversion (TAC) process. Their plants in Almeria and Seville have demonstrated the conversion of mixed plastics into TACOIL, a hydrocarbon oil that can serve as a feed-material for producing new plastics or as a fuel source [116]. These projects demonstrate that with appropriate sorting and purification strategies, pyrolysis can be scaled to commercial levels and integrated with petrochemical infrastructure.

9.5.2 Gasification Projects and Demonstrations

Industrial gasification of plastic waste is also seeing increased global interest, particularly in high-temperature processes that ensure complete breakdown of polymers. Sierra Energy's FastOx® gasification technology, currently being piloted at Fort Hunter Liggett (USA), is designed to process municipal solid waste—including plastics—at over 2000°C using oxygen and steam. This system produces clean syngas with minimal emissions, intended for power generation or chemical synthesis [117]. In Japan, engineering firms have developed commercial-scale gasification and melting plants that process MSW with high plastic content. These facilities produce syngas while converting the remaining ash into vitrified slag, which can be used as construction material [118, 119]. The successful operation of such plants demonstrates the adaptability of gasification technologies for urban waste treatment in densely populated regions.

Plasma gasification, though still in early stages of commercialization, has shown potential through high-profile demonstrations. The Tees Valley project in the UK, developed by Air Products using Westinghouse Plasma technology, was initially planned to convert 50,000 tons of waste annually into electricity. While the project was eventually halted due to technical and economic issues, it provided valuable insights into scaling challenges, cost control, and material handling in plasma environments [120, 121]. Other ongoing developments by Alter NRG and Hitachi Zosen Inova continue to explore plasma-based routes for difficult-to-treat waste including plastics and hazardous waste, with improved process controls and heat recovery mechanisms [122].

9.5.3 Integration with Petrochemical and Circular Economy Models

A promising trend is the integration of WTE processes with existing petrochemical supply chains. For instance, Shell and SABIC have entered collaborations with pyrolysis companies like Plastic Energy to develop closed-loop recycling for polyolefins. These efforts aim to "drop-in" recycled hydrocarbons into naphtha crackers, ensuring that virgin-equivalent plastics can be produced without fossil inputs [123, 124]. Likewise, Dow Chemical is partnering with Fuenix Ecogy in the Netherlands to source pyrolysis oil derived from waste plastic, which is then upgraded into polyethylene used in consumer packaging [125, 126]. These case studies illustrate that industrial uptake is feasible when WTE products meet quality and compatibility requirements of large-scale chemical processing.

Thus, the analysis of real-world applications reveals critical lessons. First, successful projects typically rely on well-sorted or pre-treated plastic streams to maintain process efficiency. Second, the proximity to refining or energy markets plays a decisive role in economic viability. Third, public-private partnerships, regulatory clarity, and funding for innovation are vital to overcoming technical and financial uncertainties. Looking forward, hybrid technologies—such as combined pyrolysis–gasification systems and modular reactor designs—offer opportunities for more flexible and decentralized waste management. As LCA tools and emissions standards mature, the environmental credibility of these industrial WTE pathways will further strengthen.

9.6 Conclusion

The sustainable valorization of plastic waste through pyrolysis and gasification presents a promising route for circular economy integration, enabling resource recovery, reduction in landfilling, and mitigating greenhouse gas emissions. Quantitative analyses from various case studies underscore the synergistic potential of co-processing biomass with plastic waste, demonstrating improved fuel quality and process efficiency. For instance, increasing plastic content (especially PS) during pyrolysis leads to reduced char yield and increased production of energy-dense hydrocarbon gases such as CH_4 and CO, while PVC significantly suppresses oil yields due to HCl formation. Gas LHV increases sharply at 100% plastic content due to enhanced hydrocarbon generation. Moreover, in co-gasification systems, rising plastic content boosts hydrogen yield, especially beyond 50%, attributed to enhanced thermal cracking and water–gas shift reactions. Conversely, CO and CO_2 contents decline as plastics replace oxygen-rich biomass. CH_4 levels increase considerably due to plastic decomposition pathways, leading to more energy-rich syngas compositions. Additionally, tar content generally decreases with plastics, reducing the need for complex gas cleaning, although excessive plastic ratios may reverse this trend due to incomplete cracking. Furthermore, the LHV of syngas and CGE improve with increased plastic ratios, supporting energy-efficient design considerations. Although carbon conversion efficiency varies based on reactor configuration and gasifying medium, overall gas yields show a consistent rise because of plastics' high volatile content and low ash levels. These findings reinforce the viability of plastic waste integration in thermochemical systems for waste-to-energy conversion. Moving forward, standardized feedstock preparation, streamlined legislative frameworks, and scalable demonstration projects are essential for commercialization. Moreover, advanced reactor designs that enable flexible feedstock usage,

AI-driven process optimization, and carbon capture integration can further enhance environmental and economic performance.

Moreover, this study also critically analyzed the techno-economic potential, industrial maturity, and operational viability of various thermal conversion technologies specifically pyrolysis and gasification highlighting their respective technology readiness levels (TRLs), feedstock compatibilities, and deployment challenges. While pyrolysis is more tailored for plastic waste conversion, slagging and non-slagging gasification demonstrate higher maturity and scalability. Plasma and SCWG offer unique pathways but face economic and technological barriers to widespread adoption. The techno-economic review revealed that pyrolysis remains economically viable under optimized process conditions, particularly when monomer recovery (e.g., styrene from PS) is achieved or when pyrolytic oil is upgraded for fuel applications. Gasification technologies, particularly non-slagging and co-gasification routes, offer higher cold gas efficiencies and are better suited for mixed or high-moisture waste streams. However, regulatory ambiguities such as the EU's exclusion of pyrolysis as a "recycling" method when used for energy pose significant hurdles to market integration and commercial financing.

Ultimately, pyrolysis and gasification of PW either individually or in co-processing with biomass represent powerful tools in the global transition toward sustainable energy systems, supporting waste reduction goals and offering valuable secondary products for energy and chemical sectors.

AI disclosure statement: The author confirms that no artificial intelligence (AI)-generated content or AI-assisted writing tools were used in the preparation of this chapter (including summaries and the abstract). All text, analysis, and conclusions are the result of the author's original work.

References

1. Alqattaf, A., Plastic waste management: Global facts, challenges and solutions. *2020 Second International Sustainability and Resilience Conference: Technology and Innovation in Building Designs (51154)*, IEEE, 2020.
2. Khoaele, K.K., *et al.*, The devastation of waste plastic on the environment and remediation processes: A critical review. *Sustainability*, 15, 6, 5233, 2023.
3. Matthews, C., Moran, F., Jaiswal, A.K., A review on European Union's strategy for plastics in a circular economy and its impact on food safety. *J. Cleaner Prod.*, 283, 125263, 2021.

4. Lin, C., Paengsri, P., Yang, Y., Impact of China's National Sword Policy on waste import: A difference-in-differences approach. *Econ. Anal. Policy*, 78, 887–903, 2023.
5. Picuno, C., *et al.*, Factors shaping the recycling systems for plastic packaging waste—A comparison between Austria, Germany and The Netherlands. *Sustainability*, 13, 12, 6772, 2021.
6. Siddique, R., Khatib, J., Kaur, I., Use of recycled plastic in concrete: A review. *Waste Manage. (Oxford)*, 28, 10, 1835–1852, 2008.
7. Bhattacharya, R., *et al.*, Challenges and opportunities: plastic waste management in India, 2018.
8. Razzak, S.A., Municipal Solid and Plastic Waste Co-pyrolysis Towards Sustainable Renewable Fuel and Carbon Materials: A Comprehensive Review. *Chem.- Asian J.*, 19, 17, e202400307, 2024.
9. Miao, Y., von Jouanne, A., Yokochi, A., Current technologies in depolymerization process and the road ahead. *Polymers*, 13, 3, 449, 2021.
10. Raj, R., Singh, D.K., Tirkey, J.V., Co-gasification of plastic waste blended with coal and biomass: a comprehensive review. *Environ. Technol. Rev.*, 12, 1, 614–642, 2023.
11. dos Santos, R.G. and Alencar, A.C., Biomass-derived syngas production *via* gasification process and its catalytic conversion into fuels by Fischer Tropsch synthesis: A review. *Int. J. Hydrogen Energy*, 45, 36, 18114–18132, 2020.
12. Uddin, N.M. and Nithe, N.A., A comprehensive exploration of biomass gasification technologies advancing United Nations sustainable development goals: Part II: Reactor types, power generation, current status and future challenges. *Johnson Matthey Technol. Rev.*, 2025.
13. Sajid, M., *et al.*, Gasification of municipal solid waste: Progress, challenges, and prospects. *Renewable Sustainable Energy Rev.*, 168, 112815, 2022.
14. Ajorloo, M., *et al.*, Experimental analysis of the effects of feedstock composition on the plastic and biomass Co-gasification process. *Renewable Energy*, 231, 120960, 2024.
15. Gupta, J., Agarwal, M., Dalai, A., An overview on the recent advancements of sustainable heterogeneous catalysts and prominent continuous reactor for biodiesel production. *J. Ind. Eng. Chem.*, 88, 58–77, 2020.
16. Adeoye, A.O., *et al.*, Pyrolysis of plastic wastes towards achieving a circular economy: an advanced chemistry and technical approach, in: *Dynamics of Transportation Ecosystem, Modeling, and Control*, pp. 215–259, Springer, 2024.
17. Clark, R.A. and Shaver, M.P., Depolymerization within a circular plastics system. *Chem. Rev.*, 124, 5, 2617–2650, 2024.
18. Hsu, C.S., *et al.*, Cracking, in: *Petroleum Science and Technology*, pp. 211–244, 2019.
19. Gajewski, J.J., *Hydrocarbon thermal isomerizations*, Elsevier, 2004.
20. Williams, P.T. and Taylor, D.T., Aromatization of tyre pyrolysis oil to yield polycyclic aromatic hydrocarbons. *Fuel*, 72, 11, 1469–1474, 1993.

21. Hahladakis, J.N., *et al.*, An overview of chemical additives present in plastics: Migration, release, fate and environmental impact during their use, disposal and recycling. *J. Hazard. Mater.*, 344, 179–199, 2018.
22. Dai, L., *et al.*, Pyrolysis technology for plastic waste recycling: A state-of-the-art review. *Prog. Energy Combust. Sci.*, 93, 101021, 2022.
23. Andrigo, P., Bagatin, R., Pagani, G., Fixed bed reactors. *Catal. Today*, 52, 2–3, 197–221, 1999.
24. Rowe, P.N. and Yates, J.G., Fluidized-bed reactors, in: *Chemical reaction and reactor engineering*, pp. 441–498, CRC Press, 2020.
25. Boateng, A.A., *Rotary kilns: transport phenomena and transport processes*, Butterworth-Heinemann, 2015.
26. Motasemi, F. and Afzal, M.T., A review on the microwave-assisted pyrolysis technique. *Renewable Sustainable Energy Rev.*, 28, 317–330, 2013.
27. Ibrahim, H.A.-H., Introductory chapter: pyrolysis, in: *Recent Advances in Pyrolysis*, p. 1, 2020.
28. Ighalo, J.O., *et al.*, Flash pyrolysis of biomass: a review of recent advances. *Clean Technol. Environ. Policy*, 24, 8, 2349–2363, 2022.
29. Klerk, A., Zeolites as catalysts for fuels refining after indirect liquefaction processes. *Molecules*, 23, 1, 115, 2018.
30. De Klerk, A., *Fischer-tropsch refining*, John Wiley & Sons, 2012.
31. Lu, M., *et al.*, Insights into Chemical Recycling and Upgrading Strategies for Polyolefin-Based Plastics. *Ind. Eng. Chem. Res.*, 64, 11, 5765–5781, 2025.
32. Raj, R., Singh, D.K., Tirkey, J., Co-gasification of Low-grade coal with Madhuca longifolia (Mahua) biomass and dual-fuelled mode engine performance: Effect of biomass blend and engine operating condition. *Energy Convers. Manage.*, 269, 116150, 2022.
33. Kumar, M., *et al.*, A review on value-addition to plastic waste towards achieving a circular economy. *Sci. Total Environ.*, 171106, 2024.
34. Walker-Franklin, I. and Jambeck, J., *Plastics*, MIT Press, 2023.
35. Shen, Y., *et al.*, Waste-to-energy: Dehalogenation of plastic-containing wastes. *Waste Manage. (Oxford)*, 49, 287–303, 2016.
36. Ferreiro, A., *et al.*, Influence of process parameters on biomass gasification: A review of experimental studies in entrained flow reactors and droptube furnaces. *Biomass Bioenergy*, 185, 107217, 2024.
37. Siedlecki, M., De Jong, W., Verkooijen, A.H., Fluidized bed gasification as a mature and reliable technology for the production of bio-syngas and applied in the production of liquid transportation fuels—a review. *Energies*, 4, 3, 389–434, 2011.
38. Havilah, P.R., *et al.*, Biomass gasification in downdraft gasifiers: A technical review on production, up-gradation and application of synthesis gas. *Energies*, 15, 11, 3938, 2022.
39. Ma, X., Wei, H., Luo, Z., Development and prospect of the acetylene production chain based on the process systems engineering: a focus on the polyvinyl chloride production. *Rev. Chem. Eng.*, 2024.

40. Liu, K., Song, C., Subramani, V., *Hydrogen and syngas production and purification technologies*, John Wiley & Sons, 2010.
41. Adeniyi, A.G., *et al.*, Thermochemical co-conversion of biomass-plastic waste to biochar: a review. *Green Chem. Eng.*, 5, 1, 31–49, 2024.
42. Ephraim, A., *et al.*, Co-pyrolysis of wood and plastics: Influence of plastic type and content on product yield, gas composition and quality. *Fuel*, 231, 110–117, 2018.
43. Jiang, X., Zhu, B., Zhu, M., An overview on the recycling of waste poly (vinyl chloride). *Green Chem.*, 25, 18, 6971–7025, 2023.
44. Yu, J., *et al.*, Thermal degradation of PVC: A review. *Waste Manage. (Oxford)*, 48, 300–314, 2016.
45. Maafa, I.M., Pyrolysis of polystyrene waste: A review. *Polymers*, 13, 2, 225, 2021.
46. Pan, R., Zan, Y., Debenest, G., Oil production from waste polyethylene and polystyrene co-pyrolysis: Interactions of temperature and carrier gas flow rate. *J. Environ. Chem. Eng.*, 10, 3, 107555, 2022.
47. Kartik, S., *et al.*, Valorization of plastic wastes for production of fuels and value-added chemicals through pyrolysis–A review. *Therm. Sci. Eng. Prog.*, 32, 101316, 2022.
48. Kida, M., Ziembowicz, S., Koszelnik, P., CH4 and CO2 emissions from the decomposition of microplastics in the bottom sediment—preliminary studies. *Environments*, 9, 7, 91, 2022.
49. Alshareef, R., Nahil, M.A., Williams, P.T., Hydrogen production by three-stage (i) pyrolysis, (ii) catalytic steam reforming, and (iii) water gas shift processing of waste plastic. *Energy Fuels*, 37, 5, 3894–3907, 2023.
50. Alshareef, R., *et al.*, Three-stage pyrolysis–steam reforming–water gas shift processing of household, commercial and industrial waste plastics for hydrogen production. *Waste Disposal Sustainable Energy*, 6, 1, 25–37, 2024.
51. Uzoejinwa, B.B., *et al.*, Co-pyrolysis of biomass and waste plastics as a thermochemical conversion technology for high-grade biofuel production: Recent progress and future directions elsewhere worldwide. *Energy Convers. Manage.*, 163, 468–492, 2018.
52. Deng, J., *et al.*, Co-pyrolysis of biomass and plastic waste into carbon materials with environmental applications: a critical review. *Green Chem.*, 2025.
53. Cudjoe, D., Brahim, T., Zhu, B., Assessing the economic and ecological viability of generating electricity from oil derived from pyrolysis of plastic waste in China. *Waste Manage. (Oxford)*, 168, 354–365, 2023.
54. KT, V., Tom, A.K., Achar, R.R., Pyrolysis of Plastic Waste to Plastic Oil: A Future Source of Fuel. *J. Sustainable Dev. Energy Water Environ. Syst.*, 13, 1, 1–22, 2025.
55. Parku, G.K., Collard, F.-X., Görgens, J.F., Pyrolysis of Waste Polypropylene Plastics for Energy Recovery: Optimization of Product Yield and Quality on a Commissioned Pilot-Scale Setup. *Energy Fuels*, 2025.

56. Lopez, G., *et al.*, Effect of polyethylene co-feeding in the steam gasification of biomass in a conical spouted bed reactor. *Fuel*, 153, 393–401, 2015.
57. Baloch, H.A., *et al.*, Parametric study of co-gasification of ternary blends of rice straw, polyethylene and polyvinylchloride. *Clean Technol. Environ. Policy*, 18, 1031–1042, 2016.
58. Pinto, F., *et al.*, Co-gasification of coal and wastes in a pilot-scale installation 1: Effect of catalysts in syngas treatment to achieve tar abatement. *Fuel*, 88, 12, 2392–2402, 2009.
59. Mishra, R., Ong, H.C., Lin, C.-W., Progress on co-processing of biomass and plastic waste for hydrogen production. *Energy Convers. Manage.*, 284, 116983, 2023.
60. Wilk, V. and Hofbauer, H., Co-gasification of plastics and biomass in a dual fluidized-bed steam gasifier: possible interactions of fuels. *Energy Fuels*, 27, 6, 3261–3273, 2013.
61. Xu, D., *et al.*, Performances of syngas production and deposited coke regulation during co-gasification of biomass and plastic wastes over Ni/γ-Al2O3 catalyst: Role of biomass to plastic ratio in feedstock. *Chem. Eng. J.*, 392, 123728, 2020.
62. Yu, H., *et al.*, Experimental study on co-gasification characteristics of biomass and plastic wastes. *BioResources*, 9, 3, 5615–5626, 2014.
63. Alvarez, J., *et al.*, Hydrogen production from biomass and plastic mixtures by pyrolysis-gasification. *Int. J. Hydrogen Energy*, 39, 21, 10883–10891, 2014.
64. Wilk, V. and Hofbauer, H., Conversion of mixed plastic wastes in a dual fluidized bed steam gasifier. *Fuel*, 107, 787–799, 2013.
65. Khatibi, M., Nahil, M.A., Williams, P.T., Pyrolysis/Non-thermal Plasma/Catalysis Processing of Refuse-Derived Fuel for Upgraded Oil and Gas Production. *Waste Biomass Valor.*, 1–28, 2025.
66. Gupta, N., *et al.*, United States Energy Association: Co-utilization of Coal and Mixed Scrap Plastics *via* Syngas Production with Carbon Capture, Utilization, and Storage (CCUS), 2021.
67. Kern, S.J., Pfeifer, C., Hofbauer, H., Cogasification of polyethylene and lignite in a dual fluidized bed gasifier. *Ind. Eng. Chem. Res.*, 52, 11, 4360–4371, 2013.
68. Bashir, M.A., *et al.*, Plastic Waste Gasification for Low-Carbon Hydrogen Production: A Comprehensive Review. *Energy Adv.*, 2025.
69. Yek, P.N.Y., *et al.*, Co-processing plastics waste and biomass by pyrolysis–gasification: A review. *Environ. Chem. Lett.*, 22, 1, 171–188, 2024.
70. Shah, H.H., *et al.*, A review on gasification and pyrolysis of waste plastics. *Front. Chem.*, 10, 960894, 2023.
71. Zaccariello, L. and Mastellone, M.L., Fluidized-bed gasification of plastic waste, wood, and their blends with coal. *Energies*, 8, 8, 8052–8068, 2015.
72. Cho, M.-H., Mun, T.-Y., Kim, J.-S., Air gasification of mixed plastic wastes using calcined dolomite and activated carbon in a two-stage gasifier to reduce tar. *Energy*, 53, 299–305, 2013.

73. Brown, J.L., *Thermal oxo-degradation of plastic wastes to valuable compounds*, Iowa State University, 2024.
74. Pan, Y., *et al.*, Removal of tar by secondary air in fluidised bed gasification of residual biomass and coal. *Fuel*, 78, 14, 1703–1709, 1999.
75. Aentung, T., Patcharavorachot, Y., Wu, W., Co-Gasification of Plastic Waste Blended with Biomass: Process Modeling and Multi-Objective Optimization. *Processes*, 12, 9, 1906, 2024.
76. McKendry, P., Energy production from biomass (part 1): overview of biomass. *Bioresour. Technol.*, 83, 1, 37–46, 2002.
77. Burra, K. and Gupta, A., Synergistic effects in steam gasification of combined biomass and plastic waste mixtures. *Appl. Energy*, 211, 230–236, 2018.
78. Mariyam, S., *et al.*, A critical review on co-gasification and co-pyrolysis for gas production. *Renewable Sustainable Energy Rev.*, 161, 112349, 2022.
79. Tomić, T., Slatina, I., Schneider, D.R., Techno-economic review of pyrolysis and gasification plants for thermochemical recovery of plastic waste and economic viability assessment of small-scale implementation. *Clean Technol. Environ. Policy*, 26, 1, 171–195, 2024.
80. Ochieng, R. and Sarker, S., Energy and techno-economic analysis of integrated supercritical water gasification of sewage sludge and fast pyrolysis of wood for power, heat, and hydrogen production. *Chem. Eng. Sci.*, 121236, 2025.
81. Raj, R. and Tirkey, J.V., Techno-economic assessment of sugarcane bagasse pith-based briquette production and performance analysis of briquette feed gasifier-engine system. *J. Environ. Manage.*, 345, 118828, 2023.
82. Tan, K.Q., *et al.*, Valorization of hazardous plastic wastes into value-added resources by catalytic pyrolysis-gasification: A review of techno-economic analysis. *Renewable Sustainable Energy Rev.*, 182, 113346, 2023.
83. Soka, O. and Oyekola, O., A feasibility assessment of the production of char using the slow pyrolysis process. *Heliyon*, 6, 7, 2020.
84. Shen, Y., *et al.*, By-products recycling for syngas cleanup in biomass pyrolysis–An overview. *Renewable Sustainable Energy Rev.*, 59, 1246–1268, 2016.
85. Lubongo, C., *et al.*, Economic feasibility of plastic waste conversion to fuel using pyrolysis. *Sustainable Chem. Pharm.*, 27, 100683, 2022.
86. Fulgencio-Medrano, L., *et al.*, Oil production by pyrolysis of real plastic waste. *Polymers*, 14, 3, 553, 2022.
87. Olomo, E., *et al.*, Thermal and catalytic pyrolysis of automotive plastic wastes to diesel range fuel. *Heliyon*, 10, 20, 2024.
88. Faizal, M., *et al.*, Purification of Synthetic Gas from Fine Coal Waste Gasification as a Clean Fuel. *J. Ecol. Eng.*, 22, 5, 114–120, 2021.
89. Braimakis, K., *et al.*, Economic evaluation of decentralized pyrolysis for the production of bio-oil as an energy carrier for improved logistics towards a large centralized gasification plant. *Renewable Sustainable Energy Rev.*, 35, 57–72, 2014.

90. McCallum, D. and Eng, P., Re: Waste to Energy Business Case Analysis–Technical Report (Final), 2011.
91. Dong, J., *et al.*, Key factors influencing the environmental performance of pyrolysis, gasification and incineration Waste-to-Energy technologies. *Energy Convers. Manage.*, 196, 497–512, 2019.
92. Porshnov, D., Evolution of pyrolysis and gasification as waste to energy tools for low carbon economy. *Wiley Interdiscip. Rev.: Energy Environ.*, 11, 1, e421, 2022.
93. Kirk, E.A., The Montreal protocol or the Paris agreement as a model for a plastics treaty?, 2020.
94. Kubiczek, J., *et al.*, Chemical recycling of plastic waste as a mean to implement the circular economy model in the European Union. *J. Cleaner Prod.*, 406, 136951, 2023.
95. van Der Marel, E.R., Trading plastic waste in a global economy: soundly regulated by the Basel Convention? *J. Environ. Law*, 34, 3, 477–497, 2022.
96. Li, S., Reviewing air pollutants generated during the pyrolysis of solid waste for biofuel and biochar production: toward cleaner production practices. *Sustainability*, 16, 3, 1169, 2024.
97. Safavi, A., Richter, C., Unnthorsson, R., Dioxin formation in biomass gasification: a review. *Energies*, 15, 3, 700, 2022.
98. Xayachak, T., *et al.*, Assessing the environmental footprint of plastic pyrolysis and gasification: a life cycle inventory study. *Process Saf. Environ. Prot.*, 173, 592–603, 2023.
99. Dong, J., *et al.*, Life cycle assessment of pyrolysis, gasification and incineration waste-to-energy technologies: Theoretical analysis and case study of commercial plants. *Sci. Total Environ.*, 626, 744–753, 2018.
100. Ostadi, M., *et al.*, CH4 and CO2 Reductions from Methanol Production Using Municipal Solid Waste Gasification with Hydrogen Enhancement. *Sustainability*, 16, 8649, 20242024.
101. Hossain, M.M., *et al.*, Techno-economic Impact of Incineration-Based Waste to Energy Generation Technology in Bangladesh: A Short Review and A Case Study, 2025.
102. Steenmans, K., Extended producer responsibility: An assessment of recent amendments to the European Union Waste Frameworks directive. *Law Env't. Dev. J.*, 15, 108, 2019.
103. Lorang, S., *et al.*, Achievements and policy trends of extended producer responsibility for plastic packaging waste in Europe. *Waste Disposal Sustainable Energy*, 4, 2, 91–103, 2022.
104. Abate, T.G. and Elofsson, K., Environmental taxation of plastic bags and substitutes: Balancing marine pollution and climate change. *J. Environ. Manage.*, 359, 120868, 2024.
105. Zhao, X., *et al.*, Plastic waste upcycling toward a circular economy. *Chem. Eng. J.*, 428, 131928, 2022.

106. Qureshi, M.S., *et al.*, Pyrolysis of plastic waste: Opportunities and challenges. *J. Anal. Appl. Pyrolysis*, 152, 104804, 2020.
107. Dodd, D., Hazard, J., Energy, S., Gasification Technology Developed for the Production of Renewable Hydrogen. *Oil Shale*, 59, 0.05.
108. Heberlein, S., *et al.*, High temperature slagging gasification of municipal solid waste with biomass charcoal as a greener auxiliary fuel. *J. Hazard. Mater.*, 423, 127057, 2022.
109. Okada, T., *et al.*, Recovery of zinc and lead from fly ash from ash-melting and gasification-melting processes of MSW–Comparison and applicability of chemical leaching methods. *Waste Manage. (Oxford)*, 27, 1, 69–80, 2007.
110. Zhuang, Q., Modern Gasification Process, in: *From Coal to Hydrogen: A Long Journey of Energy Transition*, pp. 193–223, Springer, 2024.
111. Ciuta, S., Tsiamis, D., Castaldi, M.J., *Gasification of waste materials: technologies for generating energy, gas, and chemicals from municipal solid waste, biomass, nonrecycled plastics, sludges, and wet solid wastes*, Academic Press, 2017.
112. Cudjoe, D. and Wang, H., Plasma gasification versus incineration of plastic waste: Energy, economic and environmental analysis. *Fuel Process. Technol.*, 237, 107470, 2022.
113. Sharina, I., Perepechko, L., Domarov, P., Development of technology of plasma processing of technogenic wastes (brief review). *J. Phys. Conf. Ser.*, IOP Publishing, 2019.
114. Okolie, J.A., *et al.*, A review on subcritical and supercritical water gasification of biogenic, polymeric and petroleum wastes to hydrogen-rich synthesis gas. *Renewable Sustainable Energy Rev.*, 119, 109546, 2020.
115. Hassan, W., Technical, chemical and economical aspects of plastic recycling process, Politecnico di Torino, 2022.
116. Antelava, A., *et al.*, Plastic solid waste (PSW) in the context of life cycle assessment (LCA) and sustainable management. *Environ. Manage.*, 64, 230–244, 2019.
117. Energy, S., ESTCP Project EW-201334, 2020.
118. Heberlein, S., *et al.*, Converting ash into reusable slag at lower carbon footprint: Vitrification of incineration bottom ash in MSW-fueled demonstration-scale slagging gasifier. *J. Environ. Manage.*, 352, 119967, 2024.
119. Sanito, R.C., *et al.*, A review on vitrification technologies of hazardous waste. *J. Environ. Manage.*, 316, 115243, 2022.
120. Boulos, M.I., *et al.*, Plasma In The Waste Treatment Industry, in: *Handbook Of Thermal Plasmas*, pp. 1739–1804, Springer, 2023.
121. Ducharme, C., Technical and economic analysis of Plasma-assisted Waste-to-Energy processes, Earth Engineering Center, Columbia University, 2010.
122. Themelis, N.J., Energy and materials recovery from post-recycling wastes: WTE. *Waste Dispos. Sustain. Energy*, 5, 3, 249–257, 2023.
123. Galan-Sanchez, L., *et al.*, Chemical Recycling of Mixed Plastic Waste for the Production of Aromatics and Olefins as Monomers for Circular Polyolefins

and Polycarbonate. SABIC TRUCIRCLE™ (Case Study), in: *Industrial Arene Chemistry: Markets, Technologies, Sustainable Processes and Cases Studies of Aromatic Commodities*, vol. 4, pp. 2143–2162, 2023.
124. Negri, A., Current state and decarbonisation options for the dutch polyolefins industry, 2020.
125. Eerens, H. and van Dam, D., Decarbonisation options for large volume organic chemicals production, dow chemical terneuzen, 2022.
126. Negri, A. and Ligthart, T., Decarbonisation options for the Dutch polyolefins industry, PBL Netherlands Environmental Assessment Agency, 2021, https://www.pbl.nl/en/publications/decarbonisation-options-for-the-dutch-polyolefins-industry.

10

Bioplastics as an Alternative: Sustainability and Environmental Impacts

Retno Aliyatul Fikroh and Asih Widi Wisudawati*

Department of Chemistry Education, UIN Sunan Kalijaga Yogyakarta, Yogyakarta, Indonesia

Abstract

As the third most commonly used petroleum derivative, plastics have provided numerous benefits to humankind but have also left negative impacts on the environment. In terms of smaller plastics, microplastics (MPs) have contaminated the food chain. Concerns about plastic pollution have spurred extensive research into sustainable biodegradable alternatives like bioplastics (BPs). As a polymer material derived from renewable biomass, BPs have the potential to replace fossil-based plastics. They can be made from natural polymers such as agricultural waste. In this context, banana peels and coconut husks show promise as raw materials for BPs. An investigation is necessary to differentiate the BPs obtained from the aforementioned novel agricultural waste materials. Uli (the Indonesian name for banana peels) contain starch and natural polymers that aid in the formation of a flexible and biodegradable matrix, while coconut shells, composed of lignocellulose, enhance the BPs' durability and mechanical strength. By comparing these two materials, this study evaluates their physicochemical properties, biodegradability rates, and potential applications. It emphasizes the promise of agricultural waste in producing environmentally friendly BPs, contributing to sustainable waste management and environmental protection. The framework of this chapter is also linked to harnessing the potential of BPs within the context of a circular economy.

Keywords: BPs, banana peel, sustainability, tropical materials, circular economy

**Corresponding author*: asih.wisudawati@uin-suka.ac.id

Neha Agarwal, Brijesh Pare, Vijendra Singh Solanki and Sankara Rao Miditana (eds.) Plastic Remediation: Transformative Pathways for a Rejuvenated Sustainable Environment, (245–274)

10.1 Introduction

Pollution is one of the major challenges of the modern era, contributing significantly to millions of deaths each year. Generally, pollution refers to the introduction of harmful substances or energy into the environment with the potential to degrade quality, disrupt ecosystems, and endanger health. Plastic pollution is an issue that is gaining increasing attention due to its extensive and persistent environmental impact. Plastic is a highly versatile synthetic material widely used in household packaging, electronic devices, the pharmaceutical industry, and the automotive industry. Its modifiable chemical structure allows a variety of mechanical and chemical properties, such as strength, flexibility, stability, reactivity, and durability, to be achieved. Additionally, its relatively low production costs make plastic an essential material in modern life.

The first successful plastic developed was Parkesine, a cellulose-based material that has the ability to be molded into various designs [1]. As technology developed, these natural-based plastics began to be replaced by petroleum-based synthetic polymers, such as silicone, polyethylene, polystyrene, and nylon [2]. About 140 million tons of petroleum-based polymers, such as crude oil, natural gas, and coal, are produced and used to make plastics and their derivatives each year [3]. Plastics are classified into two types based on their physical and chemical characteristics: thermoplastics and thermosets. Thermoplastics melt and reshape when heated, while thermosets maintain their shape and cannot be reshaped [4]. The high plasticity of the thermoplastic type causes it to be widely used compared to the thermoset type [5]. The Society of the Plastics Industry (SPI in Indonesia) has classified the types of plastics through a coding system that aims to facilitate the recycling process.

Petroleum-based plastics are not biodegradable and will remain in the environment for a very long time. The repercussions are twofold: environmental and health-related. When plastic is subjected to heat, it has the potential to release carcinogenic particles. Environmental impacts include accumulation in landfills, water pollution with increased biochemical oxygen demand (BOD) levels, toxicity, and other negative impacts on biodiversity. Currently, more than a third of global plastic production is used in the packaging sector, most of which is not recycled, and eventually ends up as waste. The long-term accumulation of plastic waste has significant ecological impacts. It is estimated that global plastic production exceeds 400 million tons per year, yet only about 7% is successfully recycled. The remaining 93% end up in landfills or pollute aquatic ecosystems [6].

In the process, plastics can also degrade into microscopic particles called MPs [7]. MPs in the form of PS, PP, and PE are commonly used as microbeads, which are very small plastic particles commonly found in care products such as facial soaps, toothpastes, scrubs, and cosmetics (Figure 10.1). Exposure to MPs has been linked to health disorders such as intestinal inflammation, oxidative stress, metabolic disorders, organ dysfunction, and cancer [7, 8].

Increased consumption of plastics is directly correlated with increased accumulation of waste in both terrestrial and aquatic environments [9]. In the marine environment, the degradation process is slower due to low temperatures and limited exposure to ultraviolet light. The presence of toxic chemicals in plastics exacerbates soil, water, and air pollution and can hinder groundwater movement [10]. Considering these negative impacts, plastic pollution poses a serious threat to human health and ecosystem sustainability. Therefore, the development of environmentally friendly alternative plastics from renewable resources, known as BPs, is becoming a major focus of researchers in an effort to reduce dependence on petroleum-based plastics.

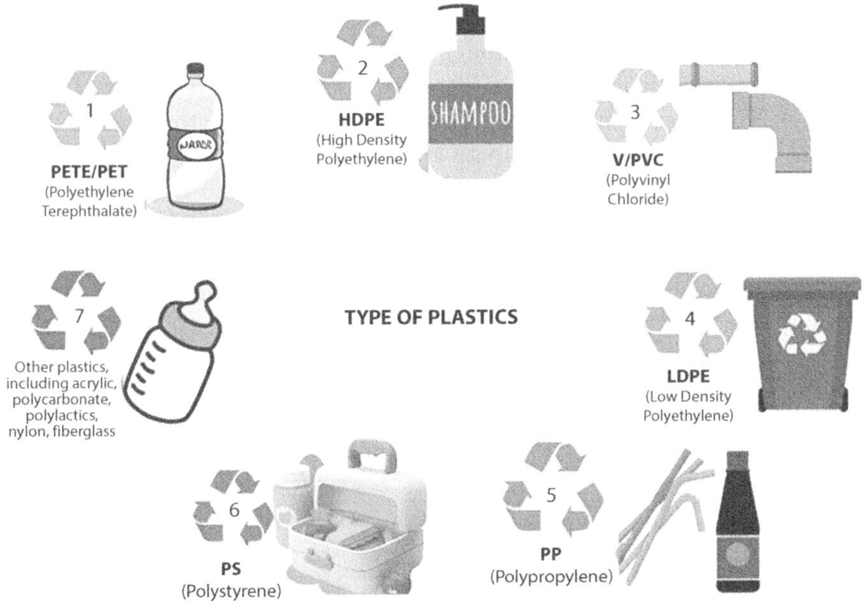

Figure 10.1 Types of plastics.

10.2 BPS

10.2.1 Characteristic of BPs

Plastics produced from plant and animal materials are commonly known as BPs [11]. Before the 20th century, people had used cellulose, rubber, and lignin from agricultural waste, plant fibers, and extracts from natural materials. Bakelite is the first synthetic polymer made from natural sources, namely the result of condensation of phenol and formaldehyde, and it has been widely used in various applications [12]. In the 20th century, polylactic acid (PLA) and polyhydroxyalkanoate (PHA) were in great demand because they were cheap, easy to use, and had long durability [13, 14]. The sources of biodegradable polymers are depicted in Figure 10.2.

BPs, known as biodegradable plastics or biodegradable polymers, are a type of plastic that is environmentally friendly, inexpensive, degradable, and recyclable. BPs are generally categorized into three types: petroleum-based BPs, bio-based BPs, and a combination of renewable materials with petroleum (bio-/petro-) [15].

The classification of biodegradable polymers is as follows:

a) Petroleum-based BPs are synthetic polymers derived from petroleum that are biodegradable. They are designed to break down after their intended use has been completed. Examples of such materials include polycaprolactone (PCL) and poly (butylene adipate-co-terephthalate) (PBAT).
b) BPs derived from renewable resources (i.e., bio-based) are biopolymers that are naturally synthesized from plants and animals. Examples of such materials include poly lactic acid (PLA), lignin, protein, cellulose, starch, and chitosan.

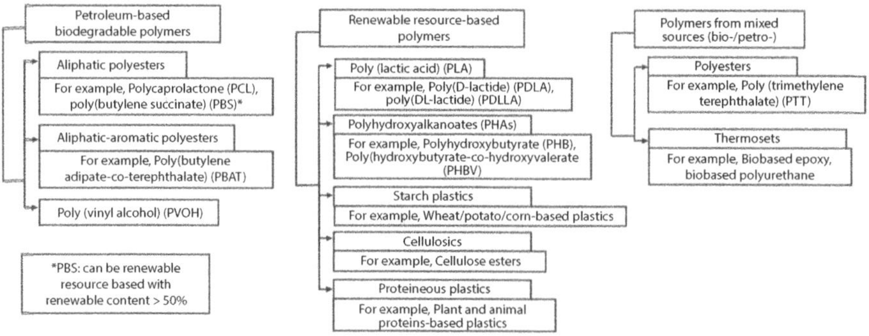

Figure 10.2 Biodegradable polymer sources [13].

c) BPs, which are derived from a blend of renewable materials and petroleum (i.e., bio-/petro-), are biodegradable materials that are produced from a combination of natural and petroleum-based polymers. Examples of such materials include poly (trimethylene terephthalate) (PTT) and biobased blends.

BPs have important properties that make them an environmentally friendly and attractive alternative to conventional plastics. These properties include the following:

a) Biodegradability: BPs can be easily decomposed through chemical and enzymatic reactions. Chemical reactions in the BP degradation process, such as hydrolysis, oxidation, and even photodegradation reactions, are important to consider. While enzymatic action involves microorganisms such as bacteria and fungi that produce enzymes such as lipase, protease, and esterase, these microorganisms are able to break down chemical bonds in BP polymers [16]. Therefore, BPs are the best alternative to petroleum-based plastics, which are difficult to decompose in the environment.
b) Transparency: BPs made from starch, cellulose, and PHA have a high level of transparency, so they are widely used in the food, pharmaceutical, and electrical industries [17].
c) Flexible and durable: BPs have flexible and durable properties, easily bending, reshaping, and deforming without breaking when exposed to mechanical stress, environmental influences, chemical exposure, UV radiation, pressure, temperature, and aging factors. These properties are very important as they expand their use in various industries, such as packaging, medical devices, textiles, and agriculture, to increase product shelf life and reduce environmental impact.
d) Heat resistance: BPs have the ability to withstand high temperatures without deforming, breaking, or changing shape. This property is important for use in packaging and electronic materials. Starch-based BPs from mango seeds are more heat-resistant than PLA [18].

This chapter generally focuses on BPs that are synthesized from renewable sources and are naturally degradable. Renewable source-based BPs can be synthesized using biomass consisting of biodegradable materials such as rice straw, banana peel, corn starch, agricultural waste, wood waste, and food waste. BPs are synthesized from plant-based components, such as cellulose and starch, which can be renewed through agricultural activities. These renewable materials will reduce dependence on fossil fuels and thereby reduce the exploitation of natural resources. BPs are biodegradable, meaning they decompose through biological processes. Although BPs are biodegradable, durability is still an important consideration so that they have similar properties to petroleum-based plastics. The use of BPs can reduce greenhouse gas emissions that would be generated in the production of petroleum-based plastics [19]. In addition, it also encourages a circular economy by replacing linear models, such as take–make–dispose, with cyclical models [20]. By adopting BPs, it can address some of the environmental concerns associated with petroleum-based plastics. Therefore, ensuring that BPs are appropriately managed throughout their life cycle, including disposal and recycling, is critical [21].

10.2.2 Types of BPs

BPs are alternatives to conventional plastics. They are divided into several types depending on the source of raw materials. These include polysaccharide-based BPs, protein-based BPs, and some aliphatic polyesters.

a) Polysaccharide-Based BPs

Polysaccharide-based BPs are increasingly being used as an alternative to petroleum-based plastics with low environmental impact. Polysaccharide-based BPs are usually made from natural materials such as cellulose, starch, chitin, and others. About 50% of BP production comes from starch annually [22]. Starch, a polysaccharide found in various plant sources such as wheat, potato, corn, cassava, and others, can be categorized as an abundant and renewable resource for producing BPs [23]. The production of starch-based BPs starts with extracting starch from plants through grinding or mashing, after which the starch granules are separated from other components, such as fibers, proteins, and lipids. The starch granules are then heated with water and plasticizers, such as glycerol or sorbitol. This stage is called gelatinization. The starch granules absorb water, causing the starch molecules to expand, break, and release amylose and amylopectin, which then interact with the plasticizer [24]. Plasticizer serves to improve the flexibility and mechanical properties of the resulting film.

The ratio of amylose to amylopectin greatly influences the properties of BPs from starch materials. Starches with high amylose content typically exhibit greater mechanical strength. The results of a recent research support this, showing that cassava starch-based BPs have high hardness due to their high amylose content, which encourages stronger intermolecular interactions [25]. However, the resulting material is more rigid. Currently, starch BPs are widely used for packaging due to their low production cost [26]. These BPs are used in various fields, including packaging, disposable cutlery, agricultural films, and everyday products. Since they are made from renewable sources and are degradable and more environmentally friendly than conventional plastics, they are a sustainable solution. Researchers continue to improve the strength, heat resistance, and water resistance of these BPs, as well as to make them easier to process. Technological innovations and new material formulations make starch-based BPs increasingly customizable to meet industry needs and support the creation of a more environmentally friendly circular economy.

Cellulose-based BPs are also of interest in the plastics industry. Worldwide, cellulose production is estimated to reach 180 billion tons [27], and the production of cellulose BPs is estimated to reach 5.33 billion tons by 2026 [28]. Cellulose-based BPs can be produced using softwood trees as the primary source of raw materials. Cellulose is a promising alternative because it is the most abundant complex carbohydrate in nature and a renewable resource for BP production. Additionally, tree bark can be used as an energy source in the production process, making the entire cycle more efficient [29]. The manufacturing process starts with extracting cellulose from wood pulp, agricultural waste, or specialized plants, such as switchgrass and bamboo. Then, the cellulose is purified to remove lignin and other impurities. Lignin makes plant cell walls strong but is undesirable in plastics because it can interfere with the final product's quality. Next, the cellulose is broken down into smaller molecules through hydrolysis or enzyme treatment to make it easier to process. The processed cellulose can then be chemically modified or blended with other polymers to improve its characteristics.

One commonly used method is esterification, which uses chemicals such as acetic acid to convert cellulose into cellulose esters, such as cellulose acetate. Cellulose acetate is widely used in packaging, filmmaking, and coatings because it has good mechanical strength; is resistant to water; and is stable at high temperatures [30, 31]. Because it is derived from natural materials, cellulose BPs are easily decomposed by microorganisms into carbon dioxide, water, and biomass when in a suitable environment, making it more environmentally friendly than ordinary plastics.

Cellulose-based BPs can also be composted, producing nutrient-rich compost, thus supporting the concept of a circular economy where products used will return to natural resources [32]. By utilizing renewable and abundant cellulose, this BP is a real alternative that can help reduce dependence on fossil-based plastics. Its biodegradable and compostable nature supports efforts to create a more sustainable environment, which is beneficial for the planet.

b) Protein-Based BPs

Protein-based BPs are made from proteins that are derived from various natural resources. These BPs are then used to produce biodegradable plastics. Proteins consist of amino acid chains, and if they are processed into BPs, they will have distinctive properties and high flexibility. The proteins used are generally sourced from plant proteins, animal proteins, and proteins from microorganisms. Protein-based BP synthesis begins by extracting proteins from the initial material and then processing or modifying them to form plastic-like materials [33]. One protein that is widely used to produce BPs is wheat gluten. Gluten in wheat has the ability to form films with mechanical strength and high flexibility, so it is widely used for packaging, coatings, and edible films [34].

Soy protein is widely used in the production of BPs, including automotive parts. However, soy protein-based films tend to be sensitive to water and have low mechanical strength. Polyester or PLA is often incorporated in soy protein to improve these weaknesses. Meanwhile, casein, a milk-derived protein, has also been investigated for BP applications. Casein-based BPs are water-resistant, have strong barrier properties, and are biodegradable [35]. They are frequently utilized in the production of food packaging and edible films. The main advantage of protein-based BPs is that they can degrade naturally. Proteins can be easily broken down by microorganism enzymes into simple compounds in a suitable environment. This reduces the accumulation of plastic waste and supports the natural carbon cycle. Research on protein-based BPs continues to grow, with a focus on improving mechanical strength, heat stability, and water resistance. Due to their renewable, biodegradable, and flexible nature, protein-based BPs are an appealing alternative to fossil-based plastics. Continued development in this field is expected to increase the use of protein-based BPs and create a more sustainable, environmentally friendly future.

c) Aliphatic Polyester-Based BPs

This type is often used due to its flexibility and biocompatibility, and it can be produced from plant material or through microbial fermentation. The aliphatic polyesters used are derivatives of diols and aliphatic dicarboxylic acids. Some examples of these acids include polyhydroxyalkanoates (PHAs), polylactic acid (PLA), polyethylene terephthalate (PET), polybutylene succinate (PBS), poly-3-hydroxybutyrate (PHB), and polyhydroxyurethanes (PHU). PHA is produced naturally by microorganisms through the fermentation of carbohydrates or fats, whereby the sugars or fats are converted into a linear polyester. The advantages of this BP are easily degradable. It is widely used in agriculture, food, and biomedical fields. PLA is a BP synthesized through the polymerization of lactic acid obtained from the fermentation of corn, cassava, starch, or sugarcane. PLA is widely used in textiles, medical devices, packaging, and 3D printing due to its strong, environmentally friendly, and degradable properties [36]. PET is made through the polymerization of ethylene glycol and terephthalic acid at a ratio of 70:30 [37]. Although PET can be synthesized from plant-based materials, most PET comes from fossil materials, making it difficult to degrade. PET is widely used for food containers, straws, and bottles because of its low production cost. PBS is produced through the esterification of succinic acid and 1,4-butanediol, with succinic acid widely produced *via* fermentation from renewable resources. Due to its high flexibility, good thermal stability, and low cost, PBS is suitable for packaging, shopping bags, and automotive components [38]. PHU is often considered a synthetic plastic, but because it has a polyol structure derived from vegetable oils, it is included in the BP category [39]. The superiority of PHU over conventional plastics lies in its properties of biodegradability, flexibility, elasticity, heat resistance, chemical resistance, water absorption, biocompatibility, and low production costs. PHB is a type of BP derived from starch or plant sugars *via* microbial fermentation [40]. It has good thermal stability, is non-toxic and biodegradable, and is easy to recycle [41].

The three types of BPs are a more environmentally friendly alternative to fossil-based plastics. These BPs decompose into carbon dioxide, water, and biomass, ensuring they do not pollute the environment. This chapter focuses on starch-based BPs made from uli, a typical Indonesian plant, and coconut shells combined with corn starch. Due to their versatility, renewability, and biodegradability, these BPs have great potential for a variety of applications, ranging from packaging to biomedical products. Continued research and innovation drive the development of BPs, supporting the circular economy and creating a more environmentally friendly and sustainable future.

10.2.3 BP and Circular Economy

The circular economy is defined as an approach that addresses environmental threats while supporting sustainable development. It is a cyclical approach in which materials and energy are used repeatedly without creating waste [42]. The circular economy is the most recent attempt to conceptualize the integration of economic activities with environmental well-being. The transition to a circular economy is also a structural change that builds long-term resilience, creates business and economic opportunities, and benefits the environment and society [43].

In the context of plastic materials, a circular economy should ideally enable the production of plastic from renewable or recycled resources. In reality, however, the current plastic cycle is linear. Approximately 79% of the plastic that is produced is sent to landfills or ends up polluting the environment. The rest is either incinerated or recycled. Recycling has been around since the 1980s. However, only about 18% of plastics from non-fiber materials are recycled. Almost no fiber can be recycled [44]. This lack of progress is largely due to the limitations of the dominant mechanical recycling method, which converts plastic waste into a new form through mechanical and thermal processes. This method is highly dependent on the quality of the input material. It requires plastic waste that has been sorted. The plastic waste must be free of contaminants.

Chemical recycling is a more robust and flexible solution, especially for mixed, contaminated, or multilayered plastic waste [45]. The process involves depolymerizing plastics into monomers that can be reprocessed into high-quality materials through repolymerization [46]. However, the complexity and high cost of initial implementation mean that this approach needs economic incentives to compete with conventional methods. Besides recycling, biodegradation is also one of the end-of-life (EOL) options, especially for easily hydrolyzable polymers such as polylactic acid (PLA). Nevertheless, the biodegradation process must be conducted in a controlled industrial setting to ensure complete degradation without producing MPs or other harmful contaminants.

Equally important, innovation in the early stages of product life (beginning-of-life) also needs to be strengthened to minimize overall environmental impacts. Educating consumers and industry on the importance of life cycle thinking will encourage a more holistic perspective, focusing not only on the impacts of use and disposal, but also on the environmental footprint of raw material sourcing, processing, and distribution [47]. In this regard, using local, nonfood renewable resources and biological waste to produce bio-based plastics through catalytic processing offers greater sustainability potential than conventional fossil fuel extraction practices [48]. Other hallmarks of the circular economy include the ability to facilitate recycling, protect critical resources, create jobs, and drive economic growth through the development of innovative business models based on eco-design and industrial symbiosis [49]. It also aims to change social and

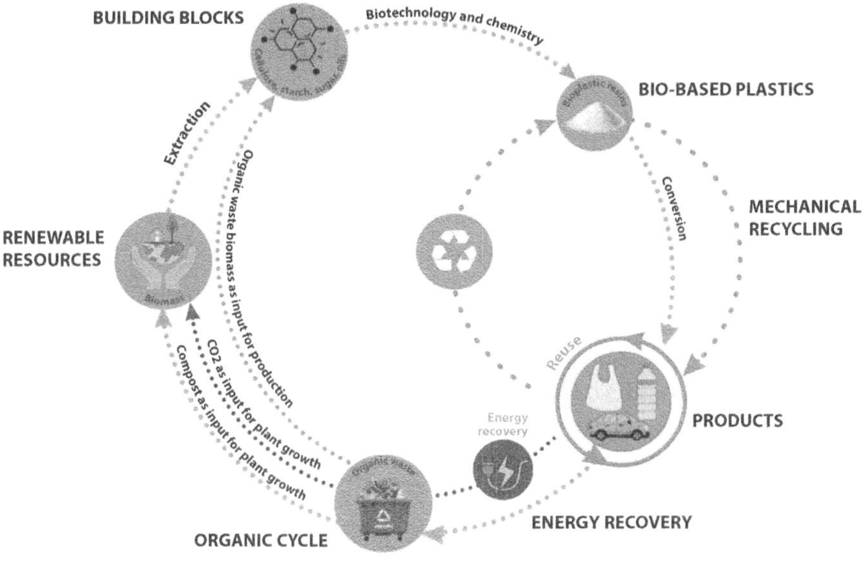

Figure 10.3 The loop of BPs.

economic dynamics by decoupling economic growth from natural resource exploitation, while reducing greenhouse gas emissions and overall negative environmental impacts [50].

Therefore, the application of circular economy principles and technologies throughout the plastic supply chain is very important. Efforts that need to be made include drastically reducing plastic consumption, designing products that can be reused and recycled, increasing energy efficiency in plastic and BP manufacturing processes supported by renewable energy, improving waste collection systems, and developing and deploying resilient and circular recycling and upcycling methods [51]. This approach is believed to be able to encourage the creation of a sustainable and environmentally friendly plastic system in the future (Figure 10.3).

10.2.4 BP Applications

In recent years, BPs and their applications in various industrial sectors have become a topic of intensive research, along with increasing concerns about plastic waste and its environmental impact. Research and innovation efforts continue to be directed towards developing environmentally friendly alternatives that can compete with conventional plastics in terms of strength and function, but with a smaller environmental footprint.

BPs, which are known to be environmentally friendly and biodegradable, are now used in many industrial fields, including the following:

a) Food packaging: Packaging is a major contributor to the use of BPs, and it is reported that in 2019, the amount reached 1.14 million tons, covering more than 53% of the total BP products [52]. BPs are an environmentally friendly packaging alternative that can reduce the negative impacts of single-use plastics. Products such as cutlery, containers, glasses, and packaging are widely made from this material.
b) Medical application: In the biomedical field, edible biopolymers are used for various applications such as wound dressings, regenerative medicine, 3D frames for tissue engineering, biomedical devices, and others [53]. Biodegradable plastics have also been specifically used as surgical sutures, surgical masks, bioresorbable implants, cosmetic cotton, and others [54].
c) Textiles: The textile industry is starting to use BP fibers for clothing, shoes, and furniture. This fiber is more easily biodegradable than conventional synthetic fibers such as polyester.
d) Agriculture: In agricultural applications, biopolymers have been widely reported to have great potential. Biodegradable polymers with superabsorbent properties have been studied for use in the gradual release of micronutrients or pesticides in plants [55]. Due to its biodegradable nature, it provides plant protection, controls weeds, and can maintain soil moisture.
e) Other applications: In the field of 3D printing, PLA as a BP material has been widely used in providing environmentally friendly filaments for the manufacture of various products and prototypes. In addition, it is also used to develop nanotubes combined with composites to improve mechanical properties for automotive product applications [56]. In automotive, BPs are used in interior components such as dashboards, door panels, and seat cushions, helping to reduce carbon emissions from vehicles.

The use of BPs for these various applications depends on the material characteristics, technical requirements, costs, and sustainability factors after their useful life. Research and innovation are ongoing to improve the quality and expand the use of BPs as a step towards a more environmentally friendly future.

10.3 Novel Research on Starch-Based BPS

As people's welfare and awareness of environmental issues increase, the demand for plastic-based products also increases. The surge in plastic use has led to a significant increase in plastic waste production. Starch-based BPs contain polysaccharide chains such as amylose, which are a potential solution to replace petroleum-based plastics. The amylose content functions to determine the plastic properties of the starch biomass. Meanwhile, to increase its flexibility and moldability, plasticizers are added. Indonesia, as a tropical country, has rich biodiversity with great potential in developing BP raw materials, especially from agricultural waste. The use of starch from banana peels, especially uli bananas, and coconut shells as raw materials for BPs offers a promising alternative when compared to petroleum-based plastics. A number of recent studies related to starch BPs have been widely reported. The synthesis of biodegradable plastics from banana peels shows that BPs from banana peels have the potential to be used commercially for packaging, shopping bags, and others [57].

Research by Beevi *et al.* reported that they had successfully synthesized BPs from banana peels and potato starch [58]. The resulting BPs were insoluble in water, but soluble in acetic acid, sulfuric acid, and acetone. Water absorption tests showed that there was an increase in weight but not significant. Research by Veena and Rani has synthesized BPs from banana peels and degraded banana peels using natural microorganisms such as bacteria, algae, and fungi [59]. The plasticizer used is glycerol, which aims to increase the strength, flexibility, and ease of use of BPs. Degradation tests were carried out using the soil burial method to ensure environmentally friendly products.

Banana peel varieties *Musa acuminata* (AAA) cv Berangan and Musa acuminata × balbisiana (ABB) cv. Awak have been synthesized by adding glycerol, sorbitol, and rice starch as plasticizers. Biodegradation tests showed that both degraded rapidly in sandy soil compared to clay soil [60]. Synthesis and characterization of BPs from banana peel flour have been carried out resulting in a maximum tensile strength of 15 minutes with increased flexibility with the addition of glycerol [61].

Latest researches have studied the physical properties and biodegradability of corn starch-based BPs [62]. The results of the study showed that corn starch-based BPs have very high biodegradability, although their tensile strength still needs to be improved by adding reinforcing fibers or nanoparticles. Another study reported that they had synthesized corn starch-based BPs with variations in the addition of silver metal to improve the mechanical and antimicrobial properties for food packaging applications [63].

The results of the study showed that the biodegradability of BPs in the soil decomposed more slowly in samples with the addition of silver, stable thermal strength, less water absorption, and reduced the number of *E. coli* bacteria by 44% and *S. aureus* by 50%. Z. Muchtar *et al.* reported that the addition of glycerol to coconut fiber BPs will produce BPs that are thinner, more flexible, more easily decomposed in the soil but less strong, and more easily absorb water while the addition of chitosan can increase tensile strength and water resistance but BPs become thick, less elastic, and wavy decompose [64].

Synthesis of BPs by utilizing a combination of corn starch and coconut fiber to produce BPs has been reported [65]. The results showed that water absorption in BPs combined with corn starch and coconut fiber was lower than BPs without the addition of coconut fiber. In addition, the addition of coconut fiber also increased the tensile strength by around 10.70 N/mm^2 but reduced the flexibility of the resulting BP. Coconut fiber has the potential as a natural reinforcing material so that it can improve the mechanical performance of BPs.

In general, the use of plastic in industry, especially packaging, is an important material. The negative impacts on the environment and the difficulty of the degradation process encourage efforts to replace it with BPs. Various solutions have been offered, one of which is the use of BPs. BP from uli banana peels is an alternative to reducing dependence on petroleum-based plastics. The starch content in banana peels is the basic ingredient for making BPs. In addition, the combination of coconut shells and corn can also be used as an alternative for making bioactives. Coconut shells contain cellulose, lignin, and hemicellulose, which are the basic ingredients for making cellulose-based BPs. Cellulose in coconut shells is expected to function to increase mechanical strength, thermal stability, reduce water absorption, and increase biodegradability and is an environmentally friendly renewable material as a sustainable alternative. Indonesia is the second largest coconut producer in the world with a production volume of 2.87 million tons or around 24 percent of world coconut production. Of course, the coconut industry will produce large amounts of waste such as coconut fiber and shells. With the availability of abundant raw materials and supporting chemical characteristics, the use of coconut shell waste as a raw material for BPs is one of the sustainable solutions in managing agricultural waste. Likewise, corn containing almost 70% starch can be directly modified to make starch-based BPs. This study focuses on the preparation of BPs from uli banana peels and BPs from a combination of corn and coconut shells to see their advantages as environmentally friendly alternatives, especially in the field of food packaging.

10.4 Experimental Section

10.4.1 Materials

a) BP-based uli banana peels: The main ingredient used is uli banana peel obtained from the Sleman Yogyakarta area. The chemicals used are AR-grade types such as acetic acid, citric acid, aquadest, glycerol, and chitosan.
b) BP-based cornstarch and cellulose coconut shells: The main ingredients used are corn starch with a combination of coconut shells obtained from traditional markets in Bantul, Yogyakarta. The materials used are AR-grade including coconut shells, corn starch, sodium hydroxide, aquadest, chitosan, glycerol, and litmus paper.

10.4.2 Methods and Evaluation

a) Preparation of BP-Based Uli Banana Peels
Uli banana peels were cut into small pieces and soaked in a 50% (w/v) citric acid solution for 60 minutes to help soften the cellular tissue. Furthermore, the banana peels were mashed with the addition of water to form a homogeneous slurry, then filtered using a cloth until a clear and colorless filtrate was obtained. The filtrate was left for 24 hours to optimally precipitate the starch. The starch precipitate formed was dried, then gravimetric water content analysis was carried out. The BP manufacturing process adopted the modified research of Pulungan *et al.* [66]. The manufacture of BPs begins with preparing two types of solutions, namely starch solution and chitosan solution. The starch solution was made by dissolving 3.5 grams of starch in 44 grams of water, then stirred using a magnetic stirrer for 10 minutes until a suspension was formed. Meanwhile, 1.5 grams of chitosan was dissolved in 50 grams of 1% acetic acid and heated at 100°C for 15 minutes to form a chitosan gel. The starch suspension was then heated at 65°C for 5 minutes, then mixed with chitosan solution and reheated at the same temperature. Furthermore, 1 gram of glycerol was added as a plasticizer, and the mixture was stirred for 10 minutes until homogeneous. The formed BP solution was poured into a mold, then dried in an oven at 50°C for 2 hours until a solid BP film was obtained. Evaluation of the quality of BPs includes tensile strength, elongation percentage, water absorption, and biodegradability tests.

b) Preparation of BP-Based Cornstarch and Cellulose Coconut Shells
The process begins with the isolation of cellulose from coconut shells which is carried out by adopting a modified method [67]. A total of 20 grams of coconut shell powder is soaked in a solution of sodium hydroxide and hydrogen peroxide. Soaking and stirring are carried out at a temperature of 55°C for 2 hours. The mixture is filtered and rinsed using distilled water then dried and finely ground using a mortar and pestle. The resulting powder is sieved with a size of 140 mesh. Two main solutions are prepared as the basic ingredients for making BPs. The first solution is a starch suspension, which is obtained by dissolving 15 grams of cornstarch in water and stirring using a magnetic stirrer for 10 minutes until a homogeneous suspension is formed. The second solution is a chitosan gel, prepared by dissolving 2.5 grams of chitosan in 50 grams of 1% acetic acid solution, then heated to a temperature of 100°C until a consistent gel is formed. The starch suspension was then heated at 65°C for five minutes, then mixed with the chitosan solution. This mixture was then added with coconut shell powder with various masses (2.5 g, 5 g, and 7.5 g) and 0.5 g of montmorillonite as a reinforcing agent. After that, the mixture was heated again at 90°C for five minutes. As a final stage, 5 g of glycerol was added as a plasticizer and stirred at 65°C for 10 minutes until perfect homogeneity was achieved. The mixed solution was then poured into a mold and dried in an oven at 50°C for approximately 1.5 hours to produce a dense and uniform BP film. Evaluation of the quality of BPs includes mechanical properties and water resistance.

10.4.3 Result and Discussion

a) BP-Based Uli Banana Peels
Evaluation of BPs from uli banana peels was carried out by testing tensile strength, percentage elongation, water absorption, and biodegradability. Tensile strength is the ability of a material to withstand external tensile force or pressure without cracking, which is measured in megapascals (MPa) [68]. Tensile strength testing is one of the initial tests carried out on samples to analyze resistance to cracking. Biodegradable films with high tensile strength can protect packaging products from mechanical damage. Figure 10.4 presents the picture of BP from uli banana peels.

The addition of 1 gram of glycerol or about 30% of the weight of the uli banana peel starch produces a tensile strength of 2.9744 Mpa. The results of this study are better than previous studies, which reported that the tensile strength value of BP from banana peel with 10% glycerol was 0.120 Mpa [69]. According to the Japanese industrial standard, the minimum

Figure 10.4 BPs from the uli banana peel starch.

tensile strength value for edible packaging is 0.392 Mpa. So, it can be said that BP from uli banana peel exceeds the standard which can be recommended as packaging. Table 10.1 shows the tensile strength and elongation percentage tests.

The elongation test shows the elasticity of BPs, which is expressed in the percentage increase in length before and after the BP is torn [70]. The higher the percentage of elongation, the greater the ability of the material to withstand loads without tearing. The addition of glycerol as a plasticizer aims to reduce polymer stiffness and increase flexibility, thereby increasing the elongation of BPs. In this study, the addition of glycerol as much as 1 gram or 30% of the weight of uli banana peel starch had an elongation percentage of 38.86%. When compared to previous research on the manufacture of BPs from Mulu bebe (Indonesian name) banana starch,

Table 10.1 Data for the tensile strength and elongation percentage test.

Treatment	a0 (mm)	b0 (mm)	Lc (mm)	F max N (MPa)	Tensile strength (MPa)	Strain at F max (elongation) (%)
First threat	0,181	5	50	2,7258	3,0119	42,102
Second threat	0,214	5	50	3,1425	2,9369	35,63
Average	0,1975	5	50	2,93415	2,9744	38,866

it produced an elongation percentage of 1.40%–13.317% [70]. A recent study reported that banana peels had an elongation value of 13.97% [3]. Combination of banana peel, glycerol, and corn starch has been reported to have an elongation value between 5.2 and 9.77% [69]. According to the Japanese industrial standard, elongation is categorized as good if it exceeds 10% and very good if it exceeds 50%. Commercial plastic bags have an elongation of 222.5%. The elongation value of this uli banana peel may not be used as a commercial plastic bag but can be used for food, cosmetic, and pharmaceutical packaging with an elongation level of around 20% [71]. Figure 10.5 is a graph of the relationship between % elongation and tensile strength as follows.

Water absorption test on BP samples was carried out to determine how much water absorption can be done by the resulting BP. The characteristics of good BP are low water absorption so that water resistance is high [72]. In this study, water absorption testing was carried out by soaking BP samples as wo in distilled water for 10 minutes, then dried and reweighed to produce the weight after soaking as w1. The water absorption capacity is the difference between w1 and w0 against wo. The results of the water absorption test on BP from uli banana peel showed a percentage of 31.48%. Figure 10.6 appears the process of BP immersion to test water absorption.

The results of this study are better when compared to previous studies. Abirami *et al.* reported that BP from yellow banana peels (Nendran) was 69%, red bananas (Sevvazhai) were 65%, and green bananas (Morris) were 58%. High water solubility can increase biodegradability but will limit

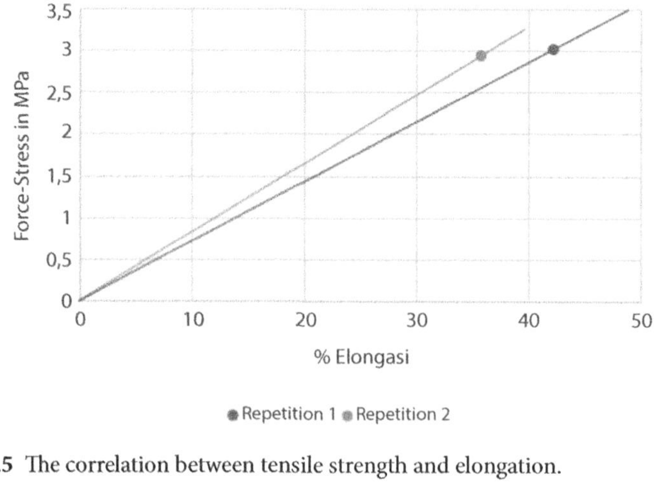

Figure 10.5 The correlation between tensile strength and elongation.

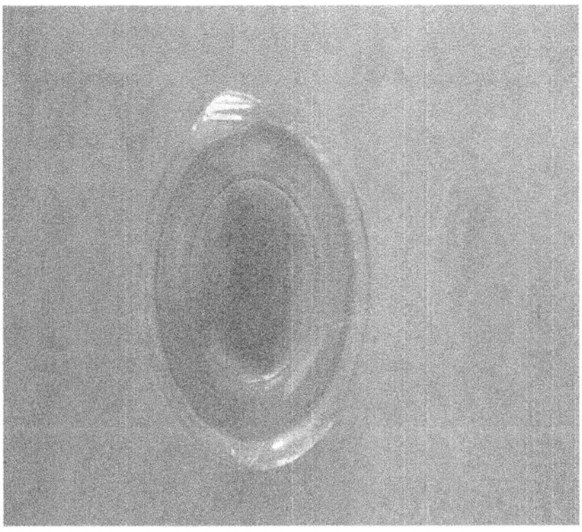

Figure 10.6 Immersion of BPs sheet in the water.

its use in applications for food packaging with high water content. Water absorption in banana peels has also been reported to have a value of 53.8% with a water solubility of 20% [73]. High starch and viscosity content usually indicate low solubility and water absorption. Research by Sánchez *et al.* reported that BP from banana peels showed water absorption of 115.23% with a solubility of 61.75%. Therefore, further modifications such as cross-linking, hydrophobic coating, or addition of nanofillers are needed to increase moisture resistance so that it can be used as a packaging plastic application [74]. BP from wak banana peel is reported to have a water absorption value of 231.5% with a concentration of 8% banana peel and 2% glycerol [75].

Water absorption can be influenced by several factors, one of which is starch. Naturally, the presence of hydroxyl groups in starch causes hydrophilic properties, thereby increasing the ability to absorb water [76]. In addition, glycerol also affects water absorption because of the nature of glycerol which is soluble in water and is naturally hygroscopic. In addition, glycerol has hydrophilic properties, which indicate that materials with hydrophilic groups can interact with water quite easily [77].

Biodegradability tests were carried out to determine whether the resulting BPs could be decomposed by various microorganisms. The test was carried out by utilizing decomposing microorganisms such as fungi and bacteria that would accelerate the degradation process [78]. In this study, the *Saccharomyces cerevisiae* fungus was used, which was found in Fermipan (Indonesia brand) for 24–72 hours. Figure 10.7 is a picture of BP soaked in Fermipan for 24 hours, 48 hours, and 72 hours.

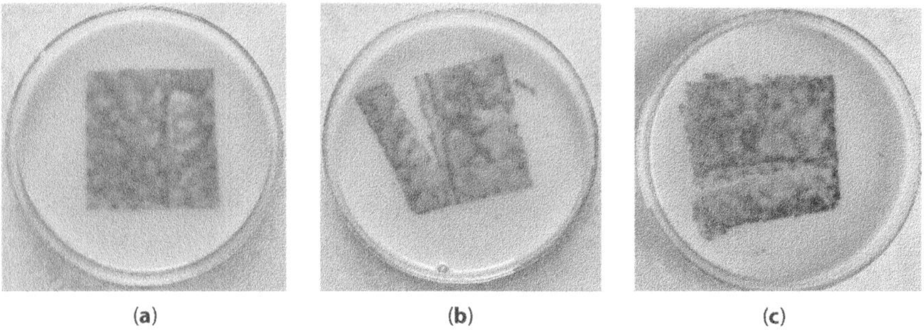

Figure 10.7 The BPs after submersion for 24 hours (a), 48 hours (b) and 72 hours (c).

The results of the biodegradability test showed that the BP made from uli banana peel starch had not undergone a degradation process in the first 24 hours, as indicated by the absence of changes in structure or microorganism activity. However, after 48 hours, changes began to appear in the structure of the BP sheet. More obvious changes were seen after 72 hours, marked by the growth of fungi on the surface of the BP during the testing process. The rate of degradation is influenced by water absorption; when water absorption is high, microorganisms can quickly degrade BP samples. In addition, the OH groups found in glycerol and starch can trigger hydrolysis reactions after absorbing water. As a result, the starch will break down into small parts until it finally breaks down and disappears. This process is highly dependent on the ability of the plastic to absorb water; the more water is absorbed, the easier the material is to degrade because water is the main medium for most microbes and bacteria [75].

b) BP-Based Cornstarch and Cellulose Coconut Shells

The making of this BP begins with the preparation of cellulose from coconut shells through the delignification and bleaching process. The purpose of delignification is to reduce the lignin content in lignocellulose. This process will open the lignocellulose structure so that the cellulose will be easier to separate. Bleaching is done to change the color of the dark brown coconut shell to light brown using H_2O_2 in an alkaline environment. The lignin content causes the dark brown color of the coconut shell so that by reducing the amount of lignin, the color of the coconut shell becomes paler (Figure 10.8).

FTIR analysis was conducted on BP samples with the aim of identifying functional groups contained in BPs, especially cellulose constituent groups. Figure 10.9 visualizes the FTIR results of coconut shell cellulose samples.

The results of the FITR analysis showed absorption at a wave number of 3339.7 cm^{-1} which is an O-H stretching vibration which is a characteristic

Figure 10.8 BP from corn extract and cellulose from coconut shells.

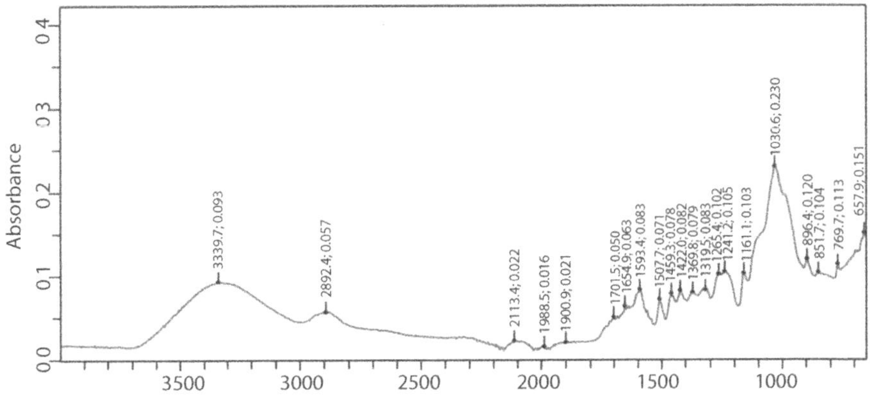

Figure 10.9 FTIR spectra for coconut shell cellulose samples.

of cellulose. This is like the research of Muhammad *et al.*, where the characteristic of cellulose films is that they have -OH absorption which represents hydroxyl groups [79]. Absorption at a wave number of 2892.4 cm^{-1} shows aliphatic C-H stretching vibrations in cellulose. The absorption peak at a wavelength of 1030.6 cm^{-1} represents the stretching vibration of the C-O bond found in the glycosidic structure of cellulose. Meanwhile, at a wavelength of 1265.4 cm^{-1}, C-O stretching vibrations originating from the aromatic lignin structure were detected [80]. The low peak intensity indicates that most of the lignin has been decomposed during the bleaching process, although there is still a little lignin residue remaining. The peak at 1507.7 cm^{-1} indicates the presence of aromatic C=C double-bond vibrations, indicating the presence of lignin in small amounts. The signal at 1701.5 cm^{-1} indicates the C=O vibrations of hemicellulose, indicating that there are still residual hemicelluloses in the material. The spectrum data shows that the cellulose obtained from coconut shells has a sufficient level of purity to be used as the main ingredient in making BPs. Remaining lignin that is still present in small amounts can have positive effects, such as increasing the

Table 10.2 The result of tensile strength test of BP from coconut shell and corn starch.

	F1 (2.5 gr)	F2 (5 gr)	F3 (7.5 gr)
Tensile strength (Mpa)	4.729	5.405	6.756

strength of the BP structure, but still needs to be monitored so as not to inhibit the natural degradation process.

The manufacture of BPs uses 3 types of formulas with the following ratio of coconut shell and corn starch 1:6 (2.5 gr: 5 gr), 1:3 (5 gr: 15 gr), and 1:2 (7.5 gr: 15 gr). Table 10.2 shows the results of the tensile strength test on the three formulas.

The tensile strength test results showed that the BP formula 3, namely the largest amount of coconut shell cellulose, had a higher tensile strength compared to the other formulas of 6756 Mpa. The bonds in starch, glycerol, and cellulose affect the tensile strength of BPs. This is in accordance with the research of Pranata *et al.*, that high cellulose content can increase bond strength [81]. The addition of cellulose from coconut fiber to the preparation of corn starch BPs significantly showed a higher tensile strength of 0.22 N/mm^2.

The increase in mechanical properties, especially tensile strength, in BPs added with cellulose is due to the linear and long structure of the cellulose polymer chain, which is able to strengthen the integrity of the BP matrix [82]. The interaction that occurs between starch and cellulose, especially through the formation of hydrogen bonds between the hydroxyl groups of each component, contributes to the internal cohesion of the material. In addition, cellulose fibers function to fill gaps or pores in the polymer matrix, resulting in a denser structure and providing higher resistance to tensile forces [83].

Water absorption tests were also carried out in this study which aimed to determine the resistance of BPs to water. The results of the water absorption test on the three types of samples are shown in Figure 10.10.

Based on the research results, the percentage of water absorption values was obtained as follows: 45.94% for F1, 40.14% for F2, and 36.06% for F3. Formula 3 with a higher amount of cellulose has the lowest absorption capacity. The low water absorption in formula 3 is due to the high amount of cellulose, which is known to have the property of reducing the hydrophilic properties of BPs. Formula 1 has a higher water absorption value due to the content of corn starch with hydrophilic properties in greater amounts than other formulas. This study is in line with previous research by Arbanah *et al.*,

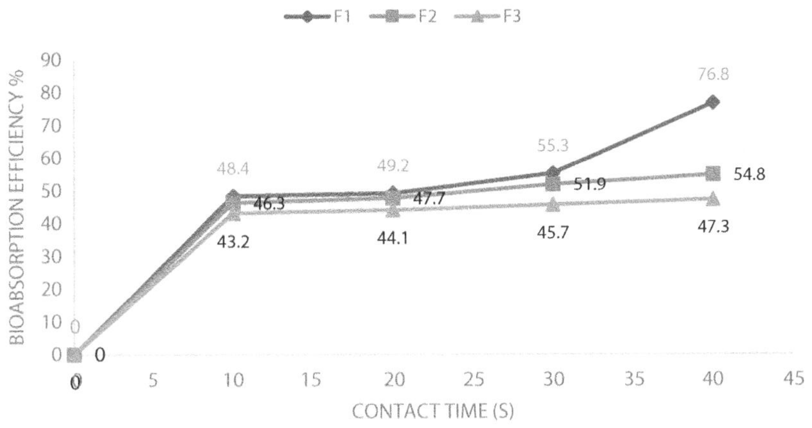

Figure 10.10 Graph of water absorption of BP each formula.

which reported that the effect of adding cellulose from coconut fiber has a lower water absorption value than without adding cellulose [65].

10.5 Conclusion

The utilization of agricultural waste, especially banana peel, coconut shell, and corn starch, shows great potential as sustainable raw materials in the production of environmentally friendly BPs that can replace petroleum-based plastics. This study confirms that banana peel, which is rich in starch content, can be efficiently processed into biodegradable BPs through a simple and sustainable process approach. Evaluation of mechanical properties revealed that the resulting BP has a tensile strength of 2.9744 MPa, exceeding the minimum threshold required for packaging applications. The elongation value of 38.86% indicates adequate flexibility, making it suitable for use in food, cosmetic, and pharmaceutical packaging applications that require a minimum elasticity level of around 20%. In addition, the water absorption value recorded at 31.48% indicates fairly good moisture resistance. Biodegradation tests showed fungal growth after 72 hours, which is an early indicator of the material's biological degradation ability. Furthermore, the BP formulation based on the combination of coconut shell cellulose and corn starch showed a significant increase in tensile strength and a decrease in water absorption, which confirmed that coconut shell cellulose functions as a structural reinforcement agent as well as a moisture absorption inhibitor. Therefore, this combination is worthy of consideration as a promising alternative in the development of BPs with superior performance.

AI disclosure statement: The author confirms that no artificial intelligence (AI)-generated content or AI-assisted writing tools were used in the preparation of this chapter (including summaries and the abstract). All text, analysis, and conclusions are the result of the author's original work.

References

1. Shah, M., Rajhans, S., Pandya, H.A., Mankad, A.U., BP for future: A review then and now. *World J. Adv. Res. Rev.*, *9*, 2, 56–67, 2021.
2. Khan, F., Ahmed, W., Najmi, A., Understanding consumers' behavior intentions towards dealing with the plastic waste: Perspective of a developing country. *Resour. Conserv. Recycl.*, *142*, November 2018, 49–58, 2019, https://doi.org/10.1016/j.resconrec.2018.11.020.
3. Azieyanti, N.A., Amirul, A., Othman, S.Z., Misran, H., Mechanical and Morphology Studies of BP-Based Banana Peels. *J. Phys. Conf. Ser.*, *1529*, 3, 2020, https://doi.org/10.1088/1742-6596/1529/3/032091.
4. Chen, R., Li, Q., Xu, X., Zhang, D., Comparative pyrolysis characteristics of representative commercial thermosetting plastic waste in inert and oxygenous atmosphere. *Fuel*, *246*, 212–221, 2019, https://doi.org/10.1016/j.fuel.2019.02.129.
5. Al-Salem, S.M., Lettieri, P., Baeyens, J., Recycling and recovery routes of plastic solid waste (PSW): A review. *Waste Manage. (Oxford)*, *29*, 10, 2625–2643, 2009, https://doi.org/10.1016/j.wasman.2009.06.004.
6. Nayanathara Thathsarani Pilapitiya, P.G.C. and Ratnayake, A.S., The world of plastic waste: A review. *Cleaner Mater.*, *11*, 100220, 2024, https://doi.org/10.1016/j.clema.2024.100220.
7. Lee, Y., Cho, J., Sohn, J., & Kim, C. Health Effects of MP Exposures: Current Issues and Perspectives in South Korea. *Yonsei Med. J.*, *64*, 5, 301–308, 2023, https://doi.org/10.3349/ymj.2023.0048.
8. Ghosh, S., Sinha, J.K., Ghosh, S., Vashisth, K., Han, S., Bhaskar, R., MPs as an Emerging Threat to the Global Environment and Human Health. *Sustainability (Switzerland)*, *15*, 14, 2023, https://doi.org/10.3390/su151410821.
9. Gonçalves de Moura, I., Vasconcelos de Sá, A., Lemos Machado Abreu, A.S., Alves Machado, A.V., BPs from agro-wastes for food packaging applications, in: *Food Packaging*, Elsevier Inc., 2017, https://doi.org/10.1016/b978-0-12-804302-8.00007-8.
10. Silva, R.V., De Brito, J., Dhir, R.K., Properties and composition of recycled aggregates from construction and demolition waste suitable for

concrete production. *Constr. Build. Mater.*, *65*, 201–217, 2014, https://doi.org/10.1016/j.conbuildmat.2014.04.117.

11. Torrejon, V.M., Fernando, S., Roy, U.K., Onwukwe, U., Anguilano, L., Transforming seaweed into BPs: a review of cultivation, harvesting and processing methods. *Appl. Phycol.*, *6*, 1, 52–73, 2025, https://doi.org/10.1080/26388081.2024.2435035.
12. Baekeland, L.H., Bakelite, a condensation product of phenols and formaldehyde, and its uses. *J. Franklin Inst.*, *169*, 1, 55–60, 1910, https://doi.org/10.1016/S0016-0032(10)90300-1.
13. Senthamaraikannan, C., Akash, K., Amanullah, S., Barath, M., Manojkumar, R., Jagadeeshwaran, J., Overview of Polylactic acid and its derivatives in medicinal applications. *IOP Conf. Ser.: Mater. Sci. Eng.*, *988*, 1, 2020, https://doi.org/10.1088/1757-899X/988/1/012003.
14. Nofar, M., Sacligil, D., Carreau, P.J., Kamal, M.R., Heuzey, M.C., Poly (lactic acid) blends: Processing, properties and applications. *Int. J. Biol. Macromol.*, *125*, 307–360, 2019, https://doi.org/10.1016/j.ijbiomac.2018.12.002.
15. Reddy, M.M., Vivekanandhan, S., Misra, M., Bhatia, S.K., Mohanty, A.K., Biobased plastics and bionanocomposites: Current status and future opportunities. *Prog. Polym. Sci.*, *38*, 10–11, 1653–1689, 2013, https://doi.org/10.1016/j.progpolymsci.2013.05.006.
16. Sun, S., Enzyme-Embedded Biodegradable Plastic for Sustainable Applications: Advances, Challenges, and Perspectives. *ACS Appl. Bio Mater.*, *8*, 1785–1796, 2025, https://doi.org/10.1021/acsabm.4c01628.
17. Gonzalez-Gutierrez, J., Partal, P., Garcia-Morales, M., Gallegos, C., Development of highly-transparent protein/starch-based BPs. *Bioresour. Technol.*, *101*, 6, 2007–2013, 2010, https://doi.org/10.1016/j.biortech.2009.10.025.
18. Blancia, G.V.V., The Potential of Mango Starch and Snake Plant Fibers as Bio-Plastic. *Pelagia Res. Lib. Asian J. Plant Sci. Res.*, *2021*, 5, 171–175, 2021, www.pelagiaresearchlibrary.com.
19. Fauziyah, S.N., Mubarak, A.S., Pujiastuti, D.Y., Application of glycerol on BP based carrageenan waste cellulose on biodegradability and mechanical properties BP. *IOP Conf. Ser.: Earth Environ. Sci*, *679*, 1, 2021, https://doi.org/10.1088/1755-1315/679/1/012005.
20. Bayer, I.S., Guzman-Puyol, S., Heredia-Guerrero, J.A., Ceseracciu, L., Pignatelli, F., Ruffilli, R., Cingolani, R., Athanassiou, A., Direct transformation of edible vegetable waste into BPs. *Macromolecules*, *47*, 15, 5135–5143, 2014, https://doi.org/10.1021/ma5008557.
21. Chung, C., Kim, J., Sovacool, B.K., Griffiths, S., Bazilian, M., Yang, M., Decarbonizing the chemical industry: A systematic review of sociotechnical systems, technological innovations, and policy options. *Energy Res. Social Sci.*, *96*, 102955, 20232023, https://doi.org/10.1016/j.erss.2023.102955.

22. Marichelvam, M.K., Jawaid, M., Asim, M., Corn and rice starch-based bio-plastics as alternative packaging materials. *Fibers*, *7*, 4, 1–14, 2019, https://doi.org/10.3390/fib7040032.
23. Gamage, A., Liyanapathiranage, A., Manamperi, A., Gunathilake, C., Mani, S., Merah, O., Madhujith, T., Applications of Starch Biopolymers for a Sustainable Modern Agriculture. *Sustainability (Switzerland)*, *14*, 10, 2022, https://doi.org/10.3390/su14106085.
24. Lee, C.L.Y. and Yeo, W.S., A Basic Characterisation Study of BPs *via* Gelatinization of Corn Starch. *J. Appl. Sci. Process Eng.*, *8*, 2, 820–833, 2021, https://doi.org/10.33736/jaspe.3445.2021.
25. Jayalath, U.T., Samaraweera, H., Samarasinghe, A., Development and characterization of gelatin-starch BPs: A comparative study of cassava, corn, and rice-based alternatives. *Sustainable Chem. Environ.*, *9*, 2025, https://doi.org/10.1016/j.scenv.2024.100190.
26. Ates, M. and Kuz, P., Starch-Based BP Materials for Packaging Industry. *J. Sustain. Construct. Mater. Technol.*, *5*, 1, 399–406, 2020, www.eds.yildiz.edu.tr/jscmt.
27. Abe, M.M., Martins, J.R., Sanvezzo, P.B., Macedo, J.V., Branciforti, M.C., Halley, P., Botaro, V.R., Brienzo, M., Advantages and disadvantages of BPs production from starch and lignocellulosic components. *Polymers*, *13*, 15, 2021, https://doi.org/10.3390/polym13152484.
28. Ali, S.S., Abdelkarim, E.A., Elsamahy, T., Al-Tohamy, R., Li, F., Kornaros, M., Zuorro, A., Zhu, D., Sun, J., BP production in terms of life cycle assessment: A state-of-the-art review. *Environ. Sci. Ecotechnol.*, *15*, 100254, 2023, https://doi.org/10.1016/j.ese.2023.100254.
29. Sultan, N.F.K. and Johari, W.L.W., The development of banana peel/corn starch BP film: a preliminary study. *Biorem. Sci. Technol. Res.*, *5*, 1, 12–17, 2017, https://doi.org/10.54987/bstr.v5i1.352.
30. Gunaseelan, V.N., Biochemical methane potential of fruits and vegetable solid waste feedstocks. *Biomass Bioenergy*, *26*, 4, 389–399, 2004, https://doi.org/10.1016/j.biombioe.2003.08.006.
31. Ilori, M.O., Adebusoye, S.A., Lawal, A.K., Awotiwon, O.A., Biogas from Banana and Plantain Peels. *Adv. Environ. Biol.*, *1*, 1, 33–38, 2007.
32. Ehiowemwenguan, G., Emoghene, A.O., Inetianbor, J.E., Antibacterial and phytochemical analysis of Banana fruit peel. *IOSR J. Pharm. (IOSRPHR)*,4, 8, 18–25, 2014, https://doi.org/10. 9790/3013-0408018025.
33. Jadhav, S.A., Kataria, P.K., Bhise, K.K., Chougule, S.A., Original Research Article Amylase Production from Potato and Banana Peel Waste. *Int.J.Curr. Microbiol.App.Sci*, *2*, 11, 410–414, 2013, http://www.ijcmas.com.
34. Umesh, M. and Preethi, K., Fermentative Utilization of Fruit Peel Waste for Lactic Acid Production by Lactobacillus plantarum. *Indian J. Appl. Res.*, *4*, 9, 449–451, 2014.

35. Hossain, A., Nano-Particle Preparation from Ligno-Cellulose Based Banana Peel Biomass as a Tool of Nano-Biotechnology. *Global J. Biol. Agric. Health Sci.*, *4*, 3, 19–21, 2015.
36. Prendiz, J., Mena, M., Vega-baudrit, J.R., Polylactic Acid (PLA) As A BP and Its Possible Applications in the Food Industry. *Food Sci. Nutr.*, *5*, 2, 1–6, 2019, https://doi.org/10.24966/fsn-1076/100048.
37. Siracusa, V., Polymers Analogous to Petroleum-Derived Ones for Packaging and Engineering Applications. *Polymers, 12*, 1641, 2020.
38. Aliotta, L., Seggiani, M., Lazzeri, A., Gigante, V., Cinelli, P., A brief review of Poly (Butylene Succinate) (PBS) and its main and applications. *Polymer, 14*, 844, 2022.
39. Mokhtari, C., Malek, F., Halila, S., Belgacem, M.N., Khiari, R., New biobased polyurethane materials from modified vegetable oil. *J. Renewable Mater.*, *9*, 7, 1213–1223, 20212021, https://doi.org/10.32604/jrm.2021.015475.
40. Adnan, M., Siddiqui, A.J., Ashraf, S.A., Snoussi, M., Badraoui, R., Alreshidi, M., Elasbali, A.M., Al-Soud, W.A., Alharethi, S.H., Sachidanandan, M., Patel, M., Polyhydroxybutyrate (PHB)-Based Biodegradable Polymer from Argomyces indicus: Enhanced Production, Characterization, and Optimization. *Polymers*, 14, 19, 2022, http://doi.org/10.3390/polym14193982.
41. dos Santos, A.J., Oliveira Dalla Valentina, L.V., Hidalgo Schulz, A.A., Tomaz Duarte, M.A., From Obtaining to Degradation of PHB:Material Properties. Part I. *Ing. Cienc.*, 13, 26, 269–298, 2017, https://doi.org/10.17230/ingciencia.13.26.10.
42. Spierling, S., Venkatachalam, V., Behnsen, H., Herrmann, C., Endres, H.J., BPs and Circular Economy Performance Indicators to Identify Optimal Pathways, in: *Sustainable Production, Life Cycle Engineering and Management*, Springer International Publishing, 2019, https://doi.org/10.1007/978-3-319-92237-9_16.
43. Geissdoerfer, M., Savaget, P., Bocken, N.M.P., Hultink, E.J., The Circular Economy – A new sustainability paradigm? *J. Cleaner Prod.*, *143*, 757–768, 2017, https://doi.org/10.1016/j.jclepro.2016.12.048.
44. Geyer, R., Jambeck, J.R., Law, K.L., Production, use, and fate of all plastics ever made, CA 93106; Bren School of Environmental Science and Management, University of California, Santa Barbara, Santa Barbara, USA. *Sci. Adv.*, *3*, July, 25–29, 2017.
45. Rahimi, A.R. and Garciá, J.M., Chemical recycling of waste plastics for new materials production. *Nat. Rev. Chem.*, *1*, 1–11, 2017, https://doi.org/10.1038/s41570-017-0046.
46. Coates, G.W. and Getzler, Y.D.Y.L., Chemical recycling to monomer for an ideal, circular polymer economy. *Nat. Rev. Mater.*, *5*, 7, 501–516, 2020, https://doi.org/10.1038/s41578-020-0190-4.
47. Miller, S.A., Five Misperceptions Surrounding the Environmental Impacts of Single-Use Plastic. *Environ. Sci. Technol.*, *54*, 22, 14143–14151, 2020, https://doi.org/10.1021/acs.est.0c05295.

48. Zheng, J. and Suh, S., Strategies to reduce the global carbon footprint of plastics. *Nat. Clim. Change*, *9*, 5, 374–378, 2019, https://doi.org/10.1038/s41558-019-0459-z.
49. van Putten, R.-J., de Vries, J.G., van der Waal, J.C., de Jong, E., Rasrendra, C.B., Heeres, H.J., Hydroxymethylfurfural, A Versatile Platform Chemical Made from. *Chem. Rev.*, *113*, 3, 1499–1597, 2013.
50. Sehnem, S., Chiappetta Jabbour, C.J., Farias Pereira, S.C., de Sousa Jabbour, A.B.L., Improving sustainable supply chains performance through operational excellence: circular economy approach. *Resour. Conserv. Recycl.*, *149*, 236–248, 2019, https://doi.org/10.1016/j.resconrec.2019.05.021.
51. Lau, W.W.Y., Shiran, Y., Bailey, R.M., Cook, E., Stuchtey, M.R., Koskella, J., Velis, C.A., Godfrey, L., Boucher, J., Evaluating scenatios toward zero plastic pollution. *Kaos GL Dergisi*, *8*, 75, 147–154, 2020, https://doi.org/10.1016/j.jnc.2020.
52. Ncube, L.K., Ude, A.U., Ogunmuyiwa, E.N., Zulkifli, R., Beas, I.N., Environmental impact of food packaging materials: A review of contemporary development from conventional plastics to polylactic acid based materials. *Materials*, *13*, 21, 1–24, 2020, https://doi.org/10.3390/ma13214994.
53. Kouhi, M., Prabhakaran, M.P., Ramakrishna, S., Edible polymers: An insight into its application in food, biomedicine and cosmetics. *Trends Food Sci. Technol.*, *103*, April, 248–263, 2020, https://doi.org/10.1016/j.tifs.2020.05.025.
54. Sikora, J., Majewski, L., Puszka, A., Modern biodegradable plastics-processing and properties: Part I. *Materials*, *13*, 8, 2020, https://doi.org/10.3390/MA13081986.
55. Álvarez-Castillo, E., Felix, M., Bengoechea, C., Guerrero, A., Proteins from agri-food industrial biowastes or co-products and their applications as green materials. *Foods*, *10*, 5, 2021, https://doi.org/10.3390/foods10050981.
56. Narancic, T., Cerrone, F., Beagan, N., O'Connor, K.E., Recent advances in BPs: Application and biodegradation. *Polymers*, *12*, 4, 2020, https://doi.org/10.3390/POLYM12040920.
57. Arjun J., Manju, R., Rajeswaran, S.R., Chandhru, M., Banana peel starch to biodegradable alternative products for commercial plastics. *GSC Biol. Pharm. Sci.*, 22, 2, 234–244, 2023, https://doi.org/10.30574/gscbps.2023.22.2.0066.
58. Beevi, R., Fathima, S., Fathima, T., Thameemunisa, N., Deepika, BP synthesis using banana peels and potato starch and characterization. *Int. J. Sci. Technol. Res.*, *9*, 1, 1809–1814, 2020.
59. Veena, S. and Rani, M.E., BPs from banana peels and biodegradation by micro organisms. *Int. J. Creat. Res. Thought (IJCRT)*, *10*, 8, 217–224, 2022.
60. Azfira, A.N., Wahida, A.R., Izzati, N.M.N., Biodegradability Study on the BP Derived from Banana Peel Fruit Waste with Various Ripening Stages. *Malays. J. Chem.*, *25*, 5, 180–189, 2023, https://doi.org/10.55373/mjchem.v25i5.180.

61. Chandarana, J. and Chandra, P.L.V.S., Production of BPs from Banana Peels. *Int. J. Sci. Res. Eng. Trends*, 7, 1, 1471–1491, 2021, https://doi.org/10.1007/978-981-99-7552-5_67.
62. Wankhade, P.K., Ahmad, F., Joshi, D., Kale, N., BP from Corn Starch; Characteristic and Biodegradable Performance. *Int. J. Sci. Res. Sci. Technol.*, 12, 1, 157–160, 2025, https://doi.org/10.32628/IJSRST2512122.
63. Kowser, M.A., Mahmud, H., Chowdhury, M.A., Hossain, N., Mim, J.J., Islam, S., Fabrication and characterization of corn starch based BP for packaging applications. *Results Mater.*, *25*, 2025, https://doi.org/10.1016/j.rinma.2025.100662.
64. Muchtar, Z., Z. M., Sari, S.A., Rahmah, S., Zubir, M., M. Z., Sarumaha, G.E., The Effect of Chitosan and Glycerol Mixture on Improving Biodegradable Plastic Properties of Young Coconut Husk (Cocos nucifera L.). *Orient. J. Chem.*, *39*, 1, 95–101, 2023, https://doi.org/10.13005/ojc/390111.
65. Arbanah, M. and Buddin, M.H.S., Effect of Coconut Fiber Reinforcement on Mechanical Properties of Corn Starch BPs. *Article Int. J. Eng. Technol.*, *7*, 4, 267–270, 2018, https://www.researchgate.net/publication/328880010.
66. Pulungan, M.H., Kapita, R.A.D., Dewi, I.A., Optimisation on the production of biodegradable plastic from starch and cassava peel flour using response surface methodology. *IOP Conf. Ser.: Earth Environ. Sci*, *475*, 1, 2020, https://doi.org/10.1088/1755-1315/475/1/012019.
67. Rojas-Valencia, M.N., Galeana-Olvera, E., Y. Fernández-Rojas, D., Mendoza-Buenrostro, C., Nájera-Aguilar, H.A., Vaca-Mier, M., Isolation of cellulose nanofibrils from coconut waste for the production of sewing thread. *Adv. Mater. Sci.*, *3*, 1, 1–4, 2018, https://doi.org/10.15761/ams.1000135.
68. Calliester, W.D. and Rethwisch, D.G., Materials Science and Eng. *Mater. Sci. Eng. A*, *42*, 1, 181, 1980.
69. Talukdar, N., Boro, R.C., Purkayastha, M.D., Nath, T., Rathi, S., Sarmah, K., Exploring banana peels as a renewable source for BP development. *Int. J. Adv. Biochem. Res.*, 9, 1, 22–30, 2025, https://doi.org/10.33545/26174693.2025.v9.ila.3407.
70. Saleh, E.R.M. and Utami, S., Characteristics of biodegradable plastic from mulu bebe banana peel starch with the addition of chitosan and glycerol plasticizer. *IOP Conf. Ser.: Earth Environ. Sci*, *1177*, 1, 2023, https://doi.org/10.1088/1755-1315/1177/1/012047.
71. Dianursanti, Gozan, M., Noviasari, C., The effect of glycerol addition as plasticizer in Spirulina platensis based BP. *E3S Web Conf.*, *67*, 11–14, 2018, https://doi.org/10.1051/e3sconf/20186703048.
72. Figaliah, F.M., Manab, A., Sawitri, M.E., Characteristics of BPs with Addition of Beeswax and Glucomannan. *BIO Web Conf.*, *88*, 1–6, 2024, https://doi.org/10.1051/bioconf/20248800021.
73. Maria Abel, D., De Castro Ruas, J., De Castro Ruas, A., Kok, T., Characterization Properties of Banana Peel as a Promising Alternative for BP. *E3S Web Conf.*, *374*, 1–7, 2023, https://doi.org/10.1051/e3sconf/202337400008.

74. Sánchez, A.A., Guamán, A., Castillo, D., Carrión, J., Riofrío, G., Padilla-Martínez, J.P., Lakshminarayanan, V., Developing and Characterization of a Biopolymeric Membrane Derived from Mature Banana Peel Biomass. *Polymers*, *17*, 6, 1–14, 2025, https://doi.org/10.3390/polym17060775.
75. Irmayanti, and Anwar, C., Synthesis of Biodegradable Plastic from Wak Banana Peel Starch with Glycerol as Plasticizer. *Food Process. Tech. Technol.*, *54*, 3, 546–557, 2024, https://doi.org/10.21603/2074-9414-2024-3-2527.
76. Santana, R.F., Bonomo, R.C.F., Gandolfi, O.R.R., Rodrigues, L.B., Santos, L.S., dos Santos Pires, A.C., de Oliveira, C.P., da Costa Ilhéu Fontan, R., Veloso, C.M., Characterization of starch-based BPs from jackfruit seed plasticized with glycerol. *J. Food Sci. Technol.*, *55*, 1, 278–286, 2018, https://doi.org/10.1007/s13197-017-2936-6.
77. Vijayakumar, V. and Nam, S.Y., A Review of Recent Chitosan Anion Exchange Membranes for Polymer Electrolyte Membrane Fuel Cells. *Membranes*, *12*, 12, 1–12, 2022, https://doi.org/10.3390/membranes12121265.
78. Lv, S., Li, Y., Zhao, S., Shao, Z., Biodegradation of Typical Plastics: From Microbial Diversity to Metabolic Mechanisms. *Int. J. Mol. Sci.*, *25*, 1, 2024, https://doi.org/10.3390/ijms25010593.
79. Muhammad, A., Roslan, A., Sanusi, S.N.A., Shahimi, M.Q., Nazari, N.Z., Mechanical properties of BP form cellulose nanocrystal (CNC) mangosteen peel using glycerol as plasticizer. *J. Phys. Conf. Ser.*, *1349*, 1, 2019, https://doi.org/10.1088/1742-6596/1349/1/012099.
80. Bhernama, B.G., Nurhayati, Saputra, S.A., Amalia, J., Karakterisasi Selulosa dan Selulosa Asetat dari Limbah Cangkang Biji Pala (Myristica Fragransi) Aceh Selatan. *J. Ris. Kim.*, *14*, 1, 81–93, 2023, https://doi.org/10.25077/jrk.v14i1.579.
81. Pranata, A., Nasution, H., Harahap, H., Yustira, A., The effect solvent type on natural fiber immersion process on tensile strength of cellulose-based BP. *IOP Conf. Ser.: Earth Environ. Sci*, *912*, 1, 2021, https://doi.org/10.1088/1755-1315/912/1/012062.
82. Rahman, R. and Putra, S.Z.F.S., Tensile properties of natural and synthetic fiber-reinforced polymer composites, in: *Mechanical and Physical Testing of Biocomposites, Fibre-Reinforced Composites and Hybrid Composites*, Elsevier Ltd, 2018, https://doi.org/10.1016/B978-0-08-102292-4.00005-9.
83. Khouaja, A., Koubaa, A., Ben, H., Chemosphere Mechanical and morphological properties of cellulose biocomposites. *Chemosphere*, *379*, 144415, 2025, https://doi.org/10.1016/j.chemosphere.2025.144415.

Part III

CHARTING FUTURE PATHWAYS: POLICY INNOVATIONS, COMMUNITY EMPOWERMENT, AND THE CIRCULAR ECONOMY

11

Grassroots to Global: Community-Led Interventions for Environmental Resilience

Amrit Krishna Mitra[1]*, Rama Saha[2], Vijendra Singh Solanki[3] and Neha Agarwal[4]

[1]*Department of Chemistry, Government General Degree College, Singur, West Bengal, India*
[2]*Department of Chemistry, Rani Rashmoni Green University, Tarakeswar, West Bengal, India*
[3]*Department of Chemistry, Institute of Science and Research, IPS Academy, Indore, Madhya Pradesh, India*
[4]*Department of Chemistry, Navyug Kanya Mahavidyalaya, Affiliated with the University of Lucknow, Lucknow, Uttar Pradesh, India*

Abstract

The widespread and often inappropriate use of plastics across sectors such as packaging, transportation, manufacturing, and agriculture both in rural and urban areas has led to a mounting global crisis of plastic waste. In 2019 alone, 370 million tons of plastic were produced worldwide, with only 9% recycled, 12% incinerated, and the rest ending up in landfills or polluting terrestrial and aquatic ecosystems. This alarming leakage of plastic waste poses serious threats to environmental and human health, prompting growing concern among governments, researchers, communities, and other stakeholders. Major barriers to effective plastic waste management include inadequate recycling infrastructure, lack of technical expertise, and limited public awareness of regulations. The environmental burden is exacerbated in both developed and developing nations, where current waste management strategies often fall short of addressing the long-term ecological and societal impacts.

This chapter explores the sources and effects of plastic waste, evaluates existing recycling methods within the framework of the circular economy, and highlights innovative approaches such as thermolysis (pyrolysis, hydrocracking, and

**Corresponding author*: ambrosia12june@gmail.com

Neha Agarwal, Brijesh Pare, Vijendra Singh Solanki and Sankara Rao Miditana (eds.) Plastic Remediation: Transformative Pathways for a Rejuvenated Sustainable Environment, (277–316)

gasification), chemolysis (alcoholysis, hydrolysis, methanolysis, and glycolysis), and biodegradation through bacteria and robotics. It also identifies knowledge gaps, policy shortcomings, and practical challenges in implementation. Emphasis is placed on the role of community-led initiatives, public awareness campaigns, and policy interventions—such as plastic bans—as vital strategies for achieving sustainable development goals through effective plastic waste remediation.

***Keywords*:** Plastic waste, microplastic, climate change, marine biodiversity, waste recycling, management solutions, challenges, policy

11.1 Introduction

Plastics are remarkable materials that have significantly contributed to economic growth and synthetic modernity. However, their indiscriminate use and improper disposal have led to widespread environmental degradation. The first synthetic plastic, Bakelite, was developed in 1907, marking the inception of the global plastics industry. Since the 1930s, plastic packaging has become increasingly prevalent in commercial development. Over the past 50 years, global plastic production has surged from 15 million tons to over 311 million tons. Owing to its extensive applications across various sectors, plastic production is expected to nearly double within the next two decades.

The increasing consumption of plastic and the resulting waste have become a major threat to both the global environment and public health. Following climate change, plastic waste is now considered the second most pressing environmental issue worldwide [1, 2]. Significant policy interventions such as China's National Sword Policy introduced in 2018 [3]; amendments to Annexes II, VIII, and IX of the Basel Convention; and the European Union's ban on the transboundary shipment of plastic waste have been implemented to curb plastic pollution [3–5].

Plastics are often formulated with various chemical additives, including dyes, antioxidants, plasticizers, and flame retardants. These additives can leach toxic metals such as lead, zinc, cadmium, and copper into the environment [6]. The majority of plastic waste originates from single-use items, with nearly half of all plastic products being used only once before disposal. Of particular concern are MPs, which result from the fragmentation of larger plastic items due to exposure to UV light, temperature fluctuations, and mechanical weathering. These MPs infiltrate ecosystems, entering food chains through ingestion by microorganisms. Recent studies have revealed the widespread presence of microplastics (MPs) in human bodies, though their long-term health implications remain uncertain. Alarmingly, countless animals continue to suffer and die from ingesting or becoming entangled in plastic bags, bottles, and packaging materials [7, 8].

To mitigate the accumulation of unmanaged plastic waste in the environment, reducing plastic consumption must be the foremost priority. Recycling remains one of the most effective strategies for managing plastic waste. Recycling techniques are broadly categorized into primary, secondary, and tertiary processes. Additionally, non-recyclable plastic waste can be repurposed through industrial co-processing for energy recovery. This chapter aims to present a strategic framework to address the multifaceted challenges of plastic waste management [9].

11.2 Outcomes of Plastic Waste

Plastics can be categorized based on their size into several types, including MPs, NPs, mesoplastics, MPs, and megaplastics, each with distinct characteristics and environmental implications. Their diverse applications span across critical sectors such as drug delivery systems, infection control, food packaging, infrastructure (e.g., plastic roads), and refuse-derived fuels (RDFs), as illustrated in Figure 11.1.

According to the World Economic Forum (WEF) report published in 2016, plastics are produced as by-products during the oil-refining process. Globally, around 300 million tons of plastic is produced each year. Of this, approximately 26% is used in packaging applications. However, only 14% of plastic packaging is recycled, while 40% ends up in landfills and 32% is released into the environment as litter [10].

Polymers derived from diverse monomers such as propylene, ethylene, vinyl chloride, styrene, bisphenol A (BPA) and various functionalized compounds (for example urethane, acrylonitrile, methyl methacrylate) have become integral to industrial and domestic applications. Their polymeric forms including polypropylene, polyethylene, poly vinyl chloride (PVC), polystyrene, polycarbonate, nylon, polyurethane, and others are widely employed in packaging, textiles, household items, construction, and medical or electronic devices. The durability, stability, and versatile physicochemical properties of these polymers account for their extensive utility; however, these same attributes contribute significantly to environmental persistence. Documented impacts include MP contamination, carcinogenic and neurotoxic effects, endocrine disruption, and bioaccumulation across trophic levels. Thus, while synthetic polymers underpin modern technological progress, their long-term environmental and biological

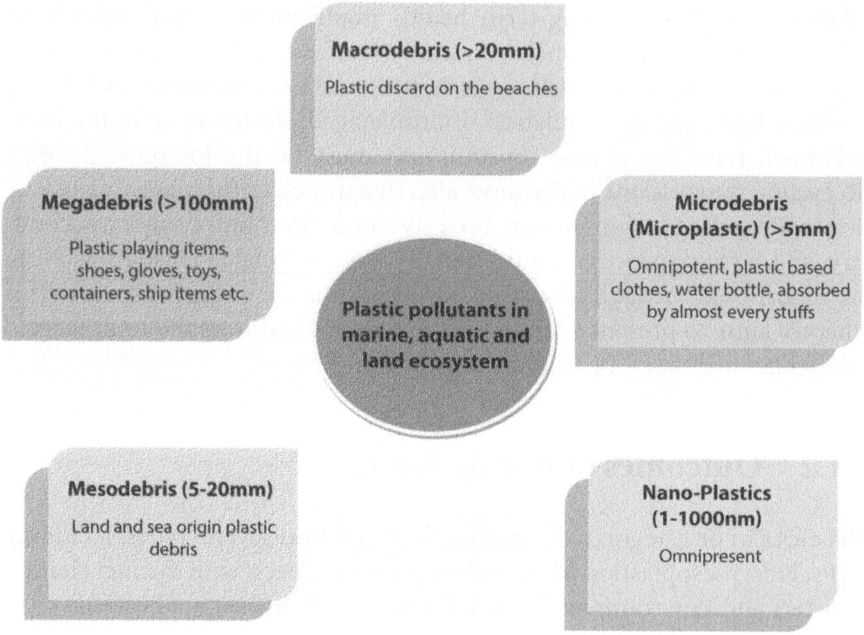

Figure 11.1 Examples of various sizes of plastic debris, highlighting their application in environmental and biological toxicity.

implications necessitate urgent attention toward sustainable alternatives, effective waste management, and policy interventions [11].

The production and degradation of plastics release a variety of hazardous substances, including BPA, polyfluorinated compounds, phthalates, and polyhalogenated flame retardants [12]. Among these, BPA is a particularly toxic compound that can leach into the environment during plastic degradation. BPA exposure has been associated with numerous adverse health effects in humans, including chromosomal abnormalities, cardiovascular disorders, immune system dysfunction, and various forms of cancer.

Phthalates and vinyl compounds, which are key constituents in the manufacturing of PVC, have also been linked to significant health risks. These include birth defects, vision impairment, throat irritation, skin diseases, respiratory conditions such as bronchitis, and several other health complications [13–15].

11.2.1 Plastic Waste in Landfills

A significant proportion of municipal solid waste (MSW) is composed of plastics, most of which are disposed of in landfills without any prior treatment. Landfilling has emerged as a major environmental concern due to factors such as inadequate infrastructure, limited regulatory enforcement, greenhouse gas emissions, and the poor biodegradability of commonly used polymers [16].

In many urban areas, scientifically engineered landfills for the systematic disposal of MSW are lacking. Instead, solid waste is often dumped in open dump yards, where it is either buried or left untreated. In India, this unsystematic dumping is widespread, largely driven by a lack of public awareness and the scarcity of land for waste management. It is estimated that at the current rate of solid waste generation, approximately 1240 hectares of land would be required annually for landfill purposes [17].

The per-capita generation of solid waste in India ranges from 0.17 kg/day in small towns to 0.62 kg/day in megacities [18]. Plastics disposed of in landfills gradually degrade into MPs, which can leach into the soil and groundwater, ultimately entering the food chain and posing serious health hazards. Landfill leachate contains a complex mixture of substances, including pesticides, pharmaceutical residues, disinfectants, organic pollutants, and heavy metals such as lead, mercury, and cadmium. These contaminants degrade soil quality and pose long-term risks to both terrestrial and aquatic ecosystems.

The adverse effects extend to surrounding vegetation, which may either cease to grow or undergo irreversible ecological changes, ultimately affecting biodiversity. Additionally, landfills are a major source of greenhouse gas emissions, predominantly carbon dioxide (CO_2) and methane (CH_4), accounting for 90% and 98%, respectively, of total emissions. Methane, generated through the anaerobic decomposition of organic matter, is highly flammable and can lead to spontaneous landfill fires. Moreover, these gases contribute to the greenhouse effect by trapping solar radiation, thereby exacerbating global warming.

11.2.2 Waste Burning

The open burning of waste remains a persistent issue affecting towns and cities across the country. It is now a common sight to observe dense black smoke emanating from landfills, indicating the combustion of mixed municipal waste. This harmful practice has proliferated primarily due to a lack of public awareness regarding its consequences and inadequate waste collection and management infrastructure.

Open waste burning is a major source of climate pollutants. It releases short-lived climate pollutants such as methane (CH_4), which, although it persists in the atmosphere for a much shorter duration than carbon dioxide (CO_2), has a global warming potential up to 5000 times greater. Therefore, curbing emissions from open burning can lead to significant and rapid climate benefits [19].

Air pollution resulting from waste burning has contributed heavily to smog events and declining air quality indices in several Indian cities, including Delhi and Varanasi. In response, the National Green Tribunal (NGT) has imposed a ban on the open burning of solid waste. Penalties have been instituted to deter this practice, with fines of ₹25,000 for burning bulk waste in open areas and ₹5,000 for backyard burning [11, 20].

Strict enforcement of this ban, along with proactive waste management strategies, is essential for reducing air pollution and mitigating its harmful effects on public health and the environment.

11.2.3 Single-Use Plastic

Single-use plastics, also known as disposable plastics, are designed to be used only once before being discarded. These include items such as food packaging, disposable utensils, coffee cups, stirrers, water and soda bottles, and lightweight plastic bags. In India, approximately 43% of the total plastic produced is used for packaging purposes, the majority of which comprises single-use products [21].

Significant quantities of single-use plastics and multi-layered plastics (MLPs) ultimately accumulate in landfills and the natural environment. During degradation, these materials release harmful substances that can enter food chains and contaminate water bodies. This is particularly concerning because many of these plastics are used in food and beverage packaging.

Improper disposal of MLPs and single-use plastics results in the leaching of toxic chemicals such as phthalates and BPA into the environment. Drinking water bottles are one of the most common contributors to this contamination. Phthalates, which can accumulate in human fatty tissues, act as anti-androgens and have been linked to various health disorders including testicular cancer, gynecomastia (male breast growth), and male reproductive dysfunction [22].

BPA, on the other hand, is known to affect human placental tissue adversely. Studies have reported that BPA exposure may result in stillbirth, intrauterine growth restriction, and premature birth [23]. These findings highlight the urgent need for regulatory action and public awareness regarding the hazards posed by single-use plastics.

11.3 Plastics and MPS: Impact on the Marine Environment

According to a study, approximately 4 to 12 million tons of plastic waste enters the world's oceans annually. It is estimated that around 80% of this plastic pollution originates from land-based sources, while the remaining 20% is attributed to ocean-based activities such as fishing and maritime industries [24, 25]. In many coastal regions, plastic waste is not effectively managed, leading to substantial leakage into marine ecosystems.

The research highlights that just five countries, China, Indonesia, the Philippines, Thailand, and Vietnam, are responsible for nearly 60% of the plastic entering the oceans. The primary drivers of marine plastic pollution are twofold: first, inadequate waste collection, and second, the low residual value of certain plastic waste materials. The McKinsey Centre for Business and Environment, in collaboration with the Ocean Conservancy, estimates that approximately 75% of ocean plastic debris from land-based sources results from uncollected waste [26].

In many regions, plastic waste is either openly burned, left to decompose, or even used as animal feed. A notable study led by Dr. Christian Schmidt from the Helmholtz Centre for Environmental Research found that ten rivers primarily in Asia and Africa are responsible for transporting approximately 90% of plastic pollution into the world's oceans [27].

Moreover, in countries with poor formal recycling infrastructure, waste collection is often handled by informal sectors, including waste pickers. These individuals tend to focus on collecting high-value plastics, which constitute only about 20% of the total plastic waste stream in the five major polluting countries. Consequently, the remaining 80% of low-value plastics are more likely to be improperly disposed of, eventually making their way into marine environments [26].

11.3.1 Impacts on Marine Biodiversity

Various forms of plastic pollution, particularly MPs and nanoplastics (NPs), pose significant threats to marine biodiversity. Long-term exposure to MPs has been associated with adverse population-level effects, such as reduced survival and increased mortality rates in zooplankton species, which serve as a critical energy source in marine ecosystems [28]. Similar detrimental impacts have been observed in crustaceans, where exposure has resulted in reduced offspring growth, lower survival rates, and decreased fertility when compared to unexposed control groups.

A substantial knowledge gap persists regarding the chemical transfer of contaminants associated with plastics and their uptake by marine organisms. Further research is needed to understand the bioavailability and toxicity of micro- and NPs, as well as their long-term ecological consequences. Current scientific findings suggest that endocrine-disrupting chemicals (EDCs), often associated with MPs, can enter marine organisms through filter-feeding mechanisms in species such as baleen whales and mussels [29] or *via* biomagnification through trophic levels in top predators like swordfish [30]. Common plasticizers with endocrine-disrupting properties such as dibutyl phthalate, dimethyl phthalate, butyl benzyl phthalate, and BPA have been shown to adversely affect the development and reproductive health of marine species. Environmental concentrations of these substances, often in the low nanogram to microgram per liter range, are sufficient to harm sensitive organisms, including mollusks, crustaceans, and amphibians. Alkylphenols and other EDCs have been documented to impair male reproductive development, causing sex reversal in mollusks and feminization or demasculinization in fish. Additionally, certain tin-containing plastic stabilizers have been linked to immunological dysfunction and the condition of imposex in gastropods [31].

11.3.2 Impacts on Food Safety and Economic Activity

As previously highlighted, endocrine-disrupting chemicals (EDCs) in plastics may already be impacting marine biodiversity, which raises significant concerns regarding food security and safety in the near future. Filter-feeding shellfish, which accumulate suspended particles on their gills for later ingestion, are directly exposed to micro- and NPs in the water column. As a result, shellfish may now represent one of the most significant sources of dietary MP exposure for humans. Extensive evidence has shown that bivalves consume MPs. For example, a study conducted in Shanghai found that nine commonly sold species of bivalves, purchased from a local fishing market, were contaminated with MPs. The calculated MP intake for Chinese shellfish consumers may exceed 100,000 particles annually [32, 33].

The impact of plastic pollution on marine biodiversity, food safety, and availability is particularly severe for countries and island nations where fish is a primary food source. This could exacerbate poverty, especially in the face of climate change and increasing competition for natural resources [34, 35]. In many small island developing states, and countries like Cambodia, Indonesia, Ghana, Sierra Leone, Bangladesh, and Sri Lanka, fish provide more than 50% of the total animal protein consumed [36]. Furthermore, fish, bivalves, and crustaceans supply approximately 20% of the average per-capita animal protein intake for 3.2 billion people and 10% for 5.1 billion people globally. Additionally, the global marine fishing industry supports the livelihoods of approximately 260 million people, encompassing both direct and indirect employment, including full-time and part-time jobs [37].

11.4 Impacts of MPs on Human Health

Plastic waste, once released into the environment, undergoes various environmental factors, weathering, and fragmentation processes, leading to the formation of MPs. These MP particles pose significant risks to human health. MPs can enter the human body in several ways. A recent study demonstrated that plastic particles smaller than 150 μm can be absorbed by tissues, organs, and cells in biota [38]. These particles can infiltrate the human body through the food chain, introducing harmful chemicals that can disrupt the normal functioning of organs and tissues.

In addition to causing acute and chronic toxicity, MPs can induce immunotoxicity, developmental toxicity, cytotoxicity, carcinogenicity, and reproductive toxicity. There is also evidence suggesting that NPs may lead to long-term harmful effects such as neurotoxicity, hepatotoxicity, and cardiovascular toxicity. These particles may exhibit developmental and genotoxic effects, which further compound their risk to human health. Studies have shown that MPs are prevalent in everyday items such as takeaway containers and seafood, including nori and shrimp [39]. Moreover, MPs have been discovered in human embryos and feces, reinforcing the notion that they can enter the human body through the alimentary canal, primarily through food consumption.

Long-term exposure to these polymers may indirectly damage vital organs such as the liver, heart, and kidneys. Revel *et al.* further reported that MPs smaller than 100 nm can penetrate subcutaneous tissues through skin folds [40, 41]. Currently, the available data on human exposure to micro- and NPs, particularly those originating from wear debris or plastic prosthesis implants, is limited and inconclusive. Medical literature on the subject reports a variety of effects, including DNA damage, alterations in gene and protein expression, coagulation of cells, necrosis, apoptosis, cell proliferation, and loss of cell viability. There is an urgent need for further scientific investigation to understand the possible effects of direct or indirect intake of marine micro- and NPs on human health, particularly during developmental stages [42].

11.5 Impacts of Micro- and NPs on Ecosystem Services and Climate Change

Every minute, approximately a million plastic bottles are purchased worldwide, and this number is expected to rise in the coming years, creating an environmental challenge that could exacerbate climate change. The deposition of MPs and NPs into sea ice may have significant impacts on living biomass, as evidenced by the positive correlation found between MPs and chlorophyll in sea ice. Importantly, sea ice in the Southern Ocean could serve as a reservoir for MPs, as these particles are likely retained and periodically released from the sea ice during the formation and melting processes, rather than being transported to deeper waters [43].

MPs impact ecosystem services in three primary media: terrestrial, aquatic, and atmospheric ecosystems. A comprehensive approach is needed to balance human well-being with the services provided by nature. Although MPs are a major environmental concern, they have yet to be fully investigated in terms of their role in ecosystem services, a concept introduced in the late 1990s [44].

This section aims to connect ecosystem services, human well-being, and the effects of micro- and NPs. Ecological research suggests that the impacts of MPs and NPs extend beyond ecological concerns, also affecting economic and social aspects of human society [45]. In terrestrial ecosystems, MPs and NPs are linked to global change and have been shown to disrupt nutrient cycles, such as nitrogen and phosphorus, leading to nutrient imbalances and oxidative stress that reduce soil productivity. This phenomenon has been particularly evident in wheat production. In aquatic environments, the accumulation of MPs contributes to mineralization, nitrification, and denitrification, which release greenhouse gases like CO_2, CH_4, and N_2O. Both resuspension of particles and algae growth increase eutrophication, leading to organic pollution and the addition of more pollutants into the food chain.

The rapid growth of the petrochemical and plastic industries is a significant factor in the global temperature rise, potentially challenging the targets set to limit the increase to 1.5 or 2°C. The increased production of plastics can destabilize the climate, contributing to environmental change [46]. As ocean temperatures rise, the influx of MPs into marine environments will likely have direct and indirect effects on global climate change, glaciers, ocean currents, and tropical cyclones.

Additionally, MPs and NPs play a detrimental role in climate change. Humans are also exposed to MPs through inhalation of polluted air, which can lead to severe health issues such as lung congestion, ulcers, cancer, and a range of nasal and olfactory diseases [47]. The growing knowledge gap regarding the toxicity of MPs and NPs poses significant ecological challenges, and the public health concerns associated with their exposure could strain healthcare systems, particularly in developing countries with overburdened medical infrastructures. This concern has the potential to lower the Human Development Index (HDI) in these regions [48].

11.6 Plastic Pollution Effects on Rural Communities

The issue of plastic pollution in rural water supplies is often underappreciated and frequently overlooked. Rural areas, characterized by lower population densities and a lack of formal or informal waste infrastructure, face widespread contamination from macroplastics (>5 mm) and MPs (<5 mm). This leads to practices such as the open burning of plastic waste and the disposal of solid debris on land and in water bodies, which are significant sources of MPs in the environment [49]. Plastic waste directly and indirectly impacts the stability and quality of water resources in these areas.

In rural regions, water resources may include treated water in tanks or bottles, as well as surface and groundwater, which can vary depending on the region. The acquisition, storage, and quality of water supplies are closely linked to the frequency of plastic contamination. As of 2017, approximately 2.2 billion people worldwide relied on poorly managed water resources [50], including untreated surface water from lakes, ponds, rivers, and streams, as well as unprotected wells and springs. These sources are vulnerable to plastic pollution due to the lack of proper waste management infrastructure. In the United States, at least 15% of the population depends on untested water supplies, such as shallow wells, which are particularly susceptible to contamination by nano- or MPs [51].

As temperatures rise, the use of plastic-bottled water is expected to increase in rural areas, especially where tap water supplies are unreliable. Depending on the level of water treatment, plastics in drinking water are becoming more prevalent. A study on the presence of MPs in raw and treated drinking water from treatment plants in the Czech Republic found that treatment was effective in removing MPs larger than 100 μm but ineffective for those smaller than 10 μm [52]. Following treatment, the average concentration of MPs in river water decreased from 330.2 particles L^{-1} to 105.8 particles L^{-1}. A global study on bottled water found that MPs larger than 100 μm were most commonly found in the form of fragments (66%) at an average concentration of 10.4 MPs L^{-1}. Open-source water, particularly in rural areas, is highly susceptible to contamination from poor waste management practices in nearby communities, as well as plastic pollution from upstream urban sources. A study by Treilles *et al.* [53] estimated that various sewer systems in Greater Paris release 8–33 tons of plastic annually. In low- and middle-income countries, only 33% of the population has access to sewer systems capable of filtering macro- and MPs before wastewater is released into the environment.

Further research is required to understand the full extent of rural plastic pollution's impact on drinking and irrigation water supplies. MP contamination poses a serious threat to agricultural soil, which is essential for food security and biodiversity. Agroecosystems, already stressed by population growth and climate change, are increasingly affected by plastic pollution [54, 55]. Sources of MPs in rural soils include the use of sewage sludge, wastewater irrigation, and compost products in farming [56].

Degradation of macroplastic items through farming activities, such as the use of greenhouse coverings, polytunnels, and plastic mulching films. Other practices, such as the open burning of plastic waste and washing synthetic clothing, contribute to MP pollution. Transfer of MPs from other environmental compartments, such as flooded freshwater bodies and particles from road surfaces [57], also contributes significantly to their redistribution and accumulation in the environment. For example, Ohtake *et al.* [58] estimated that a 60-μm layer of low-density polyethylene (LDPE) would take approximately 300 years to degrade. Over time, this could encourage soil degradation and water contamination, with potential economic impacts on rural communities due to reduced crop yields and lower-quality crops.

Inadequate waste management infrastructure contributes to the environmental damage caused by plastic waste, exacerbating public health issues related to plastic pollution. Despite some efforts to improve waste management in densely populated urban areas in Asia and Africa, awareness of the dangers of plastic contamination in food, particularly from plastic packaging, remains limited. Furthermore, the number of glass recycling facilities in former Soviet nations has decreased since the 1990s, highlighting the lack of progress in addressing this issue. Significant knowledge gaps remain regarding the potential hazards of plastic use in food packaging and how plastic pollution affects rural populations [59].

11.7 The Global Challenges of Plastic Waste

Plastic waste has emerged as a critical global issue, with growing awareness surrounding its detrimental effects, particularly regarding air and ocean pollution. These environmental challenges disrupt ecosystems and natural processes, undermining the resilience of ecosystems and the livelihoods, social well-being, and food security of millions of people. Plastics are popular for their affordability, durability, flexibility, and lightweight nature, making them widely used in households. However, the resistance of plastics to degradation leads them to persist in the environment for decades, causing long-lasting negative impacts on both ecosystems and human health [60].

Household-generated plastics contribute to human health risks. Improper disposal of plastics can foster conditions for pests like mosquitoes, which are vectors for diseases such as malaria. Plastics used in food packaging pose significant risks to human health, including cardiovascular diseases, obesity, cancer, and various other health issues. All forms of life, both aquatic and terrestrial, are affected by the careless disposal of plastics, with livestock suffering from digestive disruptions due to their exposure to plastics [61].

A major challenge in community development is managing plastic waste. This issue is complex due to the involvement of various stakeholders, including local populations, governmental agencies, and waste management systems. Effective waste management requires the engagement of all relevant parties, as the absence of any key actor can undermine the entire process. It is essential to recognize each individual's role in the process of change during community development efforts.

In the context of environmental justice, green social work aims to enhance both environmental conditions and the quality of life for people. Community development should foster the integration of ecological sustainability, social welfare, and economic well-being, promoting a human-centered approach that harmonizes human activities with the environment [62].

The rapid expansion of the plastics manufacturing industry raises significant concerns due to the harmful impact of plastics. Single-use plastics, such as shopping bags, packaging materials, and bottles, exemplify the disposable culture that underpins consumerism. A 2023 report by Mesoe *et al.*

highlighted that in Europe, 29.1 million tons of post-consumer plastic waste was collected, of which only 32.5% was recycled. In contrast, 42.6% was incinerated, and 24.9% ended up in landfills [63]. Since 2015, the average plastic consumption per person has been around 16 kg, with Africa alone producing approximately 19 million tons of plastic waste annually. In both Africa and globally, plastic waste management has become a pressing issue, with economic, social, and environmental challenges. In Africa, the poor management of plastic waste is compounded by factors such as a lack of qualified personnel, low public awareness, political instability, and weak governance. Efforts to address these challenges have led to a variety of technological and societal innovations aimed at improving waste management systems. According to Godfrey *et al.*, the growth of plastic waste in Africa is outpacing that of other types of waste, primarily due to a lack of awareness about proper waste management practices [64].

11.8 Sustainable Plastic Waste Management Solutions

Plastics are among the least regulated products, despite their ubiquitous use across various sectors. Typically, the specific polymers and additives used in plastics are not disclosed, making it difficult to fully understand their impact on human health and the environment. This lack of transparency contributes to the difficulty in recycling plastic waste. The diverse nature and composition of plastics present significant challenges and costs in the recycling process. As a result, alternative strategies for managing plastic waste have been proposed, focusing on creating value-added products.

11.8.1 Landfilling

A significant portion of municipal solid waste consists of plastics, much of which is sent to landfills without proper treatment. Factors such as inadequate infrastructure, poor maintenance, greenhouse gas emissions, legislative pressures, and the low biodegradability of many common polymers make landfilling a major environmental concern. In response to the environmental risks posed by landfill disposal, the Environmental Protection Agency has introduced improved state and federal regulations, including the establishment of landfill bed usage standards, standardized groundwater testing, and enhanced post-landfilling care for hazardous waste management [16].

Plastic waste presents a unique challenge for disposal due to its high volume-to-weight ratio, which complicates the identification of suitable

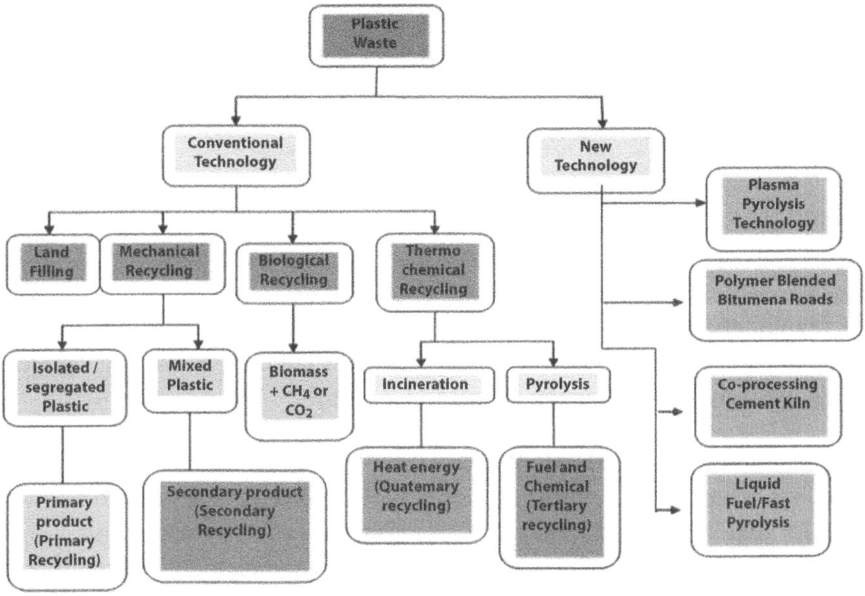

Figure 11.2 Different pathways for plastic waste management.

disposal sites. As a result, various alternative methods to traditional landfilling for waste management have been proposed, as illustrated in Figure 11.2.

11.8.2 Recycling of Plastic Waste

Plastic recycling holds significant potential in reducing environmental impacts. It helps conserve petroleum-based resources and contributes to addressing the global plastic waste crisis. Economically, recycling is also beneficial. According to Delattre *et al.*, in the early 2000s, approximately 65–70% of plastic waste was disposed of in landfills, 20–25% was incinerated, and only about 10% was recycled. However, this distribution varies considerably across countries due to differences in population size and lifestyle. While plastic recycling was minimal in the past, there has been a growing trend toward increased recycling in recent years [65].

A. Mechanical Recycling

Mechanical recycling involves converting discarded thermoplastics into new or similar plastic products. This process, which includes both primary and secondary recycling, begins with the collection of used thermoplastic materials. The main objective is to reuse homogeneous thermoplastics

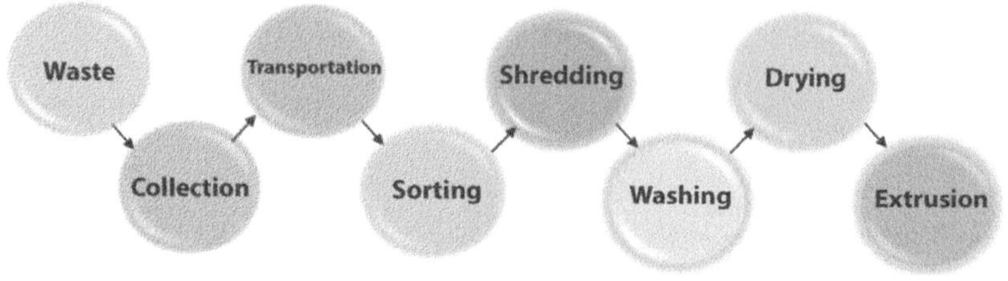

Figure 11.3 Plastic mechanical recycling process including steps.

to manufacture items that are nearly identical to the original products. Mechanical recycling involves several stages: collection, sorting, shredding, washing, and extrusion (Figure 11.3) [66].

Step 1: Collection of Plastic Waste

The first stage in mechanical recycling is the collection of post-consumer plastic materials from households, commercial establishments, and healthcare facilities. Effective plastic waste management depends on well-structured collection systems tailored to waste types and population density [67]. In densely populated urban areas, daily residential waste collection using specialized vehicles equipped with compacting and transport mechanisms has proven effective. These vehicles transport the waste to authorized recycling or waste management facilities. An efficient recycling process begins with the separation of solid waste at the source either at home or at municipal pickup points. In developed countries like the USA, Germany, and Japan, recyclable materials are typically separated and sent to designated drop-off points. In contrast, in developing countries such as Brazil, selective collection is often done door-to-door, which requires significant energy for transport [68].

Since 1993, China has emerged as the leading importer of global waste, including plastics, due to its limited availability of domestic raw materials needed for manufacturing and export. It has been estimated that China has imported between 45 and 56% of the world's plastic waste to support its plastics industry.

Step 2: Sorting of Plastic Waste

Both mechanical and chemical recycling techniques rely on the thorough sorting of waste materials before further processing. Efficient sorting is critical to ensure the quality and reusability of the recycled output. For instance, separating PVC and PET from shredded bottle waste is essential, as even small amounts of PVC can degrade the quality of an entire PET batch. Common plastic types such as PVC, PET, and PE must be identified and sorted appropriately. Color-based sorting is also carried out to improve the visual consistency of final products [69].

Sorting methods include both manual and automated techniques [70]. Automated systems, although capital-intensive, offer high throughput and require minimal human intervention. These systems employ advanced technologies like infrared spectroscopy and robotics. On the other hand, manual sorting based on visual identification and physical separation is widely used in developing countries due to its low cost and the availability of labor. However, it can be less efficient and more time-consuming. Improved manual sorting processes can significantly enhance cost-effectiveness in these settings [71].

Step 3: Shredding and Resizing

Once sorted, plastic waste is shredded to reduce its size, increasing bulk density and making it easier to handle, store, and transport. A typical shredder includes a hopper, rotating blades powered by an electric motor, and a sizing screen. As the waste passes through the system, it is chopped into flakes or granules. This stage also aids in the removal of contaminants such as metal pieces that might not have been identified earlier [72].

Step 4: Washing to Remove Impurities

After shredding, the plastic flakes undergo washing to eliminate impurities like food residue, adhesives, and dirt. Although most contaminants are removed during washing, certain materials such as agglomerated plastics may bypass this stage and be processed directly [73]. Washing can involve the use of hot water (up to 60°C) or cold water combined with mechanical agitation and chemical agents such as sodium hydroxide. The wastewater generated is treated internally. Finally, the clean plastic flakes are dried to achieve a moisture content of about 0.1% by weight, making them suitable for further processing [74].

Step 5: Extrusion

Extrusion is one of the most widely used techniques in plastic recycling. In this process, the cleaned and dried plastic flakes are fed through a hopper into an extruder. A rotating screw pushes the plastic through a heated barrel where it is melted, homogenized, and pressurized. The molten plastic is then forced through a die, forming a continuous strand which is cooled and cut into pellets of desired size and shape.

However, the repeated thermal and mechanical stress during extrusion can reduce the average molecular weight and degrade the mechanical properties of the polymers. Advances in extrusion technologies are aimed at minimizing these effects, thereby producing recycled plastics of higher quality suitable for various applications [66, 74].

Industrial recycling methods can be broadly classified into primary, secondary, and tertiary recycling pathways (outlined in Figure 11.4):

- Primary recycling involves the direct reuse of clean, homogeneous plastic scrap in similar applications. This method is cost-effective but limited by contamination and the number of possible reuse cycles [75].
- Secondary recycling entails cleaning, melting, and remolding plastics into new products, such as packaging materials that require less structural integrity. Although virgin resins can be added to improve properties, this blending often reduces material uniformity and limits end-use applications [76].
- Tertiary recycling accounts for approximately 10% of global recycling operations. It includes chemical depolymerization processes that break down long polymer chains into monomers or oligomers. These products can be reused as fuels or as feedstock for the synthesis of new polymers. Tertiary recycling is complex but allows broader and higher-value applications [77, 78].

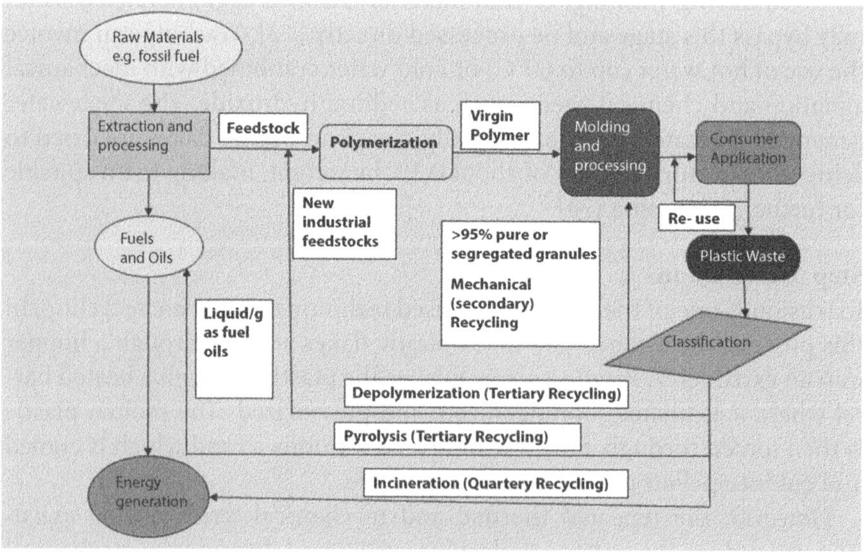

Figure 11.4 Plastic lifecycle for industrial ecology of plastic and waste management.

B. Biological Recycling

To meet the growing demand for organic recovery, biodegradable polymers have been extensively researched. The development and implementation of biological infrastructure is essential for future waste management, supported by promising technological advancements. In biological recycling, aerobic composting—using aerobic bacteria, fungi, and other microorganisms—facilitates the breakdown of biopolymers. Through this process, organic carbon resources are converted into biomass, thereby contributing to energy generation and environmental sustainability. However, degradable plastics that fail to decompose under specific conditions may be misclassified and treated as non-degradable, limiting the effectiveness of this method [79].

C. Thermal Recycling or Incineration

Thermal recycling, or incineration, involves the continuous operation of incineration plants, typically halting only for scheduled maintenance. This method is especially suited for processing composite materials, such as carbon fiber-reinforced polymers, at high temperatures ranging from 450 to 700°C [80]. Tominaga *et al.* reported that thermal recycling accounts for more than 54% of global plastic recycling efforts [81].

Due to its cost-effective processing of mixed plastic waste and reduced space requirements, incineration is widely adopted. It also contributes to energy recovery, thereby decreasing dependency on fossil fuels. For example, Japan operates 1,172 municipal solid waste (MSW) incinerators, which collectively generate about 1,770 MW of electricity. Globally, incineration of MSW yields approximately 200 kWh per ton of waste [73].

D. Chemical Recycling

Chemical recycling offers a more sustainable and versatile alternative to mechanical recycling by prioritizing energy recovery and resource circularity [82]. Unlike mechanical methods, chemical recycling is capable of processing contaminated or mixed polymer streams that vary widely in chemical composition and melting points. Chemical recycling technologies can be broadly classified into three main categories based on their decomposition mechanisms:

a) **Solvent-based purification** — Plastic waste is dissolved in solvents and purified without breaking the polymer chains, allowing the polymer to be recovered and reused.

b) **Chemical depolymerization** — Polymers are broken down through chemical reactions into their original monomer units, enabling re-polymerization for high-quality material recovery.
c) **Thermal depolymerization (pyrolysis)** — Through high-temperature cracking, hydrocarbons are produced from plastics, and monomers or other valuable chemical feedstocks are recovered.

This approach supports closed-loop recycling systems and contributes significantly to the circular economy by enabling the reuse of plastic materials in various high-value applications.

11.8.3 New Technologies Used for Plastic Waste Management

Conventional methods such as incineration, landfilling, and open dumping are inadequate for achieving sustainable plastic waste management, especially for non-biodegradable plastics. Consequently, extensive research is ongoing to develop innovative and efficient alternatives that can address the mounting global plastic waste crisis (Figure 11.5).

Plastic disposal strategies are broadly classified into two categories: degradation and recycling [83]. Among these, recycling is further subdivided based on the physicochemical characteristics of the plastic waste. According to the American Society for Testing and Materials (ASTM), plastic recycling can be categorized into four types:

a. **Primary recycling (ASTM 1):** Also referred to as physical recycling, this involves the mechanical reprocessing of clean, uncontaminated plastic waste into products that retain similar physical properties.
b. **Secondary recycling (ASTM 2):** In this method, plastic waste is mechanically recycled to produce lower-grade materials or products with reduced performance specifications.
c. **Tertiary recycling (ASTM 3):** This process breaks down plastic polymers into chemical intermediates (solids, liquids, or gases) through methods such as hydrocracking or pyrolysis.
d. **Quaternary recycling (ASTM 4):** Focused on energy recovery, this method converts plastic waste into heat, electricity, or fuel through processes like gasification or incineration [84].

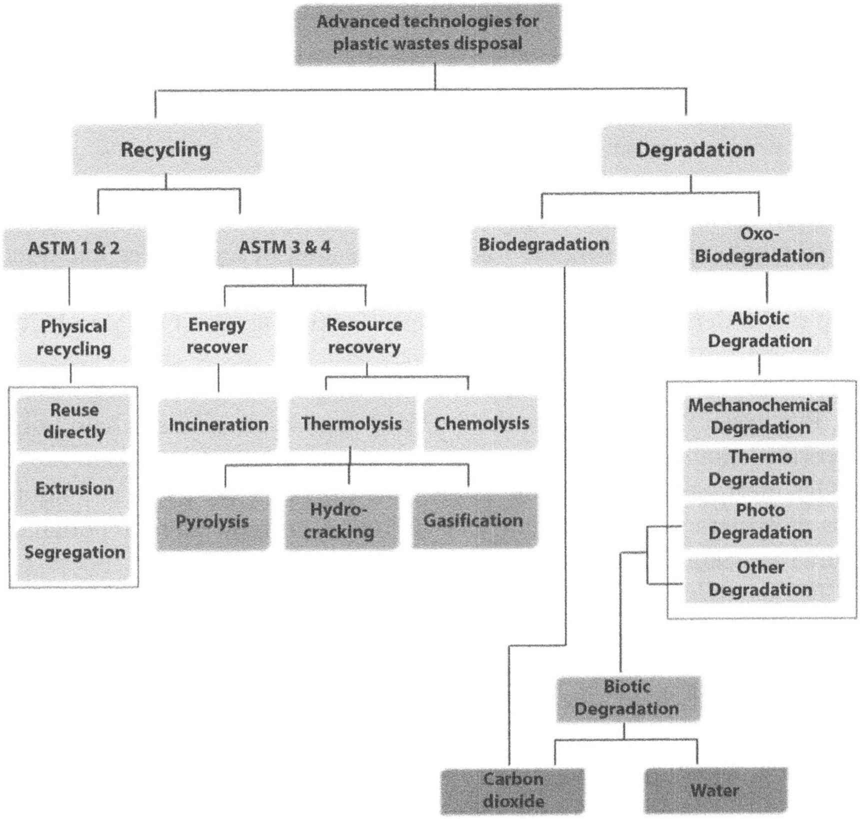

Figure 11.5 Plastic waste management using advanced methods.

While ASTM 1 and ASTM 2 primarily involve extrusion, segregation, and direct reuse, ASTM 3 and ASTM 4 encompass more advanced techniques such as pyrolysis, gasification, and hydrocracking.

This classification framework underscores the importance of selecting appropriate recycling technologies based on the type, quality, and contamination level of the plastic waste. The integration of these new technologies is essential for the transition toward a circular economy and the reduction of plastic pollution.

A. Thermolysis

Tertiary and quaternary recycling focus on product and energy recovery, respectively. A prominent approach in this domain is the thermolysis of thermoplastics, wherein high-molecular-weight polymers are broken down into smaller units [84].

Thermolysis technologies are broadly classified into the following:

a) Pyrolysis
b) Hydrocracking
c) Gasification

These techniques decompose polymers into gas, oil, and char at elevated temperatures using inert atmospheres and, in some cases, catalysts. Hydrogenation occurs in hydrogen environments, while gasification is carried out under sub-stoichiometric oxygen for producing CO, H_2, and CO_2 [85].

a) Pyrolysis: Pyrolysis is the thermal decomposition of large-chain polymers into smaller hydrocarbons at temperatures ranging from 300 to 900°C under inert gas [86]. It produces valuable products like gases and liquid oils. Polymers such as polyethylene (PE) and polyethylene terephthalate (PET) are suitable for pyrolysis [87].

The key thermal decomposition mechanisms in pyrolysis are as follows:

I. Random chain scission (fragmentation)
II. End-chain scission (depolymerization)
III. Cross-linking (formation of a network)
IV. Chain stripping (removal of side chains)

The degradation pathway depends on the polymer's chemical structure.

b) Hydrocracking

Hydrocracking or hydrogenation involves breaking down long-chain hydrocarbons into lighter fractions such as gasoline and kerosene under high pressure and in the presence of catalysts [88]. Compared to pyrolysis, hydrocracking offers the following:

I. Superior selectivity for C_6–C_{12} hydrocarbons (gasoline range) [89]
II. Reduced heteroatom and aromatic content in the products [90]
III. Better mass and heat transfer due to lower viscosity

However, it is less commonly used due to higher operational costs. For example, hydrogen can cost ~\$2800/ton compared to nitrogen's ~\$200/ton [82].

Moreover, the need for high-pressure equipment makes hydrocracking more capital-intensive than pyrolysis.

c) Gasification

Gasification is a competitive thermolysis process involving the partial oxidation of plastic waste to produce syngas (a mixture of CO and H_2) at high temperatures (1200–1500°C), often in the presence of oxygen and steam [83, 91]. Various technologies, such as Battelle and Texaco gasifiers, are used industrially. After gasification, the gases are cleaned to remove HF and HCl, resulting in dry, clean syngas containing H_2; CO; and trace amounts of CO_2, CH_4, and H_2O [79].

B. Chemolysis

Chemolysis (also known as solvolysis or chemical depolymerization) involves breaking down plastics into monomers using solvents at 80–280°C [92]. It is particularly suitable for plastics like PET, polyamides, polyesters, polycarbonate, and polyurethane [93]. Due to variations in plastic types, chemolysis is not widely adopted as a standard recycling method but offers unique advantages in closed-loop recycling [94]. The four main chemolytic pathways are as follows:

a) Alcoholysis

Plastics are depolymerized in alcohols to yield monomers [95]. For polyurethanes, alcoholysis *via* transesterification yields polyhydroxy alcohols and carbamate fragments, without CO_2 generation [96].

b) Hydrolysis

Involves depolymerization using water (acidic, neutral, or alkaline) to recover raw materials from polymers [97]. Monomers recovered can be recombined with new materials, reducing environmental impact.

c) Methanolysis

Uses methanol to depolymerize PET into dimethyl terephthalate (DMT) and ethylene glycol, key precursors for resins and polymers. This occurs at 270–300°C and 0.1–15 MPa [98].

d) Glycolysis

A widely used PET recycling method, involving depolymerization in glycol to produce specialized oligomers such as α,ω-dihydroxy compounds. Companies like DuPont and Dow use glycolysis for PET waste valorization [82].

11.9 Plastic Degradation by Bacteria

Several studies have demonstrated the potential of microorganisms to degrade various types of plastics. *Desulfovibrio nigrificans* and *Pseudomonas alcaligenes* have been isolated from plastic-contaminated soil, while *Pseudomonas putida* MTCC 2475 was recovered from garden soil. Both *P. putida* MTCC 2475 and *Enterobacter* sp. D1 isolated from the gut of *Galleria mellonella* have shown the ability to degrade polyethylene (PE). Notably, *P. putida* MTCC 2475 was capable of reducing the weight of milk packaging film by approximately 63.1–73.3% within one month of incubation. Treatment with *Enterobacter* sp. D1 led to the formation of degradation products including alcohols, esters, carboxylic acids, ethyl *n*-propanol, and 6-methyl-5-hepten-2-ol, which serve as metabolic indicators of PE biodegradation.

Four PE-degrading bacterial strains *P. putida, Bacillus amylolyticus, Bacillus subtilis,* and *Pseudomonas fluorescens* were isolated from soil. These strains were cultured individually in broth media containing polyethylene films for one month. Among them, *B. amylolyticus* achieved the highest degradation, reducing film weight by 32%, followed by *P. fluorescens* (22%), *P. putida* (18%), and *B. subtilis* (14%). Other PE-degrading microbes include *Pseudomonas aeruginosa* and *Pseudomonas stutzeri*. In *Pseudomonas* species, an inactivating operon regulates the expression of enzymes that enable the utilization of unconventional carbon sources, such as polyethylene. These enzymes include serine hydrolases and other catalytic proteins.

Furthermore, five additional PE-degrading bacterial isolates were recovered from plastic waste-contaminated soils. *P. aeruginosa* was found to reduce low-density polyethylene (LDPE) film weight by 18.75% after one month. Other bacteria such as *P. fluorescens* and *Actinobacter ursingii* also exhibit LDPE-degrading activity. Strains of *Serratia* sp. KCI-MRL, *Stenotrophomonas* sp. KC4-MRL, *Bacillus* sp. KC3-MRL, and *Bacillus licheniformis* KC2-MRL isolated from soil demonstrated surface damage to LDPE films during a one-month incubation period.

Microorganisms secrete bioemulsifiers during the degradation of polymeric materials, and biosurfactants play a pivotal role in enhancing plastic biodegradation. In particular, actinomycetes produce a wide range of metabolites that contribute significantly to plastic decomposition. *Ochrobactrum anthropi* is capable of degrading high-density polyethylene (HDPE), achieving a 20% reduction in film weight within 45 days. As for polypropylene (PP), *Bacillus cereus* and *Sporosarcina globispora* have been reported as effective degraders, with degradation rates of 0.003 g/day and 0.002 g/day, respectively [99–102].

11.10 Robotic Technologies for Plastic Waste Mitigation

The transition from traditional waste management strategies to robotic technologies represents a significant advancement, driven by the need for improved precision, operational efficiency, and adaptability. Conventional approaches often rely on manual labor and static systems, which can be time-consuming, error-prone, and ineffective in dynamic environments. In contrast, robotic systems integrate real-time sensing, computation, and actuation, enabling autonomous decision-making and adaptive responses to environmental changes [103].

Robotic solutions equipped with artificial intelligence (AI), machine learning algorithms, sensor technologies, and autonomous navigation systems are emerging as powerful tools in addressing plastic pollution. These technologies enable robots to detect, collect, and classify plastic debris with high accuracy. Aquatic robots such as Seabins and Ocean Cleanup Interceptors are designed to autonomously navigate and collect plastic waste from rivers, lakes, and oceans [104]. These systems utilize advanced imaging and sensing technologies to distinguish plastics from other debris, ensuring effective retrieval.

At Delft University of Technology, researchers have developed robots such as BeachBot, specifically designed to detect and collect MPs from sandy beaches using AI-driven algorithms for targeted waste removal [105]. The autonomous nature of these robots allows them to operate in hazardous or contaminated environments without posing risks to human health and safety.

In 2020, Sun *et al.* developed a pollen-based magnetic "hedgehog" microrobot using sunflower pollen grains (SPGs) that efficiently removes MPs and oil pollutants. This microrobot operates under a rotating magnetic field and exhibits three motion modes—rolling, revolving, and swinging. In rolling mode, it can transport 100-μm polystyrene microspheres and form chain-like structures. By combining rolling and swinging, the microrobots achieved removal efficiencies of 75% in contaminated seawater and 70% in facial cleanser samples [106].

In 2024, Velikov *et al.* introduced a novel approach using magnetic nanorobots functionalized with hydrophobic fluorescent dyes for the detection and quantification of nanoparticles (NPs) in aqueous media. These robots removed over 90% of NPs within 120 minutes [107]. Similarly, in 2023, Li *et al.* developed a self-propelled magnetorobot composed of a magnetizable ion-exchange resin sphere, capable of extracting micro- and NPs (MNPs) from diverse water sources, including river water, lake water, sewage, and deionized water. This robot consistently achieved removal efficiencies above 90% across five cycles, leveraging long-range electrophoretic attraction, and demonstrated broad applicability across varying plastic types, sizes, and water chemistries [108, 109].

11.11 Natural and Sustainable Substitutes for Plastic Products

To mitigate the environmental challenges posed by plastic pollution, a range of natural and sustainable alternatives to conventional plastic products have emerged as promising solutions. As reported by latest studies, biodegradable polymers derived from renewable biological sources such as potato, sugarcane, maize starch, and algae possess the inherent ability to decompose through natural biological processes [110, 111]. Similarly, biodegradable and compostable plastics, often produced from organic materials like bamboo or sugarcane bagasse, break down into non-toxic natural components under appropriate environmental conditions [112].

The adoption of reusable alternatives represents a practical strategy to curb the consumption of single-use plastics. Items such as shopping bags, water bottles, and cutlery made from glass, bamboo, silicone, or stainless-steel offer durable, sustainable replacements. Plant-based packaging materials including cellulose-based films and molded pulp offer environmentally responsible substitutes for traditional plastic packaging [113]. Natural fibers such as cotton, jute, sisal, and hemp also serve as eco-friendly materials in the production of textiles and carry bags. Beeswax wraps provide a biodegradable and natural alternative to conventional plastic cling films for food storage. Additionally, innovative solutions such as edible packaging produced from seaweed or other edible plant-derived substances present novel approaches to reducing packaging waste. Mycelium-based packaging, developed from the root-like structure of fungi, offers a lightweight, durable, and compostable alternative to synthetic plastic packaging [114].

11.12 Future of Bioplastics

In 2019, approximately 2.11 million tons of bioplastics was produced globally, and this figure is projected to rise to around 2.43 million tons by 2024. Asian countries account for more than 58% of the global bioplastics output, with 1.22 million tons produced in 2019 alone. India possesses significant potential for bioplastic production, owing to the abundance of agricultural biomass. A DuPont survey conducted in India revealed that approximately 63% of consumers are aware of bioplastics and their environmental advantages [115].

Common feedstocks for bioplastic production include maize starch, sugarcane, and castor oil, while materials such as wood chips, corn starch, sawdust, and vegetable fats and oils are also extensively utilized. Bioplastics can be processed using conventional polymer-processing techniques such as injection molding, blow molding, thermoforming, extrusion, and blow film extrusion. Compared to virgin plastics, bio-based alternatives exhibit approximately 70% greater mechanical and thermal stability. They are particularly well-suited for the development of multi-layered films and food packaging materials. The bioplastics market is witnessing steady expansion, driven by continuous advancements in material science, increased consumer awareness, and a growing preference for environmentally sustainable products [116].

11.13 Presenting Incentives

The implementation of incentives and financial benefits for returning discarded plastic products plays a crucial role in enhancing collection efficiency and streamlining recycling and material recovery processes. A major challenge is that people often refrain from recycling due to the perceived time and cost burden. However, interest in recycling can increase significantly when appropriate incentives, such as buy-back schemes and take-back policies, are introduced. The establishment of frequent and conveniently located take-back stations can further encourage public participation.

Nevertheless, encouraging individuals to dispose of plastic waste at designated collection points remains a persistent challenge. A viable solution is to embed collection and recycling costs into the product price, thereby raising consumer awareness about the importance of waste management and environmental conservation.

Figure 11.6 A section tool based on problems to handle plastic waste.

To address these challenges effectively, administrative bodies must integrate incentive-based approaches with comprehensive policy frameworks. These frameworks may include pricing-based, rights-based, regulatory, and behavioral strategies, as outlined in Figure 11.6 [117].

A well-structured and balanced waste management system, supported by strong government policy, is essential to achieving a significant reduction in plastic waste. Across the globe, many governments have implemented regulations that restrict the use of plastic bags based on their thickness, chemical composition, or environmental impact. Global responses to plastic pollution have taken the form of diverse policies aimed at reducing production, consumption, and environmental leakage of plastics. Several countries have targeted specific high-impact items such as plastic bags and microbeads. For instance, the United States pioneered measures at state levels, including California's ban on plastic shopping bags and restrictions on cosmetics containing plastic microbeads. Ireland introduced a levy on plastic bags as early as 2002, significantly reducing their use, while France legislated an ambitious nationwide plastic ban in 2016. Sweden has become a model recycling nation, diverting nearly all household waste from landfills, whereas China's 2008 regulation requiring consumers to pay for bags cut their consumption by half within two years. South Asian countries such as India and Bangladesh have enacted bans on ultra-thin bags and single-use plastics, with Bangladesh extending restrictions to coastal zones and the hospitality sector. Rwanda has enforced one of the strictest policies by banning the sale and import of synthetic microbeads. Likewise, Canada, Denmark, the United Kingdom, Italy, and Australia have prohibited microbeads in personal care products, with Australia also targeting lightweight shopping bags. Collectively, these measures highlight the increasing global recognition of plastic waste as a trans-boundary environmental challenge and underscore the importance of regulatory intervention in shaping sustainable consumption practices.

11.14 Knowledge Gaps and Future Scope

The rapid and widespread increase in MPs (MPs) poses a severe threat to human and animal health, as well as to the stability of global ecosystems. In aquatic environments, MPs interfere with ecological processes and disrupt microbial habitats, which can compromise water quality and biodiversity. Comprehensive methods for the identification, monitoring, and quantification of MPs in environmental matrices such as water, soil, food products, and personal care items are urgently needed to support the formulation of effective policies and regulatory frameworks [118].

Research indicates that a significant proportion of aquatic organisms, especially fish, ingest MPs both intentionally and unintentionally. This not only threatens aquatic food chains but also introduces MPs into the human diet through seafood consumption, raising serious concerns about food security and public health. Additionally, plastic pollution contributes indirectly to climate change; the degradation of plastic materials releases carbon emissions, thereby intensifying global warming and increasing the frequency of droughts [119–121]. Despite growing attention to the plastic pollution crisis, there remains a lack of comprehensive studies covering the full spectrum of plastic types and their environmental impacts. Most existing research has focused on polyethylene terephthalate (PET) and high-density polyethylene (HDPE), particularly in the context of plastic bags and bottles. Other forms of plastic pollution such as single-use plastics, microbeads in cosmetics, and plastic contaminants in food and pharmaceuticals have received comparatively little attention.

To address these challenges, it is imperative to introduce and enforce stringent policies at local, national, and international levels. These should include levies, taxes, and regulatory bans on specific plastic products. International treaties must also be strengthened to establish unified frameworks for managing marine debris. A particular focus should be placed on reducing marine litter, given its severe ecological consequences. Innovative solutions such as the use of biosurfactants may offer promising routes for mitigating plastic pollution. Moving forward, a global commitment is needed to safeguard the environment through sustainable practices and continued research. Our shared responsibility is to maintain a clean, healthy, and resilient planet for the benefit of present and future generations [122, 123].

11.15 Conclusion

This chapter primarily aimed to identify the sources of plastic waste generation and assess the extent of pollution in both terrestrial and marine environments. A comprehensive analysis of current plastic waste management strategies has also been presented, with a focus on developing sustainable policies. Plastic waste has become a global concern due to its adverse environmental impacts, including the release of toxic gases and greenhouse emissions, which are major contributors to climate change and global warming. Furthermore, plastic pollutants contaminate groundwater, reduce soil fertility, and significantly damage marine ecosystems and surrounding biodiversity.

Conventional plastic waste management methods—such as landfilling, incineration, and open dumping—not only fail to offer long-term solutions but also exacerbate environmental degradation. These approaches have been critically evaluated in this study, with particular attention given to their shortcomings in terms of recycling efficiency and energy recovery.

Raising public awareness and fostering environmental education are essential for mitigating the plastic pollution crisis. Effective interventions from both governmental and non-governmental organizations are necessary to ensure clean and safe habitats for all living beings. The promotion and adoption of eco-friendly alternatives—such as plant-based packaging, reusable products, compostable polymers, and biodegradable materials—are pivotal in reducing dependency on conventional plastics.

This chapter also examined the regulatory challenges and policy implications associated with plastic waste management, emphasizing the urgent need to reinforce existing laws and expand public engagement across all sectors of society. To achieve a sustainable future, continuous research and innovation must be directed toward the development of environmentally benign alternatives to plastic products. Only through a coordinated, multidimensional approach can the global community hope to effectively curb plastic pollution and protect ecological integrity for future generations.

AI disclosure statement: The author confirms that no artificial intelligence (AI)-generated content or AI-assisted writing tools were used in the preparation of this chapter (including summaries and the abstract). All text, analysis, and conclusions are the result of the author's original work.

References

1. Cimpan, C., Bjelle, E.L., Strømman, A.H., Plastic packaging flows in Europe: A hybrid input-output approach. *J. Ind. Ecol.*, 25, 6, 1572–1587, 2021, https://doi.org/10.1111/jiec.13175.
2. Geyer, R., Jambeck, J.R., Law, K.L., Production, use, and fate of all plastics ever made. *Sci. Adv.*, 3, 7, e1700782, 2017, Doi:10.1126/sciadv.1700782.
3. Seay, J.R., The global plastic waste challenge and how we can address it. *Clean Technol. Environ. Policy*, 24, 3, 729–730, 2022, https://doi.org/10.1007/s10098-021-02271-0.
4. Agovino, M., Cerciello, M., Musella, G., Garofalo, A., European waste management regulations and the transition towards circular economy A shift-and-share analysis. *J. Environ. Manage.*, 354, 120423, 2024, https://doi.org/10.1016/j.jenvman.2024.120423.
5. Campanale, C., Massarelli, C., Savino, I., Locaputo, V., Uricchio, V.F., A detailed review study on potential effects of MPs and additives of concern on human health. *Int. J. Environ. Res. Public Health*, 17, 4, 1212, 2020, Doi:10.3390/ijerph17041212.
6. Khan, S.A., Clearly hazardous, obscurely regulated: lessons from the Basel convention on waste trade, 2020, https://www.cambridge.org/core/journals/american-journal-of-international-law/article/clearly-hazardous-obscurely-regulated-lessons-from-the-basel-convention-on-wastetrade/E250702AE51D72AC342AA811D3003579.
7. Vethaak, A.D. and Legler, J., MPs and human health. *Science*, 371, 6530, 672–674, 2021, DOI: 10.1126/science.abe504.
8. De Silva, L., Doremus, J., Taylor, R., The Plastic Economy, Environmental Defense Fund Economics Discussion Paper Series, EDF EDP, 21-05, 2021, http://dx.doi.org/10.2139/ssrn.3859528.
9. Wiesinger, H., Wang, Z., Hellweg, S., Deep dive into plastic monomers, additives, and processing aids. *Environ. Sci. Technol.*, 55, 13, 9339–9351, 2021, https://doi.org/10.1021/acs.est.1c00976.
10. Dayrit, F.M., Circular plastics economy: Redesigning technology and reimagining society. *Trans. NAST PHL*, 44, 2022, Doi.org/10.57043/transnastphl.2022.2570.
11. Kataki, S., Nityanand, K., Chatterjee, S., Dwivedi, S.K., Kamboj, D.V., Plastic waste management practices pertaining to India with particular focus on

emerging technologies. *Environ. Sci. Pollut. Res. Int.*, 29, 17, 24478–24503, 2022, https://doi.org/10.1007/s11356-021-17974-6.

12. Das, S., Rahman, M.M., Husna, A., Akter, M., Rahaman, M.M., Islam, M.T., Adhikari, A., Respiratory, Neurological and Other Health Outcomes among Plastic Factory Workers in Gazipur, Bangladesh. *Asian J. Med. Health*, 20, 10, 109, 2022, 10.9734/AJMAH/2022/v20i1030510.
13. Almeida, S., Raposo, A., Almeida-González, M., Carrascosa, C., Bisphenol, A., Food exposure and impact on human health. *Compr. Rev. Food Sci. Food Saf.*, 17, 6, 1503–1517, 2018, https://doi.org/10.1111/1541-4337.12388.
14. Vandenberg, L.N., Hauser, R., Marcus, M., Olea, N., Welshons, W.V., Human exposure to bisphenol A (BPA). *Reprod. Toxicol.*, 24, 2, 139–177, 2007, https://doi.org/10.1016/j.reprotox.2007.07.010.
15. Rafey, A. and Siddiqui, F.Z., A review of plastic waste management in India–challenges and opportunities. *Int. J. Environ. Anal. Chem.*, 103, 16, 3971–3987, 2023, https://doi.org/10.1080/03067319.2021.1917560.
16. Garforth, A.A., Ali, S., Hernández-Martínez, J., Akah, A., Feedstock recycling of polymer wastes. *Curr. Opin. Solid State Mater. Sci.*, 8, 6, 419–425, 2004, https://doi.org/10.1016/j.cossms.2005.04.003.
17. Singh, K.D. and Mathur, A., Plastic pollution in India: An evaluation of public awareness and consumption behaviour. *OIDA IJSD*, 12, 07, 25–40, 2019, https://papers.ssrn.com/sol3/papers.cfm?abstract_id=3548729.
18. Kumar, S., Bhattacharyya, J.K., Vaidya, A.N., Chakrabarti, T., Devotta, S., Akolkar, A.B., Assessment of the status of municipal solid waste management in metro cities, state capitals, class I cities, and class II towns in India: An insight. *Waste Manag.*, 29, 2, 883–895, 2009, https://doi.org/10.1016/j.wasman.2008.04.011.
19. Skinner, J.H., International progress in solid waste management. *Environ. Sci. Stud.*, 60, 7–16, Elsevier, 1994, https://doi.org/10.1016/S0166-1116(08)71443-2.
20. Wagner, M., Scherer, C., Alvarez-Muñoz, D., Brennholt, N., Bourrain, X., Buchinger, S., Reifferscheid, G., MPs in freshwater ecosystems: what we know and what we need to know. *Environ. Sci. Eur.*, 26, 1–9, 2014, https://doi.org/10.1186/s12302-014-0012-7.
21. Madhura, R., A Study on Plastic Waste Management by Stakeholders Using Reverse Logistics. *Urban Min. Sustain. Waste Manag.*, 319–328, 2020, https://doi.org/10.1007/978-981-15-0532-4_31.
22. Graham, P.R., Phthalate ester plasticizers—why and how they are used. *Environ. Health Perspect.*, 3, 3–12, 1973, https://ehp.niehs.nih.gov/doi/pdf/10.1289/ehp.73033.
23. Benachour, N. and Aris, A., Toxic effects of low doses of Bisphenol-A on human placental cells. *Toxicol. Appl. Pharmacol.*, 241, 3, 322–328, 2009, https://doi.org/10.1016/j.taap.2009.09.005.

24. Jambeck, J.R., Geyer, R., Wilcox, C., Siegler, T.R., Perryman, M., Andrady, A., Law, K.L., Plastic waste inputs from land into the ocean. *Science*, 347, 6223, 768–771, 2015, DOI: 10.1126/science.1260352.
25. Auta, H.S., Emenike, C.U., Fauziah, S.H., Distribution and importance of MPs in the marine environment: a review of the sources, fate, effects, and potential solutions. *Environ. Int.*, 102, 165–176, 2017, https://doi.org/10.1016/j.envint.2017.02.013.
26. Vince, J. and Hardesty, B.D., Governance solutions to the tragedy of the commons that marine plastics have become. *Front. Mar. Sci.*, 5, 214, 2018, https://doi.org/10.3389/fmars.2018.00214.
27. Schmidt, C., Krauth, T., Wagner, S., Export of plastic debris by rivers into the sea. *Environ. Sci. Technol.*, 51, 21, 12246–12253, 2017, https://doi.org/10.1021/acs.est.7b02368.
28. Desforges, J.P.W., Galbraith, M., Ross, P.S., Ingestion of MPs by zooplankton in the Northeast Pacific Ocean. *Arch. Environ. Contam. Toxicol.*, 69, 3, 3, 2015, https://doi.org/10.1007/s00244-015-0172-5.
29. Fossi, M.C., Panti, C., Guerranti, C., Coppola, D., Giannetti, M., Marsili, L., Minutoli, R., Are baleen whales exposed to the threat of MPs? A case study of the Mediterranean fin whale (Balaenopteraphysalus). *Mar. Pollut. Bull.*, 64, 11, 2374–2379, 2012, https://doi.org/10.1016/j.marpolbul.2012.08.013.
30. Fossi, M.C., Casini, S., Ancora, S., Moscatelli, A., Ausili, A., Notarbartolo-di-Sciara, G., Do endocrine disrupting chemicals threaten Mediterranean swordfish? Preliminary results of vitellogenin and Zonaradiata proteins in Xiphiasgladius. *Mar. Environ. Res.*, 52, 5, 477–483, 2001, https://doi.org/10.1016/S0141-1136(01)00107-6.
31. Bergman, A., Heindel, J.J., Jobling, S., Kidd, K.A., Zoeller, R.T., Endocrine disrupting chemicals-2012, *Env. W.H.O., Geneva*, 2013, https://apps.who.int/iris/bitstream/handle/10665/78101/9789241505031_eng.pdf.
32. Avio, C.G., Gorbi, S., Milan, M., Benedetti, M., Fattorini, D., d'Errico, G., Regoli, F., Pollutants bioavailability and toxicological risk from MPs to marine mussels. *Environ. Pollut.*, 198, 211–222, 2015, https://www.sciencedirect.com/science/article/pii/S0269749114005211.
33. Li, J., Yang, D., Li, L., Jabeen, K., Shi, H., MPs in commercial bivalves from China. *Environ. Pollut.*, 207, 190–195, 2015, https://doi.org/10.1016/j.envpol.2015.09.018Get rights and content.
34. Nerland, I.L., Halsband, C., Allan, I., Thomas, K.V., MPs in marine environments: Occurrence, distribution and effects, 2014, https://www.environmentagency.no/globalassets/publikasjoner/M319/M319.pdf.
35. Johnston, E.L. and Roberts, D.A., Contaminants reduce the richness and evenness of marine communities: a review and meta-analysis. *Environ. Pollut.*, 157, 6, 1745–1752, 2009, https://doi.org/10.1016/j.envpol.2009.02.017.
36. Farsund, A.A., Daugbjerg, C., Langhelle, O., Food security and trade: Reconciling discourses in the Food and Agriculture Organization and

the World Trade Organization. *Food Secur.*, 7, 383–391, 2015, https://doi.org/10.1007/s12571-015-0428-y.
37. The, L.C. and Sumaila, U.R., Contribution of marine fisheries to worldwide employment. *Fish Fish (Oxf.)*, 14, 1, 77–88, 2013, https://doi.org/10.1111/j.1467-2979.2011.00450.x.
38. Yuan, Z., Nag, R., Cummins, E., Human health concerns regarding MPs in the aquatic environment-From marine to food systems. *Sci. Total Environ.*, 823, 153730, 2022, https://doi.org/10.1016/j.scitotenv.2022.153730.
39. Du, F., Cai, H., Zhang, Q., Chen, Q., Shi, H., MPs in take-out food containers. *J. Hazard. Mater.*, 399, 122969, 2020, https://doi.org/10.1016/j.jhazmat.2020.122969.
40. Koelmans, A.A., Nor, N.H.M., Hermsen, E., Kooi, M., Mintenig, S.M., De France, J., MPs in freshwaters and drinking water: Critical review and assessment of data quality. *Water Res.*, 155, 410–422, 2019, https://doi.org/10.1016/j.watres.2019.02.054.
41. Revel, M., Châtel, A., Mouneyrac, C., Micro (nano) plastics: a threat to human health? *Curr. Opin. Environ. Sci. Health*, 1, 17–23, 2018, https://doi.org/10.1016/j.coesh.2017.10.003.
42. Xuyang, W., Khalil, K., Zhuo, T., Keyan, C., Nazir, M.Y.M., Aqma, W.S., Hong, N.Q., Addressing the plastisphere: Sustainable approaches to combat plastic pollution. *Sci. Total Environ.*, 959, 178105, 2025, https://doi.org/10.1016/j.scitotenv.2024.178105.
43. Kelly, A., Lannuzel, D., Rodemann, T., Meiners, K.M., Auman, H.J., MP contamination in east Antarctic sea ice. *Mar. Pollut. Bull.*, 154, 111130, 2020, https://doi.org/10.1016/j.marpolbul.2020.111130.
44. Troost, T.A., Desclaux, T., Leslie, H.A., van Der Meulen, M.D., Vethaak, A.D., Do MPs affect marine ecosystem productivity? *Mar. Pollut. Bull.*, 135, 17–29, 2018, https://doi.org/10.1016/j.marpolbul.2018.05.067.
45. Beaumont, N.J., Aanesen, M., Austen, M.C., Börger, T., Clark, J.R., Cole, M., Wyles, K.J., Global ecological, social and economic impacts of marine plastic. *Mar. Pollut. Bull.*, 142, 189–195, 2019, https://doi.org/10.1016/j.marpolbul.2019.03.022.
46. Shen, M., Huang, W., Chen, M., Song, B., Zeng, G., Zhang, Y., (Micro) plastic crisis: un-ignorable contribution to global greenhouse gas emissions and climate change. *J. Clean. Prod.*, 254, 120138, 2020, https://doi.org/10.1016/j.jclepro.2020.120138.
47. Lachs, L. and Oñate-Casado, J., Fisheries and tourism: Social, economic, and ecological trade-offs in coral reef systems, in: *YOUMARES 9-the oceans: Our research, our future: Proceedings of the 2018 conference for YOUng Marine RESearcher in Oldenburg, Germany*, Springer International Publishing, pp. 243–260, 2020, https://library.oapen.org/bitstream/handle/20.500.12657/22870/1007291.pdf?sequence=1#page=253.
48. Zgheib, P.W., Ahmed, Z.U., Beldona, S., Gebara, V., The impact of population growth on Human Development Index: a comparative analysis of Middle

Eastern countries. *WRSTSD*, 3, 3, 258–269, 2006, https://doi.org/10.1504/WRSTSD.2006.010226.
49. Meijer, L.J., Van Emmerik, T., Van Der Ent, R., Schmidt, C., Lebreton, L., More than 1000 rivers account for 80% of global riverine plastic emissions into the ocean. *Sci. Adv.*, 7, 18, eaaz5803, 2021, DOI: 10.1126/sciadv.aaz5803.
50. Gomez, M., Perdiguero, J., Sanz, A., Socioeconomic factors affecting water access in rural areas of low and middle income countries. *Water*, 11, 2, 202, 2019, https://doi.org/10.3390/w11020202.
51. Mihai, F.C., Gündoğdu, S., Markley, L.A., Olivelli, A., Khan, F.R., Gwinnett, C., Molinos-Senante, M., Plastic pollution, waste management issues, and circular economy opportunities in rural communities. *Sustainability*, 14, 1, 20, 2021, https://doi.org/10.3390/su14010020.
52. Pivokonsky, M., Cermakova, L., Novotna, K., Peer, P., Cajthaml, T., Janda, V., Occurrence of MPs in raw and treated drinking water. *Sci. Total Environ.*, 643, 1644–1651, 2018, https://doi.org/10.1016/j.scitotenv.2018.08.102.
53. Treilles, R., Gasperi, J., Saad, M., Tramoy, R., Breton, J., Rabier, A., Tassin, B., Abundance, composition and fluxes of plastic debris and other macrolitter in urban runoff in a suburban catchment of Greater Paris. *Water Res.*, 192, 116847, 2021, https://doi.org/10.1016/j.watres.2021.116847.
54. He, D., Luo, Y., Lu, S., Liu, M., Song, Y., Lei, L., MPs in soils: Analytical methods, pollution characteristics and ecological risks. *TrAC, Trends Anal. Chem.*, 109, 163–172, 2018, https://doi.org/10.1016/j.trac.2018.10.006.
55. Akhtar, S., Food safety challenges—a Pakistan's perspective. *Crit. Rev. Food Sci. Nutr.*, 55, 2, 219–226, 2015, https://doi.org/10.1080/10408398.2011.650801.
56. Bläsing, M. and Amelung, W., Plastics in soil: Analytical methods and possible sources. *Sci. Total Environ.*, 612, 422–435, 2018, https://doi.org/10.1016/j.scitotenv.2017.08.086.
57. Dris, R., Gasperi, J., Saad, M., Mirande, C., Tassin, B., Synthetic fibers in atmospheric fallout: a source of MPs in the environment? *Mar. Pollut. Bull.*, 104, 1–2, 290–293, 2016, https://doi.org/10.1016/j.marpolbul.2016.01.006.
58. Ohtake, Y., Kobayashi, T., Asabe, H., Murakami, N., Studies on biodegradation of LDPE—observation of LDPE films scattered in agricultural fields or in garden soil. *Polym. Degrad. Stabil.*, 60, 1, 79–84, 1998, https://doi.org/10.1016/S0141-3910(97)00032-3.
59. Steinmetz, Z., Wollmann, C., Schaefer, M., Buchmann, C., David, J., Tröger, J., Schaumann, G.E., Plastic mulching in agriculture. Trading short-term agronomic benefits for long-term soil degradation? *Sci. Total Environ.*, 550, 690–705, 2016, https://doi.org/10.1016/j.scitotenv.2016.01.153.
60. Ngwome, G.F., Plastic waste pollution, environmental impacts and regulatory challenges: The Cameroonian example. *Envtl. L. Rev.*, 24, 2, 93–110, 2022, https://doi.org/10.1177/14614529221094254.
61. Azeez, T.O., Thermoplastic Recycling: Properties, Modifications, and. *Thermosof. Plast.*, 53, 2020, DOI: 10.5772/intechopen.81614.

62. Wang, P. and Altanbulag, A., A concern for eco-social sustainability: Background, concept, values, and perspectives of eco-social work. *Cogent Soc. Sci.*, 8, 1, 2093035, 2022, https://doi.org/10.1080/23311886.2022.2093035.
63. Mesoe, E.M., The complexity of agency and solidarity in community development: Household plastic waste management in the city of Buea, Cameroon, Master's thesis, VID Specialized University. Oslo, 2023, https://ec.europa.eu/eurostat/web/products-eurostat-news/-/EEDN-20180422-1.
64. Godfrey, L., Ahmed, M.T., Gebremedhin, K.G., Katima, J.H., Oelofse, S., Osibanjo, O., Richter, U.H., Solid Waste Management in. Reg. Dev. Afr, vol. 235, 2020, DOI: 10.5772/intechopen.86974.
65. Delattre, C., Forissier, M., Pitault, I., Schweich, D., Bernard, J.R., Improvement of the microactivity test for kinetic and deactivation studies involved in catalytic cracking. *Chem. Eng. Sci.*, 56, 4, 1337–1345, 2001, https://www.sciencedirect.com/science/article/pii/S0009250900003560.
66. Kassab, A., Al Nabhani, D., Mohanty, P., Pannier, C., Ayoub, G.Y., Advancing plastic recycling: Challenges and opportunities in the integration of 3D printing and distributed recycling for a circular economy. *Polym. Sci.*, 15, 19, 3881, 2023, https://doi.org/10.3390/polym15193881.
67. Balwada, J., Samaiya, S., Mishra, R.P., Packaging plastic waste management for a circular economy and identifying a better waste collection system using analytical hierarchy process (AHP). *Procedia CIRP*, 98, 270–275, 2021, https://doi.org/10.1016/j.procir.2021.01.102.
68. Lino, F.A. and Ismail, K.A., Incineration and recycling for MSW treatment: Case study of Campinas, Brazil. *Sustain. Cities Soc.*, 35, 752–757, 2017, https://doi.org/10.1016/j.scs.2017.09.028.
69. Ruj, B., Pandey, V., Jash, P., Srivastava, V.K., Sorting of plastic waste for effective recycling. *Int. J. Appl. Sci. Eng. Res.*, 4, 4, 564–571, 2015, DOI: 10.6088.ijaser.04058.
70. Al-Salem, S.M., Lettieri, P., Baeyens, J., Recycling and recovery routes of plastic solid waste (PSW): A review. *Waste Manage. (Oxford)*, 29, 10, 2625–2643, 2009, https://doi.org/10.1016/j.wasman.2009.06.004.
71. Jimoh, K., Ajayi, A., Ayilara, O., Intelligent model for manual sorting of plastic wastes. *Int. J. Comput. Appl.*, 101, 7, 20–26, 2014, https://www.academia.edu/download/95119784/pxc3898675.pdf.
72. Siwal, S.S., Chaudhary, G., Saini, A.K., Kaur, H., Saini, V., Mokhta, S.K., Thakur, V.K., Key ingredients and recycling strategy of personal protective equipment (PPE): Towards sustainable solution for the COVID-19 like pandemics. *J. Environ. Chem. Eng.*, 9, 5, 106284, 2021, https://doi.org/10.1016/j.jece.2021.106284.
73. Mousavi, M., Kowsari, E., Gheibi, M., Cheshmeh, Z.A., Teymoorian, T., Ramakrishna, S., Assessing bioplastics' economic, commercial, political, and energy potential with circular economy modeling: a sustainable solution to plastic waste management. *Mater. Circ. Econ.*, 6, 1, 6, 2024, https://doi.org/10.1007/s42824-023-00098-2.

74. Worrell, E. and Reuter, M.A. (Eds.), *Handbook of Recycling: State-of-the-art for Practitioners, Analysts, and Scientists*, Newnes, Elsevier, Oxford, United Kingdom, 2014.
75. Francis, R. (Ed.), *Recycling of Polymers: Methods, Characterization and Applications*, John Wiley & Sons, Hoboken, NJ, USA, 2016.
76. Singh, N., Hui, D., Singh, R., Ahuja, I.P.S., Feo, L., Fraternali, F., Recycling of plastic solid waste: A state of art review and future applications. *Compos. Part B Eng.*, 115, 409–422, 2017, https://doi.org/10.1016/j.compositesb.2016.09.013.
77. Li, H., Aguirre-Villegas, H.A., Allen, R.D., Bai, X., Benson, C.H., Beckham, G.T., Huber, G.W., Expanding plastics recycling technologies: chemical aspects, technology status and challenges. *Green Chem.*, 24, 23, 8899–9002, 2022, DOI: 10.1039/D2GC02588D.
78. Lee, A. and Liew, M.S., Tertiary recycling of plastics waste: an analysis of feedstock, chemical and biological degradation methods. *J. Mater. Cycles Waste Manag.*, 23, 1, 32–43, 2021, https://doi.org/10.1007/s10163-020-01106-2.
79. Kibria, M.G., Masuk, N.I., Safayet, R., Nguyen, H.Q., Mourshed, M., Plastic waste: challenges and opportunities to mitigate pollution and effective management. *Int. J. Environ. Res.*, 17, 1, 20, 2023, https://doi.org/10.1007/s41742-023-00507-z.
80. Verma, S., Balasubramaniam, B., Gupta, R.K., Recycling, reclamation and re-manufacturing of carbon fibres. *Curr. Opin. Green Sustain. Chem.*, 13, 86–90, 2018, https://doi.org/10.1016/j.cogsc.2018.05.011.
81. Tominaga, A., Sekiguchi, H., Nakano, R., Yao, S., Takatori, E., Thermal process-dependence of the mechanical properties and inner structures of pre-consumer recycled polypropylene, in: *AIP Conference Proceedings*, vol. 1664, no. 1, AIP Publishing, 2015, https://doi.org/10.1063/1.4918507.
82. Ragaert, K., Delva, L., Van Geem, K., Mechanical and chemical recycling of solid plastic waste. *Waste Manag.*, 69, 24–58, 2017, https://doi.org/10.1016/j.wasman.2017.07.044.
83. Berms, A., Dewil, R., Baeyens, J., Zhang, R., Gasification of plastic waste as waste-to-energy or waste-to-syngas recovery route. *Nat. Sci.*, 5, 6, 695–704, 2013, https://api.taylorfrancis.com/content/chapters/edit/download?identifierName=doi&identifierValue=10.1201/b18712-23&type=chapterpdf.
84. Okan, M., Aydin, H.M., Barsbay, M., Current approaches to waste polymer utilization and minimization: a review. *J. Chem. Technol. Biotechnol.*, 94, 1, 8–21, 2019, https://doi.org/10.1002/jctb.5778.
85. Zhang, F., Zhao, Y., Wang, D., Yan, M., Zhang, J., Zhang, P., Chen, C., Current technologies for plastic waste treatment: A review. *J. Clean. Prod.*, 282, 124523, 2021, https://doi.org/10.1016/j.jclepro.2020.124523.
86. Chen, D., Yin, L., Wang, H., He, P., Pyrolysis technologies for municipal solid waste: a review. *Waste Manag.*, 34, 12, 2466–2486, 2014, https://doi.org/10.1016/j.wasman.2014.08.004.

87. Wong, S.L., Ngadi, N., Abdullah, T.A.T., Inuwa, I.M., Current state and future prospects of plastic waste as source of fuel: A review. *Renew. Sustain. Energy Rev.*, 50, 1167–1180, 2015, https://doi.org/10.1016/j.rser.2015.04.063.
88. Al-Salem, S.M., Antelava, A., Constantinou, A., Manos, G., Dutta, A., A review on thermal and catalytic pyrolysis of plastic solid waste (PSW). *J. Environ. Manage.*, 197, 177–198, 2017, https://doi.org/10.1016/j.jenvman.2017.03.084.
89. Munir, D., Irfan, M.F., Usman, M.R., Hydrocracking of virgin and waste plastics: A detailed review. *Renew. Sustain. Energy Rev.*, 90, 490–515, 2018, https://doi.org/10.1016/j.rser.2018.03.034.
90. Fuentes-Ordonez, E.G., Salbidegoitia, J.A., Gonzalez-Marcos, M.P., Gonzalez-Velasco, J.R., Transport phenomena in catalytic hydrocracking of polystyrene in solution. *Ind. Eng. Chem. Res.*, 52, 42, 14798–14807, 2013, https://doi.org/10.1021/ie401968r.
91. Krapivin, V.F., Varotsos, C.A., Soldatov, V.Y., Simulation results from a coupled model of carbon dioxide and methane global cycles. *Ecol. Model.*, 359, 69–79, 2017, https://doi.org/10.1016/j.ecolmodel.2017.05.023.
92. Payne, J., McKeown, P., Jones, M.D., A circular economy approach to plastic waste. *Polym. Degrad. Stabil.*, 165, 170–181, 2019, https://doi.org/10.1016/j.polymdegradstab.2019.05.014.
93. Panda, A.K., Singh, R.K., Mishra, D.K., Thermolysis of waste plastics to liquid fuel: A suitable method for plastic waste management and manufacture of value added products—A world prospective. *Renew. Sustain. Energy Rev.*, 14, 1, 233–248, 2010, https://doi.org/10.1016/j.rser.2009.07.005.
94. Kumar, S., Panda, A.K., Singh, R.K., A review on tertiary recycling of high-density polyethylene to fuel. *Resour. Conserv. Recycl.*, 55, 11, 893–910, 2011, https://doi.org/10.1016/j.resconrec.2011.05.005.
95. Pardal, F. and Tersac, G., Comparative reactivity of glycols in PET glycolysis. *Polym. Degrad. Stabil.*, 91, 11, 2567–2578, 2006, https://doi.org/10.1016/j.polymdegradstab.2006.05.016.
96. Hu, B., Wang, S., Yan, J., Zhang, H., Qiu, L., Liu, W., Ge, X., Review of waste plastics treatment and utilization: Efficient conversion and high value utilization. *Process Saf. Environ. Prot.*, 183, 378–398, 2024, https://doi.org/10.1016/j.psep.2024.01.024.
97. Raheem, A.B., Noor, Z.Z., Hassan, A., Abd Hamid, M.K., Samsudin, S.A., Sabeen, A.H., Current developments in chemical recycling of post-consumer polyethylene terephthalate wastes for new materials production: A review. *J. Clean. Prod.*, 225, 1052–1064, 2019, https://doi.org/10.1016/j.jclepro.2019.04.019.
98. Genta, M., Iwaya, T., Sasaki, M., Goto, M., Hirose, T., Depolymerization mechanism of poly (ethylene terephthalate) in supercritical methanol. *Ind. Eng. Chem. Res.*, 44, 11, 3894–3900, 2005, https://doi.org/10.1021/ie0488187.
99. Urase, T., Okumura, H., Panyosaranya, S., Inamura, A., Emission of volatile organic compounds from solid waste disposal sites and importance of

heat management. *Waste Manag. Res.*, 26, 6, 534–538, 2008, https://doi.org/10.1177/0734242X07084321.

100. Xu, S.Y., Zhang, H., He, P.J., Shao, L.M., Leaching behaviour of bisphenol A from municipal solid waste under landfill environment. *Environ. Technol.*, 32, 11, 1269–1277, 2011, https://doi.org/10.1080/09593330.2010.535175.
101. Astrup, T., Møller, J., Fruergaard, T., Incineration and co-combustion of waste: accounting of greenhouse gases and global warming contributions. *Waste Manag. Res.*, 27, 8, 789–799, 2009, https://doi.org/10.1177/0734242X09343774.
102. Shen, C., Tang, X., Yao, J., Shi, D., Fang, J., Khan, M.I., Chen, Y., Levels and patterns of polycyclic aromatic hydrocarbons and polychlorinated biphenyls in municipal waste incinerator bottom ash in Zhejiang province, China. *J. Hazard. Mater.*, 179, 1–3, 197–202, 2010, https://doi.org/10.1016/j.jhazmat.2010.02.079.
103. Muaddi, H., El Hafid, M., Choi, W.J., Lillie, E., de Mestral, C., Nathens, A., Karanicolas, P.J., Clinical outcomes of robotic surgery compared to conventional surgical approaches (laparoscopic or open): a systematic overview of reviews. *Ann. Surg.*, 273, 3, 467–473, 2021, DOI :10.1097/SLA.0000000000003915.
104. Conway, C. and Gaynullin, A., Research of alternative methods of marine litter collection: Study of marine plastic pollution, 2022, https://www.diva-portal.org/smash/record.jsf?pid=diva2:1707714.
105. Frey, A.R., Mahanta, R., Maharana, A., Quevedo, F., Villa, G., Gravitational waves from high temperature strings. *J. High Energy Phys.*, 2024, 12, 1–41, 2024, https://doi.org/10.1007/JHEP12(2024)174.
106. Sun, M., Chen, W., Fan, X., Tian, C., Sun, L., Xie, H., Cooperative recyclable magnetic microsubmarines for oil and MPs removal from water. *Appl. Mater. Today*, 20, 100682, 2020, https://doi.org/10.1016/j.apmt.2020.100682.
107. Velikov, D.I., Jancik-Prochazkova, A., Pumera, M., On-the-fly monitoring of the capture and removal of NPs with nanorobots. *ACS Nanosci. Au*, 4, 4, 243–249, 2024, https://doi.org/10.1021/acsnanoscienceau.4c00002.
108. Li, W., Wu, C., Xiong, Z., Liang, C., Li, Z., Liu, B., Li, D., Self-driven magnetorobots for recyclable and scalable micro/NP removal from nonmarine waters. *Sci. Adv.*, 8, 45, eade1731, 2022, DOI: 10.1126/sciadv.ade1731.
109. Ali, S., Khan, M.H., Zuhra, Z., Wang, J., Innovative materials that behave like robots to combat plastic pollution. *Mater. Horiz.*, 2025, https://doi.org/10.1039/D4MH01772B.
110. Acquavia, M.A., Pascale, R., Martelli, G., Bondoni, M., Bianco, G., Natural polymeric materials: A solution to plastic pollution from the agro-food sector. *Polym. Sci.*, 13, 1, 158, 2021, https://doi.org/10.3390/polym13010158.
111. Moshood, T.D., Nawanir, G., Mahmud, F., Mohamad, F., Ahmad, M.H., AbdulGhani, A., Biodegradable plastic applications towards sustainability: A recent innovations in the green product. *Clean. Eng. Technol.*, 6, 100404, 2022, https://doi.org/10.1016/j.clet.2022.100404.

112. Nizamuddin, S. and Chen, C., Biobased, biodegradable and compostable plastics: chemical nature, biodegradation pathways and environmental strategy. *Environ. Sci. Pollut. Res.*, 31, 6, 8387–8399, 2024, https://doi.org/10.1007/s11356-023-31689-w.
113. Miao, W. and Bai, H., The new material science towards sustainable robotics. *J. Mater. Chem. C*, 12, 33, 12721–12733, 2024, DOI: 10.1039/D4TC01868K.
114. Ali, M.Q., Azhar, M.A., Munaim, M.S.A., Ruslan, N.F., Alsubhi, L.M., Ahmad, N., Noman, A.E., Seaweed organic compounds source of hydrocolloids and sustainable food packaging: properties, application, and future direction. *Discov. Food*, 4, 1, 101, 2024, https://doi.org/10.1007/s44187-024-00173-w.
115. Mazhandu, Z.S., Muzenda, E., Mamvura, T.A., Belaid, M., Nhubu, T., Integrated and consolidated review of plastic waste management and bio-based biodegradable plastics: Challenges and opportunities. *Sustainability*, 12, 20, 8360, 2020, https://doi.org/10.3390/su12208360.
116. Miditana, S.R., Dokka, W.T., Babu, A.R., Ampolu, S., Agarwal, N., Simma, N., Cyclodextrins for UV protection in food, beverages, and agricultural products: A comprehensive review. *N. Afr. J. Food Nutr. Res.*, 9, 20, 219–235, 2025, https://doi.org/10.51745/najfnr.9.20.219-235.
117. Alpizar, F., Carlsson, F., Lanza, G., Carney, B., Daniels, R.C., Jaime, M., Wahdera, S., A framework for selecting and designing policies to reduce marine plastic pollution in developing countries. *Environ. Sci. Policy*, 109, 25–35, 2020, https://doi.org/10.1016/j.envsci.2020.04.007.
118. Koelmans, A.A., Besseling, E., Foekema, E., Kooi, M., Mintenig, S., Ossendorp, B.C., Scheffer, M., Risks of plastic debris: unravelling fact, opinion, perception, and belief, 2017, https://doi.org/10.1021/acs.est.7b02219.
119. Fadare, O.O., Wan, B., Guo, L.H., Zhao, L., MPs from consumer plastic food containers: are we consuming it? *Chemosphere*, 253, 126787, 2020, https://doi.org/10.1016/j.chemosphere.2020.126787.
120. Miditana, S.R., Tirukkovalluri, S.R., Alim, S.A., Raju, I.M., Photocatalytic degradation of Allura Red by Mn–Ni co-doped nanotitania under visible light irradiation. *Int. J. Innov. Technol. Explor. Eng.*, 8, 12, 2019, https://doi.org/10.35940/ijitee.K2458.1081219.
121. Princiotta, F.T. and Loughlin, D.H., Global climate change: The quantifiable sustainability challenge. *J. Air Waste Manage.*, 64, 9, 979–994, 2014, https://doi.org/10.1080/10962247.2014.923351.
122. Reid, A.J., Carlson, A.K., Creed, I.F., Eliason, E.J., Gell, P.A., Johnson, P.T., Cooke, S.J., Emerging threats and persistent conservation challenges for freshwater biodiversity. *Biol. Rev.*, 94, 3, 849–873, 2019, https://doi.org/10.1111/brv.12480.
123. Yadav, P. and Mishra, V., Comprehending MP pollution in diverse environment: Assessing fate, impacts, and remediation approaches. *Int. Biodeterior. Biodegradation.*, 196, 105953, 2025, https://doi.org/10.1016/j.ibiod.2024.105953.

12

Regulatory Approaches to Plastic Mitigation: Policies for Production, Use, and Remediation

Amrit Krishna Mitra[1*] **and Rama Saha**[2]

[1]*Department of Chemistry, Government General Degree College, Singur, West Bengal, India*

[2]*Department of Chemistry, Rani Rashmoni Green University, Tarakeswar, West Bengal, India*

Abstract

Originally hailed for their versatility and durability, plastics have become one of the most pressing environmental concerns of our time. Their widespread accumulation particularly in marine ecosystems has led to serious implications for biodiversity, ecosystem functioning, and human health. This chapter delves into the full lifecycle of plastic materials, focusing on the growing threat posed by microplastics (MPs), which are now considered a potential planetary boundary issue. It provides a critical overview of evolving regulatory measures across local, national, and global levels aimed at reducing plastic production, limiting consumption, and improving post-consumer waste management. Special attention is given to policies targeting single-use plastics and emerging strategies to control MP contamination. By identifying existing gaps in implementation and enforcement, the chapter underscores the need for integrated and context-sensitive policy solutions. It calls for collaborative global efforts and sustained legislative action to reduce plastic pollution and protect environmental and public health.

Keywords: Plastic pollution, MPs, environmental regulation, single-use plastics, waste management policies, sustainable mitigation strategies

**Corresponding author*: ambrosia12june@gmail.com

Neha Agarwal, Brijesh Pare, Vijendra Singh Solanki and Sankara Rao Miditana (eds.) Plastic Remediation: Transformative Pathways for a Rejuvenated Sustainable Environment, (317–358)

12.1 Introduction

Hailed as one of the most transformative inventions of the 20th century, plastics have revolutionized nearly every aspect of human life from consumer goods and electronics to medical devices and food packaging. However, this convenience has come at a steep environmental cost. Today, plastics are ubiquitous in terrestrial environments and dominate marine debris across oceans worldwide. Among the most commonly found types in marine ecosystems are polyethylene (PE), polypropylene (PP), and polystyrene (PS) [1]. The accumulation of plastic waste in the ocean is often driven by large-scale ocean currents and gyres, where discarded plastics are trapped and broken down into smaller fragments. These plastic particles exist in diverse forms and sizes, posing threats to human health, marine biodiversity, and even the global economy causing losses in tourism, damage to fishing equipment, and disruptions in coastal livelihoods [2, 3].

Plastic pollution is now recognized as a serious planetary boundary threat. Its impacts are considered irreversible due to the persistent nature of most plastic materials, their resistance to degradation, and the impracticality of recovering all the dispersed waste. At a planetary scale, plastics disrupt Earth system processes by altering the physicochemical environment and harming ecological networks [4]. The entire life cycle of plastics from production to disposal contributes to biodiversity loss and climate change, compounding the urgency of effective regulatory interventions.

Despite the material benefits plastics offer, improper disposal and inefficient waste management practices have exacerbated their global footprint. This has drawn increasing scrutiny from stakeholders, researchers, and policymakers alike, as the ecological and socioeconomic ramifications continue to unfold [5]. One of the challenges in managing plastic pollution lies in its wide variability. Plastics are classified based on size: macroplastics (>2.5 cm), mesoplastics (5 mm–20 mm), MPs (1 micron–5 mm), and nanoplastics (<1 micron) [6, 7]. MPs, in particular, are of growing concern due to their ability to infiltrate food chains and natural systems. A widely accepted 2019 definition describes MPs as "any synthetic solid particle or polymeric matrix, with a regular or irregular shape, ranging in size from 1 micron to 5 mm, of either primary or secondary origin, and insoluble in water." Primary MPs are manufactured intentionally at small sizes, while secondary MPs result from the breakdown of larger plastic items through environmental exposure, mechanical stress, or microbial activity [8, 9].

Alarmingly, it is estimated that at least 3% of mismanaged plastic waste eventually degrades into secondary MPs [10].

Given this complexity, mitigation strategies must address both macroplastics and MPs. While community clean-up drives, citizen science campaigns, and awareness initiatives have shown promise in reducing localized plastic loads, these approaches alone are insufficient [11]. Long term, systemic solutions are needed to address the root causes of plastic pollution. A key area of focus in current regulatory frameworks is the restriction or phasing out of single-use plastics (SUPs). Governments across the globe are now introducing new policies and alternative material strategies aimed at curbing plastic production and consumption. However, the success of these efforts depends on robust implementation, enforcement, and adaptability to local contexts [12].

This chapter provides a comprehensive overview of the evolving regulatory landscape surrounding plastic mitigation. It explores the major policy strategies implemented at national and international levels, highlights existing knowledge gaps, and offers a timeline of key governmental actions. In doing so, it aims to offer a foundational perspective for both scientific and policy-driven approaches to addressing the global plastic crisis.

12.2 Origin of Plastic Waste

Plastics have become inseparable from modern life. Their versatility, durability, and low cost have made them indispensable across industries. However, the very qualities that make plastics so useful have also led to a mounting environmental crisis. As global plastic production has skyrocketed over the decades, the world now struggles to manage the massive volumes of plastic waste it generates.

From a modest 2 million tons (MT) in 1950, global plastic production surged to over 454 MT by 2018 largely derived from fossil fuels. The global source-wise plastic waste generation snapshot is displayed in Figure 12.1. According to estimates, around 9.7 billion tons of plastic was produced between 1950 and 1980 alone, and this figure is expected to double by 2025 and triple by 2050 [13]. The global annual generation of plastic waste is projected to reach 343 MT [14], with major contributions coming from different regions: 35 MT from North America, 45 MT from Europe and Central Asia, and 57 MT from East Asia and the Pacific [14].

A major portion of this waste originates from household sources, often referred to as residential waste. This includes common items such as shopping bags, food packaging, printed paper, diapers, and organic waste.

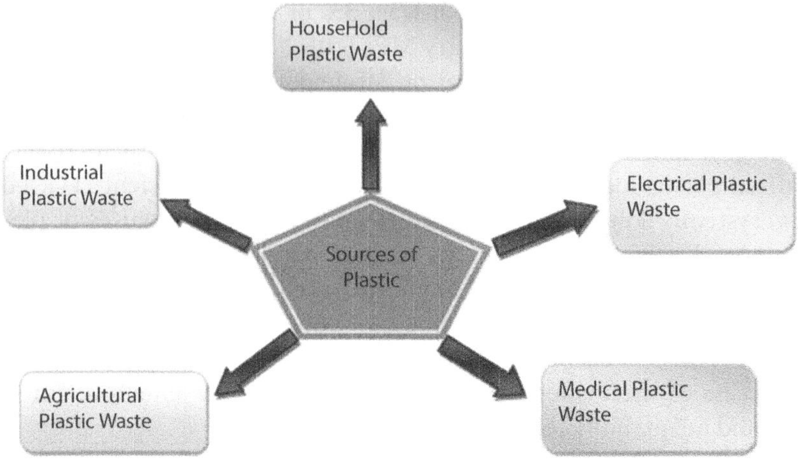

Figure 12.1 Source of different types of plastic wastes.

While some components like paper and food residues are non-hazardous, items such as used batteries, electronics, plastics, and medical waste are classified as hazardous.

Another significant source is industrial plastic waste, generated from manufacturing, packaging, and processing sectors. Industries such as automotive, electronics, construction, and demolition contribute to this stream. The positive aspect is that industrial plastic waste often remains uncontaminated and is available in large quantities offering opportunities for efficient recycling, provided robust regulatory frameworks are in place.

In the agriculture sector, plastics are used extensively for crop protection, mulching, irrigation systems, silage covers, seed packaging, and more. These materials play a critical role in enhancing productivity and crop security. However, they also result in substantial agricultural plastic waste (APW). An estimated 6.5 MT of agricultural plastics are used globally each year [15], most of which are burned, buried, or abandoned in fields, eventually making their way into waterways and causing widespread environmental harm.

Medical plastic waste (MPW) has also emerged as a pressing concern, particularly with the expansion of healthcare services and modern medical technologies. Items like syringes, PPE kits, gloves, and medication packaging are indispensable but disposable. In 2018, around 817,000 tons of MPW were generated by over 200 medium and large cities in China alone [16]. The COVID-19 pandemic significantly exacerbated this issue. In 2020, more than 200 million medical masks were distributed globally,

89 million in the U.S. and 24.37 million in the UK alone [17]. Bangladesh reported over 14,000 tons of plastic waste in a single month during the pandemic, and in countries like Sri Lanka, limited waste infrastructure led to increased burning of plastic waste to reduce contamination risks, thereby worsening air pollution and soil degradation [18].

These various sources, residential, industrial, agricultural, and medical, highlight the multifaceted nature of plastic waste generation. Each stream presents unique challenges and requires tailored strategies for reduction, management, and eventual remediation.

12.3 Impacts of Plastic Waste

Plastic waste has far-reaching consequences on environmental and human health. Typically, plastics discarded from households, commercial establishments, and workplaces are collected and sent to secondary transfer stations. Here, basic sorting is carried out. Depending on the local infrastructure and available technologies, plastics may then be recycled, reused, incinerated in open spaces, or ultimately dumped in landfills.

Once plastic waste infiltrates the soil, it begins to alter the physical and chemical properties of the land. Plastics affect soil aggregation, porosity, and water retention, ultimately disrupting natural soil dynamics. MPs due to their tiny size and high dispersion ability mix readily with the soil matrix and reduce soil fertility. They often carry hazardous substances such as polybrominated diphenyl ethers (PBDEs); perfluorinated compounds (PFOS); and heavy metals like copper (Cu), zinc (Zn), and lead (Pb) [19, 20]. Moreover, plastics hinder the activity of essential soil organisms particularly earthworms and beneficial microbes thereby contributing to the degradation of soil health. Interestingly, while MPs may stimulate the activity of certain soil enzymes, this is generally an early stress response and not indicative of improved soil quality.

In open dumping sites, environmental conditions such as heat, humidity, and pressure contribute to the fragmentation of plastics into MPs, which are easily carried by wind, water, and runoff into surrounding ecosystems. These MPs find their way into our environment contaminating soil, surface water, and even indoor air. Humans are exposed to them *via* inhalation of airborne particles, consumption of contaminated food, or drinking water. Studies show that children and adults may ingest approximately 3223 and 1063 MP particles, respectively, from indoor dust alone [21]. Figure 12.2 illustrates the primary demographic pathways of human exposure to MPs.

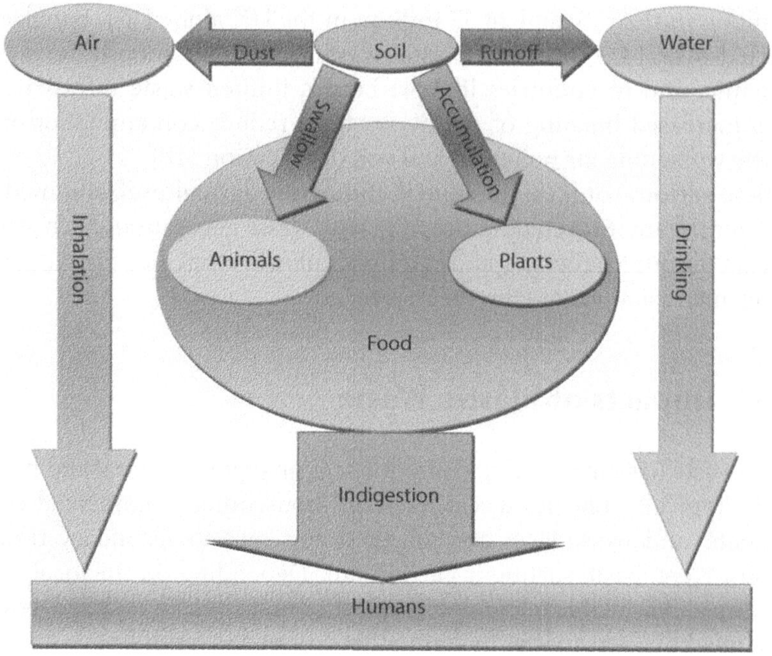

Figure 12.2 MP entering the human body through different routes.

The problem doesn't stop on land. According to the United Nations Environment Programme (UNEP), between 10 and 20 million tons (MT) of plastic waste enters aquatic ecosystems annually, adding to the estimated 150 MT already present in our oceans. The consequences for marine biodiversity and ecological balance are profound. Floating plastic debris harms marine animals through ingestion and entanglement and disrupts the food chain at multiple levels. Microorganisms like zooplankton and copepods, which help absorb around 30% of atmospheric CO_2, are especially vulnerable to plastic pollution [22]. Protecting these microscopic organisms is essential in the broader fight against climate change.

Most marine plastic pollution originates from land-based activities, particularly in densely populated coastal regions [23]. Rivers, tides, and shoreline activity act as primary pathways for transporting plastic waste into the oceans. In Asia, where several major rivers serve as lifelines for billions, high population density amplifies the flow of plastic into marine environments. These rivers have now become critical contributors to global marine plastic pollution [24].

The cumulative impact of plastic waste across terrestrial, freshwater, and marine ecosystems is alarming. From degrading soil fertility to threatening human health and destroying marine biodiversity, the consequences demand immediate and sustained regulatory action.

12.4 Improving the Disposal of Waste

Effective plastic mitigation relies heavily on a robust waste management framework, often summarized by the four Rs: reduce, reuse, recycle, and recover. These principles form the foundation of integrated waste management (IWM), which views waste not as a burden, but as a resource. While reducing production and reusing materials remain the ultimate goals, practical challenges particularly in reusing packaging mean that some items must go through recovery processes such as sorting and refilling. These practices are typically feasible only for high-value goods like automobiles and electronics [25].

When waste is unavoidable, the priority should be recycling. If recycling is not possible due to material contamination or degradation, the next best option is energy recovery or conversion into feedstock. Only the residual waste such as incineration ash should end up in landfills. Adopting a well-structured solid waste management system not only reduces visible plastic pollution but also limits the formation of MPs, which result from the gradual breakdown of plastics in the environment [26].

12.4.1 Policy-Driven Success Stories

Global examples demonstrate how strategic policies can improve waste disposal outcomes. In Taiwan, for instance, a suite of regulatory measures such as the Plastic Restriction Policy, the Recycling Act, and the Compulsory Trash-Sorting Policy significantly lowered waste generation from 0.9 to 0.48 kg per capita and reduced the presence of plastic litter on beaches. These policies included bans on plastic bags and disposable tableware and mandated household-level waste sorting.

Similarly, Australian local councils have allocated substantial budgets toward beach clean-up and waste management, showing the value of investment in local action. On a global scale, the Basel Convention, now ratified by over 180 countries, places restrictions on the transboundary movement of mixed plastic waste. This has forced many nations to develop domestic recycling and waste processing solutions, rather than depending on exports [27].

12.4.2 Collection and Sorting Strategies

Waste collection begins either at the source (households and businesses) or through post-consumer sorting at dedicated centers. Source collection is often more economical and helps reduce contamination, making downstream recycling processes more efficient. Collection methods vary depending on local infrastructure and include the following:

- Door-to-door collection, with or without user fees
- Curbside pickup
- Drop-off centers (without monetary exchange)
- Buy-back centers, which incentivize returns through small payments [28, 29]

Setting the right fee structures for collection is essential. If fees are too low, they may not incentivize waste reduction. If they are too high, they may lead to illegal dumping or burning. In this context, buy-back schemes have proven particularly effective. For example, container deposit legislation in the United States and Australia has significantly decreased the number of beverage containers found along coastlines [30]. Additionally, deposit-refund schemes and tax incentives are increasingly being used to boost recycling rates by offering positive reinforcement.

12.4.3 The Plastic Recycling Process

Recycling plastics is a complex, multi-step process that requires meticulous separation and processing:

1. Collection of segregated waste either directly from consumers or through centralized centers
2. Sorting of recyclables and removal of contaminants
3. Separation by polymer type and color, often using techniques such as Fourier-transformed near-infrared (FT-NIR) spectroscopy
4. Extrusion of clean, uniform materials into plastic pellets
5. Sale of recycled pellets to manufacturers as raw material [28, 29, 31]

One of the most challenging steps is sorting plastics by polymer type. Inconsistent or mixed plastics reduce the quality of the recycled product, which may not meet industry standards. Nonetheless, recycling remains a top priority due to its far-reaching benefits:

- Reduces dependence on landfills
- Conserves energy and natural resources
- Cuts emissions of harmful pollutants
- Creates jobs and strengthens local economies
- Decreases the need for virgin material imports

While recycling can be expensive upfront, its long-term economic and environmental benefits far outweigh the costs.

12.4.4 Challenges and Future Directions

Not all plastics are recyclable. Contaminated, degraded, or multi-layered plastics often cannot be processed efficiently and must be diverted to energy-recovery systems or used as industrial feedstock. Moreover, the plastics manufacturing industry requires a consistent supply of high-quality raw material, which is not always achievable from post-consumer recycled plastic. To address these challenges, countries are exploring strategies such as the following:

- ➢ Enhancing primary recycling techniques
- ➢ Imposing taxes on virgin plastic production
- ➢ Mandating the use of recycled plastic in certain product categories

Additionally, technological innovations in sorting, identification, and processing are making recycling more precise and economically viable. By improving recycling rates and the quality of recycled materials, we can gradually shift toward a circular economy where plastic waste is minimized and resources are kept in use for as long as possible.

12.5 Principle of Extended Producer Responsibility for Packaging Waste

The foundational concerns surrounding waste management and environmental responsibility can be traced back to the 1987 Brundtland Report, which emphasized the environmental threats posed by transboundary waste shipments particularly to the Global South and the urgent need to mitigate pollution [32]. For decades, the prevailing model for waste management has relied heavily on public funding, with taxpayers bearing the costs of collection, recycling, and landfill operations. However, by the late 20th century, many countries began facing mounting challenges: increasing volumes of solid waste, limited landfill capacity, and the environmental burden of poorly managed product disposal. These challenges necessitated a shift toward more holistic and accountable waste governance strategies.

A critical evolution in this domain came with the development of Extended Producer Responsibility (EPR), a policy principle that holds producers accountable for the environmental impacts of their products throughout the entire life cycle, particularly at the post-consumer stage. Originally articulated in a report to the Swedish Ministry of Environment in the 1990s by Thomas Lindhqvist and his team, EPR aims to shift the responsibility for waste management from municipalities to the producers themselves [33]. By doing so, it encourages manufacturers to internalize the environmental costs associated with their products, from design to disposal.

The central tenet of EPR is to promote sustainability by integrating the true environmental cost of products into business and design decisions. This includes encouraging eco-friendly product design, facilitating easier recycling or reuse, and financing the infrastructure needed for the end-of-life treatment of products. In practical terms, EPR policies require producers individually or collectively through Producer Responsibility Organizations (PROs) to organize and often finance the collection, sorting, recycling, or safe disposal of waste generated from their products.

EPR has proven to be especially effective in sectors where products present high environmental risks or are difficult to manage at their end-of-life stage. These include waste electrical and electronic equipment (WEEE), household appliances, vehicles, and increasingly, packaging materials [33, 34]. When applied to packaging waste, EPR offers dual benefits: it reduces the financial burden on public waste systems and incentivizes manufacturers to rethink packaging design, opting for materials that are easier to recycle or generate less waste. By mandating producer responsibility, EPR systems create structured networks for post-consumer material recovery and foster circular economic practices. Producers are motivated not only to meet compliance obligations but also to innovate—developing lighter, more durable, and recyclable materials, thus reducing environmental externalities at the source.

Moreover, EPR for packaging waste is gaining traction as an essential policy tool in mitigating plastic pollution. It has been successfully adopted in various regions, with positive outcomes in waste reduction, recycling rate improvements, and reduced pressure on landfills. As more governments adopt or strengthen EPR regulations, the role of industry stakeholders in environmental stewardship is becoming increasingly central to the future of sustainable packaging.

12.5.1 Procedural Responsibility Model

While fee-based Extended Producer Responsibility (EPR) programs for packaging waste remain the cornerstone of modern waste policy, they represent only one facet of a broader toolkit. A wide array of complementary strategies ranging from regulatory mandates and product take-back schemes to voluntary industry initiatives and public education campaigns are also integral to achieving the objectives of EPR. These diverse approaches, when effectively integrated, can significantly curb waste generation, facilitate material recovery, and prevent environmental pollution.

Yu, Hills, and Welford have outlined various regulatory and economic instruments that support both government and industry in the sustainable stewardship of materials [35]. Extended Producer Responsibility has emerged as a pivotal policy framework designed to shift the responsibility of waste management from consumers and municipalities to producers, thereby promoting more sustainable production and consumption systems. Different approaches to EPR are employed worldwide to address the challenges of plastic and electronic waste. Voluntary industry practices, such as self-imposed codes of conduct, public–private partnerships, and product leasing or servicing models, encourage industries to proactively minimize waste generation. Regulatory instruments, on the other hand, impose minimum product standards, restrict hazardous substances, enforce disposal bans, and mandate recycling, ensuring compliance through legal obligations. Informative measures, including product labeling, disclosure of material composition, and reporting to authorities, play a crucial role in raising awareness among consumers and recyclers while enhancing transparency. Economic instruments, such as deposit–refund schemes, advance recycling fees, disposal charges, and material-based taxes or subsidies, provide financial incentives that influence consumer behavior and industry practices. Collectively, these complementary EPR instruments not only encourage sustainable product design and waste reduction but also create a circular economy framework where producers are held accountable for the entire life cycle of their products.

These tools help to address systemic mismanagement of resources, restrict pathways through which pollutants enter the environment, and broaden the scope of responsibility to include the post-use or end-of-life phase of products. Such expanded oversight ensures that discarded goods are not merely treated as waste, but as potential resources within a circular economy. Historical precedents illustrate the practical evolution of procedural responsibility. For example, the state of Oregon launched the United States' first deposit-refund program for beverage containers in 1972, laying the groundwork for what would become a popular waste reduction strategy. Vermont and Saskatchewan followed suit in 1973, with Michigan and Maine introducing similar initiatives in 1978 and Iowa in 1979 [36]. These programs not only increased recycling rates but also fostered greater public participation and awareness.

Over time, deposit-refund systems and other EPR-related tools have expanded to include product categories beyond packaging such as waste electrical and electronic equipment (WEEE), household electronics, used vehicle tires, lead-acid batteries, and hazardous waste. These initiatives not only promote higher recovery rates but also reduce the financial and environmental burdens traditionally placed on municipal systems. In essence, the procedural responsibility model reflects an evolving, flexible policy framework, one that combines legal mandates with economic incentives and voluntary compliance to drive more sustainable production and consumption cycles. By broadening the range of tools used to manage waste, policymakers can tailor solutions to local contexts while advancing global objectives for plastic and material mitigation.

12.5.2 Application of the EPR Principle to Packaging Waste

Governments around the world are increasingly adopting Extended Producer Responsibility (EPR) as a key strategy for managing packaging waste. One of the main motivations is to ease the financial burden on taxpayers, who traditionally fund public waste management programs. By holding producers accountable for the full life cycle of their products including disposal and recycling, EPR schemes aim to increase recycling rates, reduce landfill dependency, and foster a circular economy where better product design aligns with environmental objectives. EPR adoption for packaging waste is now a global phenomenon, complementing broader plastic waste regulations. However, implementation strategies vary widely across regions. The Ellen MacArthur Foundation (EMF) classifies countries as having "emerging," "limited," or "voluntary" EPR programs, depending on the design, scope, and legal status of their systems [37].

Africa: In 2021, South Africa took a significant step by implementing mandatory, fee-based EPR regulations under Section 18 of its National Waste Management Act. This legislation covers paper, packaging, and several single-use products. Kenya also initiated EPR program development in 2021 under the Environmental Management and Coordination Act. While Namibia and Tunisia have formulated EPR frameworks, they have not yet operationalized them. Nigeria currently maintains a voluntary EPR model covering limited packaging types, whereas Ghana and Cameroon have restricted EPR systems in place, addressing only a narrow range of materials. Across the continent, momentum is building to expand and strengthen EPR applications for packaging [38].

Asia: The implementation of EPR in Asia varies widely across countries. Japan pioneered its EPR scheme for packaging with the 1995 Act on the Promotion of Sorted Collection and Recycling of Containers and Packaging, achieving a 27% increase in recycling rates from 1997 to 2010. South Korea followed with the 2003 Act on the Promotion of Saving and Recycling Resources, while Taiwan introduced packaging-related EPR through its 1998 Waste Disposal Act. More recently, India's 2020 Plastic Waste Management Rules mandated an EPR program focused specifically on plastic packaging. These legislative frameworks highlight Asia's increasing commitment to packaging waste management through robust EPR mechanisms [39, 40].

Latin America: Several Latin American countries have embraced mandatory EPR programs for packaging waste. Chile enacted a nationwide EPR law in June 2019, followed by Venezuela in 2020 and Colombia in 2018. These initiatives establish legal obligations for producers to take responsibility for packaging waste. Other countries, including Argentina, are in the process of developing mandatory systems. Meanwhile, Brazil, Uruguay, Bolivia, and Ecuador have introduced voluntary EPR programs, typically targeting selected types of packaging materials. This mixed landscape reflects a growing recognition of the need for shared responsibility in managing packaging waste across the region [39].

Europe: Europe has long been a global leader in EPR implementation. The European Union (EU) embedded EPR within its broader waste management framework in the early 1990s. A landmark development came with the 1994 Packaging and Packaging Waste Directive, which aimed to curb the rising volumes of household packaging waste. Germany, however, had already taken the lead by introducing its Packaging Waste Ordinance in 1991 [41].

Under EU rules, member states are required to meet progressively ambitious recycling targets. By 2025, countries must ensure that 65% of all packaging waste by weight is recycled, including at least 50% for plastics. These thresholds will rise further by 2030, mandating a 70% overall recycling rate, with a minimum of 55% for plastic packaging. Most EU countries with active EPR systems are on track to meet or exceed these targets, demonstrating the effectiveness of the policy in driving environmental outcomes and industry innovation.

12.6 Bans and Policies Addressing MPs and Single-Use Plastics

Several countries have introduced bans and regulatory measures to combat the environmental harm caused by post-consumer plastic waste, particularly the more visible and persistent forms such as plastic bags and MPs. These actions often aim to curb pollution at its source by eliminating the use of problematic products. Recent policy trends, as discussed in this chapter, show a clear global shift toward targeting single-use plastics (SUPs). Governments are increasingly enacting legislation to restrict, phase out, or ban these items, reflecting a broader commitment to reducing plastic leakage into ecosystems.

Over the past two decades, numerous countries have adopted targeted national policies to curb plastic pollution, reflecting growing global recognition of its environmental and health consequences. Early interventions began in Ireland (2002) with a levy on plastic bags, which significantly reduced consumption and in Bangladesh (2002), which became one of the first countries to ban thin plastic bags, later extending restrictions in 2020 to single-use plastics (SUPs) in coastal zones and hospitality sectors. Similar measures followed worldwide, with China (2008) mandating consumer charges on plastic bags, leading to a 50% reduction in use within two years and imposing a ban on ultra-thin bags below 25 μm. Australia (2011) and New Zealand (2019) also enforced bans on lightweight plastic bags of varying thicknesses. In parallel, bans on microbeads in cosmetics and personal care products gained momentum, beginning with Rwanda (2008) and extending to countries such as the United States (2015), United Kingdom (2015), Denmark (2016), Canada (2016), Italy (2018), South Korea (2018), and California (2020). Broader legislation included France's Plastic Ban law (2016) and India's Plastic Waste Management Rules (2016), which outlawed bags below 50 μm and later extended restrictions to disposable plastics in Delhi and the National Capital Region. Other nations, such as Sri Lanka (2017), Zambia (2018), Taiwan (2018), and Colombia (2017), introduced partial bans on plastic bags or SUPs, while Finland (2016) opted for framework agreements to reduce lightweight carrier bag usage. Sweden (2017) stands out as a global leader in waste management, recycling nearly all household waste and sending less than 1% to landfills. Collectively, these measures reveal both diversity and convergence in policy approaches, from levies and bans to recycling mandates, underscoring the critical role of national regulation in tackling the plastic crisis.

12.6.1 Plastic Bag Bans Around the World

Governments across the globe have responded to the growing plastic waste crisis by enacting bans and levies on plastic bags, one of the most ubiquitous single-use plastics. These regulatory measures vary in scope, enforcement, and impact but collectively reflect a global shift toward more sustainable consumption practices.

- Africa has been a leader in banning plastic bags, with several countries implementing strict policies. South Africa, Uganda, Somalia, Rwanda, Botswana, Kenya, and Ethiopia have enforced comprehensive plastic bag bans. These measures aim to combat widespread littering and environmental degradation caused by plastic pollution.
- In Bangladesh, a turning point came after devastating floods between 1988 and 1998 submerged two-thirds of the country. Plastic bags were blamed for clogging drainage systems, prompting the government to enact one of the world's first plastic bag bans in 2002. However, enforcement remains weak, and plastic bags are still distributed by many retailers.
- Europe has adopted a mix of taxation and regulation. In Ireland, a plastic bag levy introduced in 2002 led to a dramatic 90% drop in usage. Denmark followed with a fee on retailers in 2003, resulting in a 66% reduction in plastic and paper bag consumption. Belgium imposed a similar fee in 2007, with ongoing reductions in plastic bag use. Italy, in a more ambitious step, banned the sale of non-biodegradable plastic bags in 2011.
- China implemented a nationwide ban on free plastic bags in 2008. Retailers were instructed to stop distributing them by mid-year. This bold move not only improved environmental

conditions but also aimed to conserve resources, with projections indicating a potential saving of 37 million barrels of oil annually.
- In Australia, while there is no federal ban, states like South Australia and the Northern Territory have implemented their own legislation. The movement began in Coles Bay, Tasmania, the first locality to ban plastic bags. South Australia's 2008 ban under its "Zero Waste" strategy reportedly eliminated the use of 400 million bags annually. By mid-2019, major supermarket chains nationwide voluntarily phased out single-use bags and introduced a charge for reusable alternatives.
- The United States has adopted a patchwork approach. By July 2014, around 20 million Americans were living in jurisdictions with some form of plastic bag ban. These included regulations either in effect or under consideration in 20 states and 132 cities or counties. While federal action is absent, local bans are gaining momentum.
- In England, the introduction of charges on plastic bags resulted in a staggering 85% drop in usage from nearly seven billion bags annually to around one billion. Other countries, such as Wales, Haiti, Scotland, France, and Germany, have also enacted measures to regulate plastic bags and reduce plastic waste.

These examples demonstrate how countries, despite differences in governance and economic development, are converging toward a shared goal: curbing plastic pollution through targeted bans, levies, and public awareness campaigns [42–45].

12.6.2 The UK Microbeads Ban

The United Kingdom took a significant regulatory step in its fight against plastic pollution by targeting microbeads, tiny plastic particles commonly found in personal care products. In 2017, the Department for Environment, Food, and Rural Affairs (DEFRA) released a detailed response to a public consultation initiated in 2016. This consultation had been driven by growing public and scientific concern over the environmental damage caused by MPs, particularly those used in cosmetics and personal care products.

Microbeads, typically less than 5 mm in size, are added to products like facial scrubs, toothpastes, and shower gels to provide exfoliating or cleansing properties. However, because of their small size, these plastic particles easily bypass water filtration systems and enter rivers and oceans, where they pose serious threats to marine life and ecosystems. According to DEFRA's summary, while there were discussions about extending the ban to include "leave-on" products such as makeup and sunscreens, the final regulation focused exclusively on "rinse-off" products. This decision was based on both the availability of alternatives and the relative environmental impact of different product categories.

The UK's ban on the manufacture of rinse-off cosmetics containing microbeads came into effect in January 2018, followed by a ban on sales starting June 2018. This legislation positioned the UK as a global leader, implementing what was then considered one of the strictest microbead bans in the world. The new rules prohibited the addition of microbeads to rinse-off items such as facial cleansers, body scrubs, and dental products, marking a crucial move in reducing MP pollution at its source. By eliminating the production and sale of these microbead-laden products in England and Scotland, the UK not only responded to public demand but also set a strong precedent for other countries to follow in tackling MP pollution through targeted regulatory action [46].

12.6.3 The U.S. Microbead-Free Waters Act of 2015

In a landmark move toward curbing MP pollution, the United States Congress passed the Microbead-Free Waters Act of 2015, amending the Federal Food, Drug, and Cosmetic Act. This legislation marked a significant step in the regulation of MPs, specifically targeting plastic microbeads found in rinse-off personal care products.

The Act prohibits the manufacture, packaging, and distribution of rinse-off cosmetics containing intentionally added plastic microbeads. It covers not only cosmetic products such as exfoliating scrubs and cleansers, but also over-the-counter drugs, including microbead-containing toothpastes. The inclusion of nonprescription drugs reflects the broad scope of the legislation and underscores the concern about microbeads entering waterways through common hygiene products. Before the federal law was enacted, several U.S. states had already introduced bans on microbeads. However, inconsistencies among state-level regulations created a fragmented legal landscape, complicating compliance for manufacturers and distributors. Recognizing the need for uniformity, Congress established a nationwide standard through the Microbead-Free Waters Act.

The federal law preempts state and local governments from enacting or enforcing separate legislation that would restrict the manufacture, sale, or interstate distribution of rinse-off cosmetics containing microbeads except where such laws were already in place before the Act. This provision ensures consistent regulation across the country while acknowledging early legislative action at the state level. With this act, the United States became one of the first countries to adopt a comprehensive national ban on plastic microbeads in personal care products, reflecting a growing global consensus on the urgent need to address MP pollution at the source [45].

12.6.4 Australia's Regulation of Plastics

As a federal nation, Australia's approach to regulating plastic packaging is multi-tiered, involving various levels of government. State and territorial governments are primarily responsible for enacting waste management legislation and policies, while municipal governments oversee day-to-day waste operations. At the national level, the Federal Government plays a coordinating role, particularly in fulfilling Australia's obligations under international frameworks such as the Global Instrument to end Plastic Pollution.

Despite these structures, Australia's plastic packaging regulations have faced widespread criticism from government bodies, academic researchers, industry stakeholders, and civil society. Recent studies [47, 48] point to significant shortcomings, with a 2021 independent review calling for substantial reforms. A report prepared for the Environmental Protection Agency of South Australia encapsulated the issue starkly:

"Governments have also been quite reluctant to regulate, and when they have, it has been inefficient and patchy, city by city or state by state" [49]. Australia's current regulatory structure relies on a co-regulatory framework consisting of the National Environmental Protection Measure (NEPM, 2011) and the Australian Packaging Covenant, which work together to guide industry practices and promote recycling initiatives. However, enforcement remains inconsistent, and voluntary participation often limits impact.

In 2021, the Federal Government introduced the National Plastics Plan, a policy roadmap that laid out ambitious targets primarily focusing on recycling enhancement and public awareness. One of the most notable legislative steps was the Recycling and Waste Reduction Act 2020, which prohibited the export of unsorted plastic waste and raw plastic materials [50]. This move was prompted by global outcry over plastic waste from wealthy countries including Australia being exported to Southeast Asian nations lacking adequate waste-processing infrastructure.

However, these export restrictions have unintentionally led to an increase in domestic landfilling [51]. The Recycling Modernization Fund (RMF) was launched to build national recycling capacity, but its impact remains limited thus far, and Australia continues to grapple with the shortfall in infrastructure for onshore waste processing. Recognizing these gaps, the Federal Government has recently committed to introducing mandatory regulations on packaging standards, signaling a shift away from voluntary compliance toward a more enforceable national policy framework [52]. This move is expected to address inconsistencies across jurisdictions and enhance accountability across the plastics value chain.

12.6.5 Indian Government Regulations for Plastic Waste Management

India has established a series of laws and policy guidelines aimed at managing plastic waste; however, plastic waste management (PWM) still lacks widespread awareness and effective implementation. According to estimates by the Central Pollution Control Board (CPCB), India generates approximately 9.46 million tons of plastic waste annually, out of which only about 3.78 million tons are collected. The remaining waste is often dumped in landfills or open dumpsites, posing serious environmental and health hazards [53]. Over the years, several assessments and studies have been conducted to evaluate the effectiveness of existing PWM policies. These studies consistently highlight the shortcomings in the current infrastructure, particularly the lack of organized collection centers and inefficient post-consumer recycling mechanisms. As a result, the recovery and reuse of plastic waste remain far below their potential.

To address these gaps, the Swachh Bharat Abhiyan (SBA), a flagship cleanliness initiative launched by the Government of India, incorporated plastic waste awareness and management as a key component. The SBA emphasizes public education and behavioral change, promoting responsible disposal and segregation practices among citizens [54].

The PWM strategy under SBA for rural areas outlines a structured implementation plan, which includes the following key components [55]:

a. Source segregation and systematic collection: ensuring the separation of plastic waste at the source, followed by organized collection, storage, and transportation to authorized recyclers.

b. Stakeholder awareness: conducting targeted awareness programs to inform and engage all stakeholders, including local communities, village authorities, and sanitation workers.
c. Ban on open burning: strict prohibition on the open burning of plastic waste to reduce air pollution and toxic emissions.

To strengthen the rural PWM framework, the government has proposed the development of waste storage facilities at the village level and Material Recovery Facilities (MRFs) or dedicated PWM units at the district or block level. These decentralized units aim to facilitate efficient sorting, recycling, and value recovery from plastic waste.

While these efforts mark a step in the right direction, consistent enforcement, capacity building, and incentivized participation from both urban and rural stakeholders remain crucial to achieving long-term sustainability in plastic waste management across India.

12.7 Action Plans and Regulatory Agreements to Reduce Plastic Inputs from Land- or Sea-Based Sources

Numerous international conventions and agreements have been established to prevent or minimize the discharge of plastics and MPs into the marine environment. The global challenge of plastic pollution is now increasingly recognized as a "common concern for humankind" [56]. Addressing this issue on a meaningful scale requires enhanced cooperation and the implementation of comprehensive national and international legal frameworks. While the formulation of these frameworks marks significant progress, the key challenges remain in their implementation, monitoring, and enforcement. Several regional conventions have nonetheless played a vital role in setting standardized methodologies, performance indicators, and assessment tools while also promoting joint litter reduction initiatives and evaluating their effectiveness [57].

12.7.1 The OSPAR Convention

The OSPAR Convention is one of the most significant international instruments designed to reduce plastic pollution in the marine environment. It evolved from the earlier Oslo and Paris Conventions and came into force in September 1992. The convention is supported by a wide range of parties, including Belgium, France, Germany, Denmark, the European Union, Finland, Iceland, Ireland, Luxembourg, the Netherlands, Norway, Portugal, Spain, Sweden, Switzerland, and the United Kingdom [58].

The OSPAR Convention defines marine litter as any solid material discarded, disposed of, or abandoned in the marine and coastal environment, originating from both land-based and sea-based sources. These include inputs *via* rivers, drainage systems, sewage discharges, and even wind-blown debris. Notably, plastic accounts for over 80% of marine litter found on the seabed, shorelines, and sea surface within the OSPAR region [59].

The Convention operates across six strategic areas:

1. Eutrophication and hazardous substances
2. Biodiversity and ecosystems
3. Human activities
4. Offshore industries
5. Radioactive substances
6. Cross-cutting issues

Key actions under the OSPAR Action Plan on Marine Litter include the following:

- Reduction of waste from the fishing industry
- Imposing fines for marine pollution
- Targeting single-use plastics
- Supporting "Fishing for Litter" programs
- Retrieving lost and abandoned fishing gear
- Promoting education and awareness
- Eliminating MPs from consumer products [60]

12.7.2 The HELCOM Convention

The HELCOM Convention formally known as the Helsinki Convention aims to protect the Baltic Sea, a uniquely sensitive marine ecosystem. Signed in 1992 by Denmark, Estonia, Finland, Germany, Latvia, Lithuania, Poland, Russia, and Sweden, the convention seeks to prevent marine pollution from both land- and sea-based sources within the Baltic region.

In 2007, the HELCOM Baltic Sea Action Plan (BSAP) outlined four key areas of focus:

1. Eutrophication
2. Biodiversity

3. Hazardous substances
4. Maritime activities [61]

In 2015, HELCOM introduced a Regional Action Plan (RAP) for marine litter, with the goal of reducing plastic waste in the Baltic Sea by 2025. This plan identifies consumer behavior as a major contributor to marine plastic pollution. In the Baltic region, household waste and tourism-related activities are responsible for approximately 33% and 48%, respectively, of plastic litter [62].

Under this agreement, each contracting state must

- Commit to implementing the RAP guidelines at both regional and national levels
- Introduce public education and awareness campaigns to reduce marine litter
- Promote policy reforms and behavioral change in plastic consumption and disposal practices [63]

12.8 Policies Tackling MP Contamination

Nearly seven decades after the invention of synthetic plastic, a series of international treaties and national policies began to emerge to address the growing issue of plastic pollution. While early legislation primarily targeted macroplastic waste, from the mid-2000s to 2020, the regulatory focus began to shift more explicitly toward MP contamination due to its rising environmental and health concerns. Figure 12.3 presents a timeline of major policies and regulations that directly or indirectly address MP pollution. These policy developments highlight a growing recognition of MPs as a distinct environmental hazard.

For analytical clarity, we categorize the evolution of these policies into three thematic areas:

- Establishment of Pioneer Conventions: foundational international agreements that laid the groundwork for global cooperation in plastic pollution control, often before MPs were explicitly recognized as pollutants.
- Formulation of Regulations Addressing Plastic Waste: legislation and policies developed to manage and mitigate the generation, disposal, and recycling of plastic waste, which indirectly contributed to limiting MP release through improved waste handling systems.

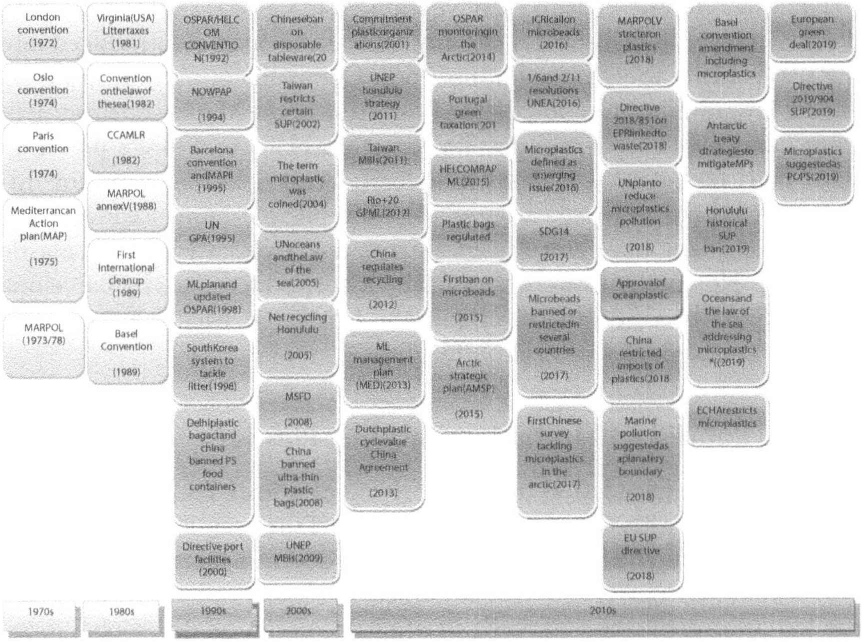

Figure 12.3 Timeline of policies targeting plastic contamination.

- ➢ Formulation of Regulations Addressing MP Contamination: recent and more targeted policies specifically crafted to limit the production, use, and environmental discharge of MPs, such as bans on microbeads in personal care products and standards for synthetic fiber emissions during washing.

12.8.1 Pioneer Conventions

Legislative initiatives to control land-based and marine pollution began in the 1970s, targeting the release of substances or energy by human activities that negatively impact ecosystems, living resources, infrastructure, and human health [64]. In the following decade, several United Nations (UN) treaties addressing maritime pollution and the transboundary movement of hazardous materials came into effect. These treaties led to the development of agendas, mandatory annexes, and refund–deposit systems for waste management [65]. These early regulatory frameworks laid the foundation for more comprehensive anthropogenic waste management strategies in the 1990s.

For example, the Helsinki Convention (HELCOM), adopted in 1992, defines "dangerous substances" as "any substance which, if introduced into the sea, is likely to cause pollution" [66]. Complementing this, the UN's Global Programme of Action for the Protection of the Marine Environment from Land-based Activities developed action plans to address pollution from both terrestrial and marine sources. While plastics were not initially singled out in these early policies, several well-known conventions and action plans have since been amended to explicitly address plastic pollution. Notably, MPs were specifically incorporated into the Barcelona Convention, the OSPAR Convention, the Northwest Pacific Action Plan, and the revised Basel Convention [67, 68].

12.8.2 Policies and Initiatives Tackling Plastics

Building on the foundation of these early conventions, regulations explicitly targeting plastics began to emerge. The European Marine Strategy Framework Directive (MSFD) took a strong stance against plastic pollution, leading Europe to implement numerous regulations addressing plastics [69]. Similarly, the UN recognized plastic pollution as a global issue, promoting stakeholder participation, funding regional sea protection programs, exploring market-based solutions, and issuing guidelines to support worldwide efforts [66].

Following these initiatives, global plastic organizations committed to addressing the economic losses caused by plastic waste [6]. At the national level, China enacted the Prevention and Control of Waste Plastic Processing and Utilization law, Taiwan introduced deposit–refund schemes, and measures were implemented to reduce littering in the Antarctic [70].

Between 2013 and 2020, policies addressing plastic pollution including bans on certain plastics expanded globally. These policies take various forms, including regional action plans, national and regional agreements, amendments to MARPOL 73/78 Annexes, monitoring programs, national laws, and taxation reforms, such as Portugal's restrictions on plastic waste imports.

These efforts have transformed plastic design, manufacture, and recycling practices, notably through the European Strategy for Plastics in a Circular Economy. Scientists also recognized marine plastic pollution as a planetary boundary threat. Early plastic bans in Asia appeared in countries such as China, India, and Taiwan [71, 72].

12.8.3 Policies and Initiatives Directly Tackling MPs

In subsequent years, several countries including the U.S., Netherlands, France, Taiwan, South Korea, Sweden, and the UK enacted bans on MPs in rinse-off cosmetics. International bodies such as the International Coral Reef Initiative and the Secretariat of the Antarctic Treaty endorsed efforts to reduce plastic microbeads. The European Chemicals Agency (ECHA) adopted similar measures, while Canada classified MPs as toxic substances in personal care products.

However, other sources of MPs, such as abrasive materials used in plastic blasting and automotive molding, have generally been overlooked in regulations [66, 73]. The COVID-19 pandemic has also had a notable impact on plastic pollution. The healthcare sector's increased demand for single-use plastics, along with a surge in takeout food packaging, has contributed to a rise in plastic waste. Additionally, some countries delayed or reversed policies aimed at reducing plastic pollution; for example, the UK postponed its ban on plastic straws. The full extent of COVID-19's impact on plastic pollution remains uncertain.

12.9 Upstream Responses to Prevent MP Pollution

Preventive interventions are generally more cost-effective than mitigation strategies and largely depend on robust governance mechanisms [74]. Figure 12.4 summarizes the key upstream actions implemented to reduce MP pollution at its source [74].

12.9.1 Circular Economy

A circular economy (CE) integrates social equity, economic development, and environmental sustainability [75]. Transitioning the design, manufacture, use, and end-of-life handling of plastic products towards CE principles can significantly minimize waste output and reduce dependency on raw materials. CE offers a long-term solution to mitigate plastic pollution and is instrumental in limiting the formation of secondary MPs in the environment.

Through eco-design and the development of alternative materials, CE enhances the mechanical properties, shelf life, and recyclability of single-use plastic (SUP) products [76]. Recycling is one of its fundamental pillars, where polymers are reprocessed into new products after sorting, treating,

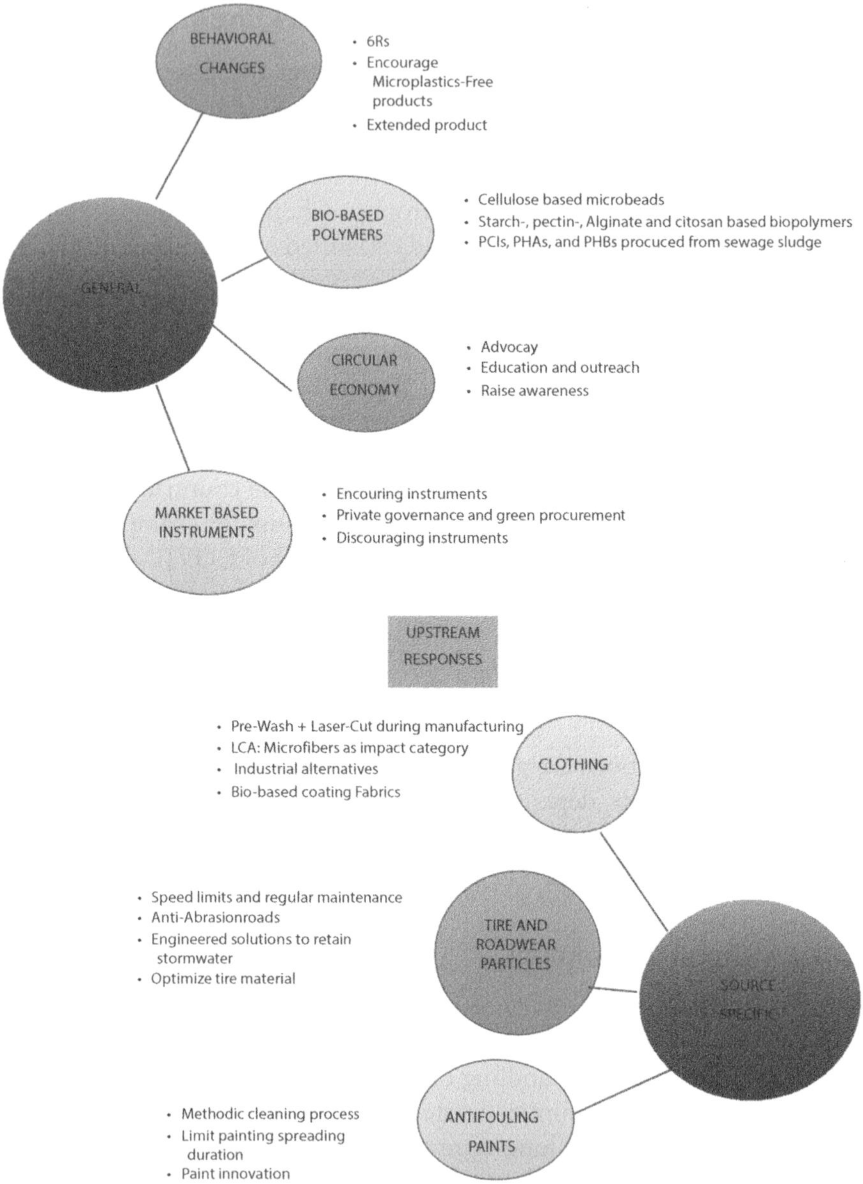

Figure 12.4 Upstream responses to prevent MP pollution.

melting, extruding, and pelletizing. However, technical and economic barriers hinder the recycling of many plastic types [63].

Several EU-funded initiatives, such as BLUENET, OCEANETS, MARELITT, and SEACYCLE, aim to upcycle marine debris and abandoned fishing gear. Some industries have started converting recycled nets into yarns. Extended Producer Responsibility (EPR) frameworks also promote the design of recyclable and reusable goods while minimizing the use of hazardous materials [77].

12.9.2 Behavioral Changes

Preventing MP pollution necessitates multi-level governance and behavioral change among governments, industries, NGOs, academic institutions, fishermen, and local communities [74]. Promoting environmental literacy among children and adults and engaging stakeholders are crucial for reducing littering and plastic consumption [78, 79].

Awareness campaigns such as "*Basuraleza*" in the Basque Country and "Keep Britain Tidy" in the UK address behavior-related issues. In 2016, the Dutch government initiated studies to raise awareness of MP pollution, while Belgium developed mechanisms encouraging industries to reduce their reliance on primary MPs. Media-driven campaigns, instructional videos, and educational programs including MOOCs on marine litter and ocean plastic webinars can further support this cause [80]. These efforts should span all ecosystems: aquatic, aerial, and terrestrial, though most currently focus on marine environments [77].

12.9.3 Bio-Based Polymers

Bio-based plastics offer a promising alternative to petroleum-derived plastics, potentially reducing fossil fuel consumption and environmental pollution. Though they comprised only about 1% of the 2.1 million tons of plastic produced in 2019, interest in these materials has been growing. Polymers such as polyhydroxyalkanoates (PHAs), polylactic acid (PLA), and polyhydroxybutyrate (PHB) derived *via* fermentation of starch and other feedstocks exhibit favorable mechanical properties. Other promising biomatrices include chitosan, pectin, lignin, starch, and adultified, which are being actively investigated for their potential to replace traditional plastics [81].

12.9.4 Market-Based Instruments (MBIs)

Market-based instruments can effectively influence behaviors linked to plastic use and MP pollution [82]. These tools internalize the external costs of plastic littering through taxation, incentives, penalties, and other economic measures [83].

Key MBIs include the following:

- Green procurement policies, such as requiring reusable materials in public settings
- Plastic taxes, targeting environmentally harmful products
- Monitoring systems to restrict MP use and trace sources especially challenging in developing countries
- Reinvestment of collected levies and penalties into upstream solutions
- Consumer education to foster environmentally conscious choices
- Governance frameworks to guide industries toward producing eco-friendly alternatives [84]

12.9.5 Primary Microfibers from Clothing

The textile industry significantly contributes to primary MP pollution, with microfibers being released during manufacturing and domestic washing [85, 86]. Factors such as fabric type, polymer content, cutting methods, washing machine design, and garment age influence microfiber shedding. Solutions include applying silicon- or bio-based polymer coatings, which reduce microfiber loss by ~30%, and using detergents, pre-wash cycles, and filter meshes, which can reduce emissions by over 50%. Despite these innovations, over 15,000 tons of microfibers is still released annually [87]. Collaboration among textile manufacturers, appliance companies, and policymakers is crucial to develop emission tracking tools and establish fiber-release reduction standards [85].

12.9.6 Tire and Road Wear Particles (TRWPs)

TRWPs contribute significantly to MP emissions estimated at 5.5 kg per capita annually. This equates to approximately 1.12 million tons per year in the U.S. and 1.33 million tons in the EU.

To mitigate TRWP emissions, key strategies include the following:

- Standardizing assessment methods
- Developing emission parameters
- Estimating tire wear per mileage
- Establishing biomarkers for environmental TRWP detection [88, 89]

Maintenance-related actions include the following:

- Proper wheel alignment and balancing
- Imposing taxes on studded tires
- Designing wear-resistant tires using materials like silica to improve durability under stress

12.9.7 Antifouling Paints

Antifouling paints are widely used in the shipping and fishing industries to reduce drag caused by biofouling. However, these coatings are a significant source of MP emissions.

Effective measures include the following:

- Regulating emissions during grazing and coating removal
- Developing innovative paints using silicon-based and other environmentally friendly antifouling agents
- Minimizing dust dispersion, excess coating, and brush contamination
- Raising awareness among ship crews about sustainable practices
- Enhancing paint durability and incorporating catalytic agents to improve degradation and reduce environmental impact [74]

12.10 Downstream Strategies to Mitigate MP Pollution

To address the increasing levels of MP contamination already present in the environment, downstream strategies play a critical role. These measures focus on the degradation, recovery, and removal of MPs from terrestrial and aquatic ecosystems. An overview of these strategies is presented in Figure 12.5.

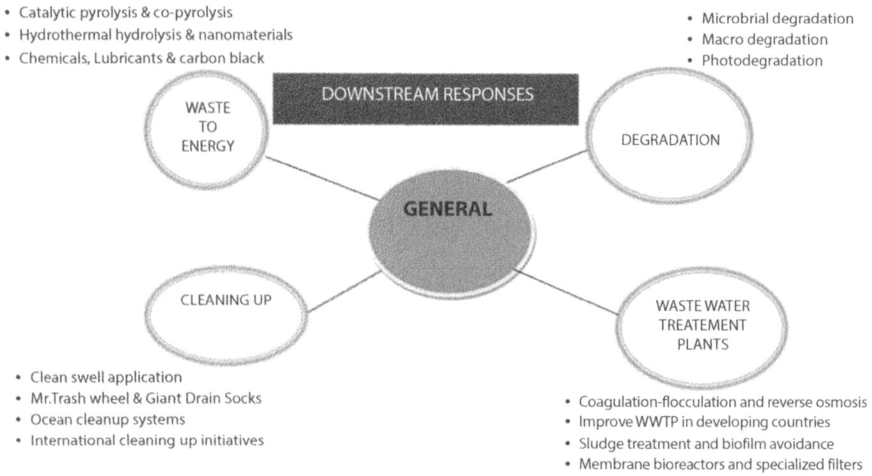

Figure 12.5 Downstream responses to mitigate MP pollution.

12.10.1 Degradation of MPs

Several techniques have been developed to degrade plastic waste. Photodegradation, which requires only oxygen and sunlight, initiates surface oxidation and causes physical breakdown of plastics. Innovative methods focus on enhancing this process using visible light-induced plasmonic photocatalysts. For instance, partial degradation of low-density polyethylene (LDPE) films has been achieved using platinum-decorated zinc oxide (ZnO) nanorods, while high-density polyethylene (HDPE) MPs have been decomposed using nitrogen-doped titanium dioxide (N–TiO_2) semiconductors [90].

Biodegradation using macro-organisms has also been explored. Certain species, including *Lumbricusterrestris* (earthworms), *Galleria mellonella* (wax moth) larvae, and *Tenebriomolitor Linnaeus* (mealworms), have been shown to host gut microbiota capable of degrading polyethylene (PE) and polystyrene (PS). Additionally, marine organisms such as *Alcanivorax borkumensis* and *Zalerion maritimum* have demonstrated the potential to degrade LDPE and ingest PE, respectively [91–94].

However, several uncertainties remain regarding the efficiency and scalability of biodegradation. Critical concerns include the management of biodegradable polymers by existing waste systems, greenhouse gas emissions from landfills, the effect of contaminants on compost quality, land use for bio-based plastics, and the actual conversion of plastic-derived carbon into microbial biomass, CO_2, or CH_4 [95].

12.10.2 Waste-to-Energy (WTE) Conversion

For polymers that are difficult to recycle, waste-to-energy conversion presents a viable alternative. Catalytic pyrolysis, co-pyrolysis, and microwave-assisted pyrolysis have been employed to convert polyesters, thermoplastics, and polyolefins into fuel and energy. Additionally, valuable byproducts such as carbon nanotubes and carbon black can be recovered and used in the production of chemicals and pitch enhancers [96].

Emerging methods include the use of peroxymonosulfate, which, when activated by carboxylic oxidation and hydrothermal hydrolysis, can break down plastics into intermediate organic compounds. These intermediates can promote the growth of algae, which can subsequently be used to produce biodiesel. Despite its sustainability advantages, this process often relies on costly materials, limiting its feasibility in resource-constrained communities.

12.10.3 Water Treatment Facilities

Water treatment infrastructure plays a significant role in capturing MPs and preventing their release into the environment. Wastewater treatment plants (WWTPs), sewage treatment plants (STPs), drinking water treatment plants (DWTPs), landfills, and integrated solid waste management (ISWM) systems contribute to MP retention.

Advanced WWTPs are capable of retaining over 90% of incoming MPs, while DWTPs also achieve high removal efficiencies [97]. Technologies employed include the following:

- **Membrane bioreactors (MBRs)** — considered the most effective, with >98% MP removal efficiency.
- **Sedimentation and coagulation-flocculation** — These methods retain nearly half of MP content.
- **Reverse osmosis, ozonation, and rapid sand filtration** — These supplementary techniques further enhance MP removal [98].
- **Microfiber recovery filters** and **air flotation + activated sludge systems** — commonly applied to recover fibers and remove MPs from sludge.

12.10.4 Cleanup Strategies

Environmental cleanups, although often localized, contribute to the reduction of macro- and MP pollution. Shoreline cleanup activities have proven effective in limiting plastic accumulation and, indirectly, reducing MP generation [99, 100].

Technological innovations in this space include the U-shaped ocean cleanup system and river interceptors developed by *The Ocean Cleanup*. These systems are already in operation in countries such as the Dominican Republic, Vietnam, Jamaica, and Indonesia. The initiative also aims to convert the collected marine debris into marketable products to fund continued operations [74].

12.11 Future Directions for Plastic Mitigation

Plastic pollution has become a critical global concern, affecting both terrestrial and aquatic ecosystems. Although several measures such as regulatory bans, governmental policies, sustainable waste management practices, and the development of bioproducts have been implemented to mitigate plastic hazards, these efforts are still insufficient. To address the magnitude of this challenge, collaborative action involving governments, industries, researchers, and local communities is essential. Furthermore, increased public awareness and focused scientific research are necessary to develop effective long-term solutions.

The following areas are recommended for future research and policy development [11]:

- ***Chemical contaminant absorption:*** Investigate how chemical pollutants adhered to MPs are absorbed into the tissues of aquatic organisms after ingestion.
- ***Ecosystem-level impact studies:*** Move beyond single-species laboratory studies to conduct research involving multiple species and trophic levels to better understand the ecological consequences of MP pollution.
- ***Biomagnification and human health:*** Study the biomagnification of chemical pollutants associated with absorbed MPs and assess their effects on higher trophic levels, including humans and key marine species.
- ***Long-term environmental monitoring:*** Establish long-term monitoring programs to characterize MPs in various

ecosystems and examine their interactions with persistent organic pollutants.

- ***Nanoplastics research:*** Explore how MPs degrade into nanoplastics and investigate the potentially more severe, size-dependent impacts of nanoplastics on aquatic life and ecosystem health.
- ***Policy and community reassessment:*** Continuously evaluate and adapt government policies, industry practices, and community initiatives to ensure the effectiveness of plastic waste reduction strategies.

12.12 Conclusion

MPs have become ubiquitous in aquatic environments across the globe. Their widespread distribution, persistence, and ability to adsorb and concentrate chemical pollutants make them a significant environmental hazard. These pollutants, once bound to MPs, are easily ingested by aquatic organisms and can bioaccumulate through the food chain, ultimately reaching humans. The ingestion of MP-bound chemical contaminants and the additives used in plastic production pose serious ecological and potential health risks. While the full extent of human health impacts from consuming seafood containing MPs remains unclear, early-stage research underscores the urgency of the issue.

In light of these concerns, increased attention from all stakeholders' governments, industries, researchers, and the public is essential. There is a pressing need for comprehensive legislation at municipal, national, regional, and international levels to regulate plastic use, minimize environmental discharge, and promote effective waste management practices. Only through coordinated efforts and robust policies can the global community mitigate the threats posed by MP pollution and safeguard both environmental and public health.

AI disclosure statement: The author confirms that no artificial intelligence (AI)-generated content or AI-assisted writing tools were used in the preparation of this chapter (including summaries and the abstract). All text, analysis, and conclusions are the result of the author's original work.

References

1. Andrady, A.L., MPs in the marine environment. *Mar. Pollut. Bull.*, 62, 8, 1596–1605, 2011, https://doi.org/10.1016/j.marpolbul.2011.05.030.
2. Karbalaei, S., Hanachi, P., Walker, T.R., Cole, M., Occurrence, sources, human health impacts and mitigation of MP pollution. *Environ. Sci. Pollut. Res. Int. ESPR*, 25, 36046–36063, 2018, https://doi.org/10.1007/s11356-018-3508-7.
3. Liu, T.K., Wang, M.W., Chen, P., Influence of waste management policy on the characteristics of beach litter in Kaohsiung, Taiwan. *Mar. Pollut. Bull.*, 72, 1, 99–106, 2013, https://doi.org/10.1016/j.marpolbul.2013.04.015.
4. Avery-Gomm, S., Walker, T.R., Mallory, M.L., Provencher, J.F., There is nothing convenient about plastic pollution. Rejoinder to Stafford and Jones "Viewpoint–Ocean plastic pollution: A convenient but distracting truth?". *Mar. Pol.*, 106, 103552, 2019, https://doi.org/10.1016/j.marpol.2019.103552.
5. Al-Salem, S.M., Uddin, S., Al-Yamani, F., An assessment of MPs threat to the marine environment: A short review in context of the Arabian/Persian Gulf. *Mar. Environ. Res.*, 159, 104961, 2020, https://doi.org/10.1016/j.marenvres.2020.104961.
6. Puskic, P.S. and Coghlan, A.R., Minimal meso-plastics detected in Australian coastal reef fish. *Mar. Pollut. Bull.*, 173, 113074, 2021, https://doi.org/10.1016/j.marpolbul.2021.113074.
7. Vandermeersch, G., Van Cauwenberghe, L., Janssen, C.R., Marques, A., Granby, K., Fait, G., Devriese, L., A critical view on MP quantification in aquatic organisms. *Environ. Res.*, 143, 46–55, 2015, https://doi.org/10.1016/j.envres.2015.07.016.
8. Frias, J.P. and Nash, R., MPs: Finding a consensus on the definition. *Mar. Pollut. Bull.*, 138, 145–147, 2019, https://doi.org/10.1016/j.marpolbul.2018.11.022.
9. Cole, M., Lindeque, P., Halsband, C., Galloway, T.S., MPs as contaminants in the marine environment: a review. *Mar. Pollut. Bull.*, 62, 12, 2588–2597, 2011, https://doi.org/10.1016/j.marpolbul.2011.09.025.
10. vanWijnen, J., Ragas, A.M., Kroeze, C., Modelling global river export of MPs to the marine environment: Sources and future trends. *Sci. Total Environ.*, 673, 392–401, 2019, https://doi.org/10.1016/j.scitotenv.2019.04.078.
11. Onyena, A.P., Aniche, D.C., Ogbolu, B.O., Rakib, M.R.J., Uddin, J., Walker, T.R., Governance strategies for mitigating MP pollution in the marine environment: a review. *MPs*, 1, 1, 15–46, 2021, https://doi.org/10.3390/MPs1010003.
12. Mitrano, D.M. and Wohlleben, W., MP regulation should be more precise to incentivize both innovation and environmental safety. *Nat. Commun.*, 11, 1, 5324, 2020, https://doi.org/10.1038/s41467-020-19069-1.
13. Lusher, A., Hollman, P., Mendoza-Hill, J., MPs in fisheries and aquaculture: status of knowledge on their occurrence and implications for aquatic

organisms and food safety, FAO, 2017, https://oceanrep.geomar.de/id/eprint/49179/1/MPs%20in%20fisheries%20and%20aquaculture.pdf.
14. Geyer, R., Production, use, and fate of synthetic polymers, in: *Plastic waste and recycling*, pp. 13–32, Academic Press, Elsevier, Amsterdam, Netherlands, 2020, https://doi.org/10.1016/B978-0-12-817880-5.00002-5.
15. Lanorte, A., De Santis, F., Nolè, G., Blanco, I., Loisi, R.V., Schettini, E., Vox, G., Agricultural plastic waste spatial estimation by Landsat 8 satellite images. *Comput. Electron. Agric.*, 141, 35–45, 2017, https://doi.org/10.1016/j.compag.2017.07.003.
16. Ding, Z., Chen, H., Liu, J., Cai, H., Evrendilek, F., Buyukada, M., Pyrolysis dynamics of two medical plastic wastes: drivers, behaviors, evolved gases, reaction mechanisms, and pathways. *J. Hazard. Mater.*, 402, 123472, 2021, https://doi.org/10.1016/j.jhazmat.2020.123472.
17. Dey, S., Anand, U., Kumar, V., Kumar, S., Ghorai, M., Ghosh, A., Dey, A., Microbial strategies for degradation of MPs generated from COVID-19 healthcare waste. *Environ. Res.*, 216, 114438, 2023, https://doi.org/10.1016/j.envres.2022.114438.
18. Kibria, M.G., Masuk, N.I., Safayet, R., Nguyen, H.Q., Mourshed, M., Plastic waste: challenges and opportunities to mitigate pollution and effective management. *Int. J. Environ. Res.*, 17, 1, 20, 2023, https://doi.org/10.1007/s41742-023-00507-z.
19. Gaylor, M.O., Harvey, E., Hale, R.C., Polybrominateddiphenyl ether (PBDE) accumulation by earthworms (Eiseniafetida) exposed to biosolids-, polyurethane foam microparticle-, and penta-BDE-amended soils. *Environ. Sci. Technol.*, 47, 23, 13831–13839, 2013, https://doi.org/10.1021/es403750a.
20. Hodson, M.E., Duffus-Hodson, C.A., Clark, A., Prendergast-Miller, M.T., Thorpe, K.L., Plastic bag derived-MPs as a vector for metal exposure in terrestrial invertebrates. *Environ. Sci. Technol.*, 51, 8, 4714–4721, 2017, https://doi.org/10.1021/acs.est.7b00635.
21. Dehghani, S., Moore, F., Akhbarizadeh, R., MP pollution in deposited urban dust, Tehran metropolis, Iran. *Environ. Sci. Pollut. Res. Int.*, 24, 20360–20371, 2017, https://doi.org/10.1007/s11356-017-9674-1.
22. Agarwal, N., Solanki, V., *et al.*, Bacterial Laccases as Biocatalysts for the Remediation of Environmental Toxic Pollutants: A Green and Eco-Friendly Approach—A Review. *Water*, 14, 4068, 2022, 10.3390/w14244068.
23. Jambeck, J.R., Geyer, R., Wilcox, C., Siegler, T.R., Perryman, M., Andrady, A., Law, K.L., Plastic waste inputs from land into the ocean. *Sci.*, 347, 6223, 768–771, 2015, DOI: 10.1126/science.1260352.
24. Eriksen, M., Lebreton, L.C., Carson, H.S., Thiel, M., Moore, C.J., Borerro, J.C., Reisser, J., Plastic pollution in the world's oceans: more than 5 trillion plastic pieces weighing over 250,000 tons afloat at sea. *PLoS One*, 9, 12, e111913, 2014, https://doi.org/10.1371/journal.pone.0111913.

25. Schneider, D.R. and Ragossnig, A., Recycling and incineration, contradiction or coexistence?. *Waste Manage. Res.*, 33, 8, 693–695, 2015, doi: 10.1177/0734242X15593421.
26. Wu, W.M., Yang, J., Criddle, C.S., MPs pollution and reduction strategies. *FESE*, 11, 1–4, 2017, https://doi.org/10.1007/s11783-017-0897-7.
27. Prata, J.C., Silva, A.L.P., Da Costa, J.P., Mouneyrac, C., Walker, T.R., Duarte, A.C., Rocha-Santos, T., Solutions and integrated strategies for the control and mitigation of plastic and MP pollution. *Int. J. Environ. Res. Public Health*, 16, 13, 2411, 2019, https://doi.org/10.3390/ijerph16132411.
28. Liu, Z., Adams, M., Walker, T.R., Are exports of recyclables from developed to developing countries waste pollution transfer or part of the global circular economy?. *Resour. Conserv. Recycl.*, 136, 22–23, 2018, https://doi.org/10.1016/j.resconrec.2018.04.005.
29. Santander, P., Sanchez, F.A.C., Boudaoud, H., Camargo, M., Closed loop supply chain network for local and distributed plastic recycling for 3D printing: a MILP-based optimization approach. *Resour. Conserv. Recycl.*, 154, 104531, 2020, https://doi.org/10.1016/j.resconrec.2019.104531.
30. Schuyler, Q., Hardesty, B.D., Lawson, T.J., Opie, K., Wilcox, C., Economic incentives reduce plastic inputs to the ocean. *Mar. Policy*, 96, 250–255, 2018, https://doi.org/10.1016/j.marpol.2018.02.009.
31. Ignatyev, I.A., Thielemans, W., Vander Beke, B., Recycling of polymers: a review. *ChemSusChem*, 7, 6, 1579–1593, 2014, https://doi.org/10.1002/cssc.201300898.
32. Lea, D.A., Löffler, E., Douglas, I., Review symposium: Our common future: The world commission on environment and development. *Aust. Geogr.*, 20, 2, 195–201, 1989, https://doi.org/10.1080/00049188908702992.
33. Gottberg, A., Morris, J., Pollard, S., Mark-Herbert, C., Cook, M., Producer responsibility, waste minimisation and the WEEE Directive: Case studies in eco-design from the European lighting sector. *Sci. Total Environ.*, 359, 1–3, 38–56, 2006, https://doi.org/10.1016/j.scitotenv.2005.07.001.
34. Kunz, N., Mayers, K., Van Wassenhove, L.N., Stakeholder views on extended producer responsibility and the circular economy. *Calif. Manag. Rev.*, 60, 3, 45–70, 2018, https://doi.org/10.1177/0008125617752694.
35. Yu, J., Hills, P., Welford, R., Extended producer responsibility and eco-design changes: perspectives from China. *Corp. Soc. Resp. Env. Manag.*, 15, 2, 111–124, 2008, https://doi.org/10.1002/csr.168.
36. Favot, M., Grassetti, L., Massarutto, A., Veit, R., Regulation and competition in the extended producer responsibility models: Results in the WEEE sector in Europe. *Waste Manage.*, 145, 60–71, 2022, https://doi.org/10.1016/j.wasman.2022.04.027.
37. Foschi, E., Bonoli, A., The commitment of packaging industry in the framework of the European strategy for plastics in a circular economy. *Adm. Sci.*, 9, 1, 18, 2019, https://doi.org/10.3390/admsci9010018.

38. De Kock, L., Sadan, Z., Arp, R., Upadhyaya, P., A circular economy response to plastic pollution: Current policy landscape and consumer perception. *S. Afr. J. Sci.*, 116, 5–6, 1–2, 2020, https://doi.org/10.17159/sajs.2020/8097.
39. Diggle, A., Walker, T.R., Environmental and economic impacts of mismanaged plastics and measures for mitigation. *Environ.*, 9, 2, 15, 2022, https://doi.org/10.3390/environments9020015.
40. Kale, R.C. and Jain, M.S., A review study on plastic waste management: An Indian perspective, in: *Solid Waste Management for Resource-Efficient Systems*, pp. 233–253, Elsevier, Amsterdam, Netherlands, 2024. https://doi.org/10.1016/B978-0-443-23775-1.00019-9.
41. Leal Filho, W., Saari, U., Fedoruk, M., Iital, A., Moora, H., Klöga, M., Voronova, V., An overview of the problems posed by plastic products and the role of extended producer responsibility in Europe. *J. Cleaner Prod.*, 214, 550–558, 2019, https://doi.org/10.1016/j.jclepro.2018.12.256.
42. Knoblauch, D. and Mederake, L., Government policies combatting plastic pollution. *Curr. Opin. Toxicol.*, 28, 87–96, 2021, https://doi.org/10.1016/j.cotox.2021.10.003.
43. Gionfra, S., Richer, C., Watkins, E., The role of policy in tackling plastic waste in the aquatic environment, in: *Plastics in the Aquatic Environment-Part II: Stakeholders' Role Against Pollution*, pp. 119–138, Springer International Publishing, Cham, 2020, https://doi.org/10.1007/698_2020_484.
44. Kish, R.J., Using legislation to reduce one-time plastic bag usage. *Econ. Aff.*, 38, 2, 224–239, 2018, https://doi.org/10.1111/ecaf.12287.
45. Kosior, E. and Crescenzi, I., Solutions to the plastic waste problem on land and in the oceans, in: *Plastic waste and recycling*, pp. 415–446, Academic Press, Elsevier, Amsterdam, Netherlands, 2020, https://doi.org/10.1016/B978-0-12-817880-5.00016-5.
46. Zhu, Q., An appraisal and analysis of the law of "Plastic-Bag Ban". *Energy Procedia, 5*, 2516–2521, 2011, https://doi.org/10.1016/j.egypro.2011.03.432.
47. Anderson, L., Plastic waste in Australia-And the recycling greenwash-By 2050, the amount of plastic consumed in, 2024, https://coilink.org/20.500.12592/sbcc78f.
48. Jones, S. and Head, B.W., The limits of voluntary measures: packaging the plastic pollution problem in Australia. *Aust. J. Pub. Admin.*, 83, 4, 647–665, 2024, https://doi.org/10.1111/1467-8500.12614.
49. Johnson, H., Keane, K., McGillivray, L., Akhtar-Khavari, A., Chambers, L., Barner-Kowollik, C., Blinco, J., Reforming plastic packaging regulation: Outcomes from stakeholder interviews and regulatory analysis. *Sustainable Prod. Consumption*, 54, 52–63, 2025, https://doi.org/10.1016/j.spc.2024.12.017.
50. Sulaiman, S.A. and Ahmad, R.K., Plastic Waste Issue in Malaysia: Where Are We?, in: *Energy and Environment in the Tropics*, pp. 119–144, Springer Nature Singapore, Singapore, 2022, https://doi.org/10.1007/978-981-19-6688-0_8.

51. Cáceres Ruiz, A.M. and Zaman, A., The current state, challenges, and opportunities of recycling plastics in Western Australia. *Recycling*, 7, 5, 64, 2022, https://doi.org/10.3390/recycling7050064.
52. Macintosh, A., Simpson, A., Neeman, T., Dickson, K., Plastic bag bans: Lessons from the Australian capital territory. *Resour. Conserv. Recycl.*, 154, 104638, 2020, https://doi.org/10.1016/j.resconrec.2019.104638.
53. Sri SasiJyothsna, T. and Chakradhar, B., Current scenario of plastic waste management in India: way forward in turning vision to reality, in: Urban Mining and Sustainable Waste Management, pp. 203–218, Springer Nature Singapore Pte Ltd., Singapore, 2020. https://doi.org/10.1007/978-981-15-0532-4_21.
54. Bura, N., An overview of plastic waste management in India, in: *Waste Management and Resource Efficiency: Proceedings of 6th IconSWM 2016*, pp. 935–943, 2018, https://doi.org/10.1007/978-981-10-7290-1_78.
55. Iyer, P., Pandey, A., Vashisht, M., Smith, D.W., Swachh Bharat Mission or the Mission to Make India Clean: Addressing Open Defecation at Massive Scale (A), Indian Institute of Management Ahmedabad, pp. 1–10, 2023, https://doi.org/10.1108/CASE.IIMA.2023.000041.
56. Agarwal, N., Solanki, V., Awasthi, R.R., Bind, V., Edison, B., Verma, M., Kumar, V., Release and accumulation of pharmaceuticals in the environment: Critical risk assessment and challenges for environment, ecosystem and human health, 2023.
57. Miditana, S.R., Rao, T.S., Alim, S.A., Study on structural and optical properties of Ni and Mn co-doped TiO_2 nanomaterials and its application in visible light photocatalytic activity of Indigo Carmine. *J. Nanosci. Technol.*, 682–687, 2019, https://doi.org/10.30799/jnst.239.19050209.
58. O'Hagan, A.M., Paterson, S., Le Tissier, M., Addressing the tangled web of governance mechanisms for land-sea interactions: Assessing implementation challenges across scales. *Mar. Policy*, 112, 103715, 2020, https://doi.org/10.1016/j.marpol.2019.103715.
59. Claussen, U., Zevenboom, W., Brockmann, U., Topcu, D., Bot, P., Assessment of the eutrophication status of transitional, coastal and marine waters within OSPAR, in: *Eutrophication in Coastal Ecosystems: Towards better understanding and management strategies Selected Papers from the Second International Symposium on Research and Management of Eutrophication in Coastal Ecosystems*, 20–23 June 2006, Springer Netherlands, Nyborg, Denmark, pp. 49–58, 2009, https://doi.org/10.1007/978-90-481-3385-7_5.
60. Matz-Lück, N. and Fuchs, J., The impact of OSPAR on protected area management beyond national jurisdiction: Effective regional cooperation or a network of paper parks?. *Mar. Policy*, 49, 155–166, 2014, https://doi.org/10.1016/j.marpol.2013.12.001.
61. Ruiz, M. and Stankiewicz, M., Five years since the 2013 HELCOM ministerial declaration, in: *Plastics in the Aquatic Environment-Part II: Stakeholders'*

Role Against Pollution, pp. 179–204, Springer International Publishing, Cham, 2019, https://doi.org/10.1007/698_2019_378.
62. Strempel, R., Regional Action, Global Benefits: Central Elements of HELCOM's Contribution to UN Sustainable Development Goals, in: *SDGs in the European Region*, pp. 1–11, Springer International Publishing, Cham, 2023, https://doi.org/10.1007/978-3-030-91261-1_79-1.
63. Ogunola, O.S., Onada, O.A., Falaye, A.E., Mitigation measures to avert the impacts of plastics and MPs in the marine environment (a review). *Environ. Sci. Pollut. Res. Int.*, 25, 9293–9310, 2018, https://doi.org/10.1007/s11356-018-1499-z.
64. Boehmer-Christiansen, S., Marine pollution control in Europe: Regional approaches, 1972–80. *Mar. Policy*, 8, 1, 44–55, 1984, https://doi.org/10.1016/0308-597X(84)90056-3.
65. Constable, A.J., de la Mare, W.K., Agnew, D.J., Everson, I., Miller, D., Managing fisheries to conserve the Antarctic marine ecosystem: practical implementation of the Convention on the Conservation of Antarctic Marine Living Resources (CCAMLR). *ICES J. Mar. Sci.*, 57, 3, 778–791, 2000, https://doi.org/10.1006/jmsc.2000.0725.
66. Pettipas, S., Bernier, M., Walker, T.R., A Canadian policy framework to mitigate plastic marine pollution. *Mar. Policy*, 68, 117–122, 2016, https://doi.org/10.1016/j.marpol.2016.02.025.
67. Fossi, M.C., Vlachogianni, T., Galgani, F., DegliInnocenti, F., Zampetti, G., Leone, G., Assessing and mitigating the harmful effects of plastic pollution: the collective multi-stakeholder driven Euro-Mediterranean response. *Ocean Coast. Manag.*, 184, 105005, 2020, https://doi.org/10.1016/j.ocecoaman.2019.105005.
68. Kuo, F.J. and Huang, H.W., Strategy for mitigation of marine debris: analysis of sources and composition of marine debris in northern Taiwan. *Mar. Pollut. Bull.*, 83, 1, 70–78, 2014, https://doi.org/10.1016/j.marpolbul.2014.04.019.
69. Bottari, T., Savoca, S., Mancuso, M., Capillo, G., GiuseppePanarello, G., MartinaBonsignore, M., Fazio, E., Plastics occurrence in the gastrointestinal tract of Zeus faber and Lepidopuscaudatus from the Tyrrhenian Sea. *Mar. Pollut. Bull.*, 146, 408–416, 2019, https://doi.org/10.1016/j.marpolbul.2019.07.003.
70. Eriksen, M., Borgogno, F., Villarrubia-Gómez, P., Anderson, E., Box, C., Trenholm, N., Mitigation strategies to reverse the rising trend of plastics in Polar Regions. *Environ. Int.*, 139, 105704, 2020, https://doi.org/10.1016/j.envint.2020.105704.
71. Villarrubia-Gómez, P., Cornell, S.E., Fabres, J., Marine plastic pollution as a planetary boundary threat–The drifting piece in the sustainability puzzle. *Mar. Policy*, 96, 213–220, 2018, https://doi.org/10.1016/j.marpol.2017.11.035.
72. Rajmohan, K.V.S., Ramya, C., Viswanathan, M.R., Varjani, S., Plastic pollutants: effective waste management for pollution control and abatement.

Curr. Opin. Environ. Sci. Health, 12, 72–84, 2019, https://doi.org/10.1016/j.coesh.2019.08.006.
73. Kokalj, A.J., Kuehnel, D., Puntar, B., Gotvajn, A.Ž., Kalčikova, G., An exploratory ecotoxicity study of primary MPs versus aged in natural waters and wastewaters. *Environ. Pollut.*, 254, 112980, 2019, https://doi.org/10.1016/j.envpol.2019.112980.
74. Rochman, C.M., Cook, A.M., Koelmans, A.A., Plastic debris and policy: Using current scientific understanding to invoke positive change. *Environ. Toxicol. Chem.*, 35, 7, 1617–1626, 2016, https://doi.org/10.1002/etc.3408.
75. Hartley, K., van Santen, R., Kirchherr, J., Policies for transitioning towards a circular economy: Expectations from the European Union (EU). *Resour. Conserv. Recycl.*, 155, 104634, 2020, https://doi.org/10.1016/j.resconrec.2019.104634.
76. Prata, J.C., Silva, A.L.P., Da Costa, J.P., Mouneyrac, C., Walker, T.R., Duarte, A.C., Rocha-Santos, T., Solutions and integrated strategies for the control and mitigation of plastic and MP pollution. *Int. J. Environ. Res. Public Health*, 16, 13, 2411, 2019, https://doi.org/10.3390/ijerph16132411.
77. Munhoz, D.R., Harkes, P., Beriot, N., Larreta, J., Basurko, O.C., MPs: a review of policies and responses. *MPs*, 2, 1, 1–26, 2022, https://doi.org/10.3390/MPs2010001.
78. Brennan, C., Ashley, M., Molloy, O., A system dynamics approach to increasing ocean literacy. *Front. Mar. Sci.*, 6, 360, 2019, https://doi.org/10.3389/fmars.2019.00360.
79. Clausen, L.P.W., Hansen, O.F.H., Oturai, N.B., Syberg, K., Hansen, S.F., Stakeholder analysis with regard to a recent European restriction proposal on MPs. *PLoS One*, 15, 6, e0235062, 2020, https://doi.org/10.1371/journal.pone.0235062.
80. Henderson, L. and Green, C., Making sense of MPs? Public understandings of plastic pollution. *Mar. Pollut. Bull.*, 152, 110908, 2020, https://doi.org/10.1016/j.marpolbul.2020.110908.
81. Khoramnejadian, S., Zavareh, J.J., Khoramnejadian, S., Bio-based plastic a way for reduce municipal solid waste. *Procedia Eng.*, 21, 489–495, 2011, https://doi.org/10.1016/j.proeng.2011.11.2042.
82. Agarwal, N., Srivastava, A., Solanki, V., Singh, N., Yadav, R., Emerging trends in the use of microbial isolates for the treatment of dye containing wastewater, in: *Emerging Innovative Trends in the Application of Biological Processes for Industrial Wastewater Treatment*, Elsevier, Amsterdam, Netherlands, 2024. https://doi.org/10.1016/B978-0-443-13561-3.00010-7.
83. Klemeš, J.J., Fan, Y.V., Jiang, P., Plastics: friends or foes? The circularity and plastic waste footprint. *Energy Sources Part A*, 43, 13, 1549–1565, 2021, https://doi.org/10.1080/15567036.2020.1801906.
84. Misund, A., Tiller, R., Canning-Clode, J., Freitas, M., Schmidt, J.O., Javidpour, J., Can we shop ourselves to a clean sea? An experimental panel approach to assess the persuasiveness of private labels as a private governance approach to

MP pollution. *Mar. Pollut. Bull.*, 153, 110927, 2020, https://doi.org/10.1016/j.marpolbul.2020.110927.
85. Henry, B., Laitala, K., Klepp, I.G., Microfibres from apparel and home textiles: Prospects for including MPs in environmental sustainability assessment. *Sci. Total Environ.*, 652, 483–494, 2019, https://doi.org/10.1016/j.scitotenv.2018.10.166.
86. Napper, I.E., Thompson, R.C., Release of synthetic MP plastic fibres from domestic washing machines: Effects of fabric type and washing conditions. *Mar. Pollut. Bull.*, 112, 1–2, 39–45, 2016, https://doi.org/10.1016/j.marpolbul.2016.09.025.
87. Hartline, N.L., Bruce, N.J., Karba, S.N., Ruff, E.O., Sonar, S.U., Holden, P.A., Microfiber masses recovered from conventional machine washing of new or aged garments. *Environ. Sci. Technol.*, 50, 21, 11532–11538, 2016, DOI:10.1021/acs.est.6b03045.
88. Baensch-Baltruschat, B., Kocher, B., Stock, F., Reifferscheid, G., Tyre and road wear particles (TRWP)-A review of generation, properties, emissions, human health risk, ecotoxicity, and fate in the environment. *Sci. Total Environ.*, 733, 137823, 2020, https://doi.org/10.1016/j.scitotenv.2020.137823.
89. Wagner, S., Hüffer, T., Klöckner, P., Wehrhahn, M., Hofmann, T., Reemtsma, T., Tire wear particles in the aquatic environment-a review on generation, analysis, occurrence, fate and effects. *Water Res.*, 139, 83–100, 2018, https://doi.org/10.1016/j.watres.2018.03.051.
90. Tofa, T.S., Ye, F., Kunjali, K.L., Dutta, J., Enhanced visible light photodegradation of MP fragments with plasmonic platinum/zinc oxide nanorodphotocatalysts. *Catal.*, 9, 10, 819, 2019, https://doi.org/10.3390/catal9100819.
91. Bombelli, P., Howe, C.J., Bertocchini, F., Polyethylene bio-degradation by caterpillars of the wax moth Galleria mellonella. *Curr. Biol.*, 27, 8, R292–R293, 2017, https://www.cell.com/current-biology/fulltext/S0960-9822(17)30231-2?xtor=AL-32280680.
92. Huerta Lwanga, E., Gertsen, H., Gooren, H., Peters, P., Salánki, T., Van Der Ploeg, M., Geissen, V., MPs in the terrestrial ecosystem: implications for Lumbricusterrestris (Oligochaeta, Lumbricidae). *Environ. Sci. Technol.*, 50, 5, 2685–2691, 2016, DOI: 10.1021/acs.est.5b05478.
93. Yang, Y., Yang, J., Wu, W.M., Zhao, J., Song, Y., Gao, L., Jiang, L., Biodegradation and mineralization of polystyrene by plastic-eating mealworms: part 2. Role of gut microorganisms. *Environ. Sci. Technol.*, 49, 20, 12087–12093, 2015, DOI: 10.1021/acs.est.5b02663.
94. Delacuvellerie, A., Cyriaque, V., Gobert, S., Benali, S., Wattiez, R., The plastisphere in marine ecosystem hosts potential specific microbial degraders including Alcanivoraxborkumensis as a key player for the low-density polyethylene degradation. *J. Hazard. Mater.*, 380, 120899, 2019, https://doi.org/10.1016/j.jhazmat.2019.120899.
95. Baytekin, B., Baytekin, H.T., Grzybowski, B.A., Retrieving and converting energy from polymers: deployable technologies and emerging concepts.

Energy Environ. Sci., 6, 12, 3467–3482, 2013, DOI-https://doi.org/10.1039/C3EE41360H.

96. Poulikakos, L.D., Papadaskalopoulou, C., Hofko, B., Gschösser, F., Falchetto, A.C., Bueno, M., Partl, M.N., Harvesting the unexplored potential of European waste materials for road construction. *Resour. Conserv. Recycl.*, 116, 32–44, 2017, https://doi.org/10.1016/j.resconrec.2016.09.008.
97. Shen, M., Song, B., Zhu, Y., Zeng, G., Zhang, Y., Yang, Y., Yi, H., Removal of MPs *via* drinking water treatment: Current knowledge and future directions. *Chemosphere*, 251, 126612, 2020, https://doi.org/10.1016/j.chemosphere.2020.126612.
98. Bayo, J., López-Castellanos, J., Olmos, S., Membrane bioreactor and rapid sand filtration for the removal of MPs in an urban wastewater treatment plant. *Mar. Pollut. Bull.*, 156, 111211, 2020, https://doi.org/10.1016/j.marpolbul.2020.111211.
99. deCarvalho, D.G. and Neto, J.A.B., MP pollution of the beaches of Guanabara Bay, Southeast Brazil. *Ocean Coast. Manag.*, 128, 10–17, 2016, https://doi.org/10.1016/j.ocecoaman.2016.04.009.
100. Andres, M., Delpey, M., Ruiz, I., Declerck, A., Sarrade, C., Bergeron, P., Basurko, O.C., Measuring and comparing solutions for floating marine litter removal: lessons learned in the south-east coast of the Bay of Biscay from an economic perspective. *Mar. Policy*, 127, 104450, 2021, https://doi.org/10.1016/j.marpol.2021.104450.

13

Plastic Waste Management: Roadmap for Circular Economy

Subrata Jana[1*], Dinkar Verma[2], Vikrant Sharma[3], Mangilal Choudhary[4], Satya Prakash Pati[5] and Michael Annor-Nyarko[6]

[1]Department of Nuclear Science and Technology, School of Engineering & Technology, Mody University of Science and Technology, Laxmangargh, Rajasthan, India
[2]Department of Basic and Applied Sciences, School of Engineering & Sciences, GD Goenka University, Gurugram, Haryana, India
[3]Department of Mechanical Engineering, School of Engineering & Technology, Mody University of Science and Technology, Laxmangargh, Rajasthan, India
[4]Department of Physics and Astrophysics, University of Delhi, New Delhi, Delhi, India
[5]Department of Electrical and Electronic Engineering, Kyushu University, Nishi-ku, Fukuoka, Japan
[6]Nuclear Installations Directorate, Nuclear Regulatory Authority, Kwabenya, Accra, Ghana

Abstract
Plastic waste has become a serious global environmental and social challenge. The increasing volume of plastic that accumulates in landfills, oceans, and natural ecosystems poses a major threat to the health of the planet and the communities that depend on it. Addressing this issue requires a shift toward a circular economy model, one that prioritizes reducing, reusing, and recycling plastic products to minimize their environmental impact. A key strategy within this model is the development of decentralized, community-based circular systems for managing plastic waste, particularly in low-income regions where recycling and waste management infrastructure is often limited. By ensuring that plastic waste is collected, processed, remanufactured, and reused within the same community, both the environmental and economic costs associated with long-distance transportation and centralized processing can be significantly reduced.

**Corresponding author*: eiesubrata@gmail.com

Neha Agarwal, Brijesh Pare, Vijendra Singh Solanki and Sankara Rao Miditana (eds.) Plastic Remediation: Transformative Pathways for a Rejuvenated Sustainable Environment, (359–394)

Furthermore, integrating AI-powered waste management technologies can enhance the efficiency of waste collection, sorting, and processing, leading to smarter, more sustainable, and cost-effective solutions. This chapter presents a comprehensive framework for applying circular economy principles to plastic waste management. It emphasizes the essential roles of governments, businesses, consumers, and civil society in enabling this transition. Ultimately, the study provides stakeholders with a roadmap for implementing sustainable, circular solutions to plastic waste, strengthening global resource efficiency and building a more resilient economy.

Keywords: Plastic waste management, circular economy, SDGs, sustainable solutions to plastic waste, net zero carbon and sustainability

13.1 Introduction

Plastic waste harms people and the environment because it is produced in large quantities, does not decompose easily, and is often disposed of improperly. Each year, millions of tons of plastic end up in landfills, oceans, and ecosystems, releasing greenhouse gases and microplastics that pollute air and water. Single-use plastics, especially packaging, make the problem worse. Wildlife is injured, and chemicals leaching from plastics can affect human health. Despite growing awareness, recycling rates remain low. To reduce the economic and environmental impacts of plastic pollution, we must change how plastics are designed, used, and discarded [1–5]. A circular economy approach is more effective than traditional recycling because it considers the full life cycle of plastic products. It emphasizes designing long-lasting, repairable, and easily recyclable materials. Developing bio-based and biodegradable alternatives can further reduce pollution. Advanced recycling technologies, including chemical recycling and pyrolysis, break plastics down into their basic components to create new, high-quality materials. Reuse and refill systems also cut reliance on single-use plastics. Additionally, digital tools such as AI and blockchain improve waste management by optimizing collection, enhancing sorting accuracy, and ensuring traceability, increasing efficiency and trust in recycling systems [6–10].

Transitioning to a circular economy for plastic waste requires a top-down approach involving stakeholders across industries. Governments play a crucial role by creating policies that support sustainable practices. For example, extended producer responsibility (EPR) policies hold manufacturers accountable for a product's entire lifecycle, while bans on single-use plastics push companies to develop reusable alternatives [11, 12].

These measures encourage businesses to adopt circular models and drive innovation throughout the value chain.

Companies are increasingly integrating circular principles by using recyclable materials, designing products for easy disassembly, and implementing closed-loop supply chains that return materials back into production [13–15]. New business models, such as product-as-a-service, shift the focus from ownership to access, promoting durability and resource efficiency. However, challenges remain, including the difficulty of recycling complex plastics, scaling circular systems, and changing consumer behavior. Continued investment in research, supportive policies, and widespread public education is essential to achieving a fully circular plastic economy that benefits the environment, the economy, and society [16, 17].

13.1.1 Overview of Plastic Waste and Its Global Impact

Plastic waste harms the environment worldwide because it is used in large quantities, does not break down easily, and is often disposed of improperly. Every year, millions of tons of plastic end up in landfills, oceans, and ecosystems. During production and as it degrades, plastic releases pollutants such as greenhouse gases and microplastics, which harm wildlife,

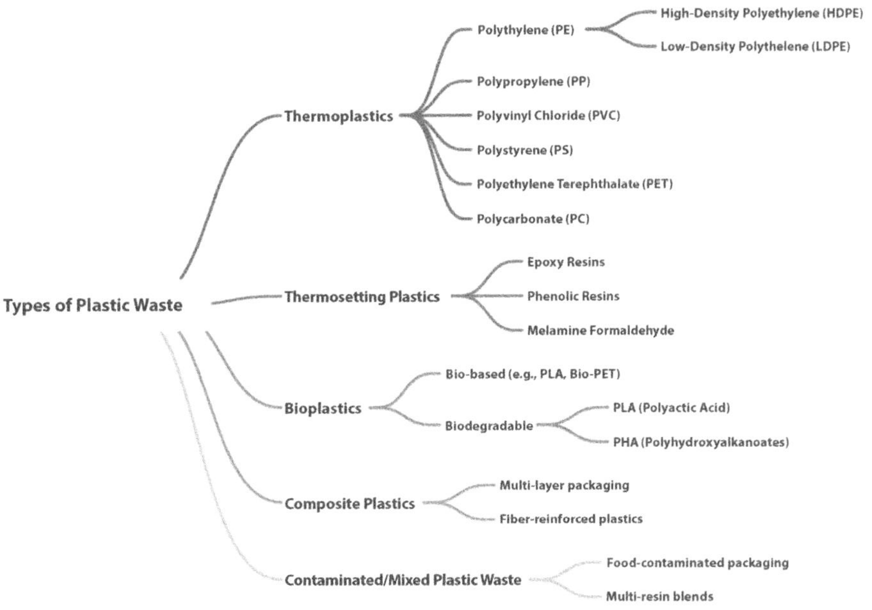

Figure 13.1 Different types of plastic waste.

contaminate water, and worsen environmental problems. Much of this waste comes from single-use plastics, especially packaging that are often recyclable only once, further adding to the issue.

Marine animals can be injured or killed when they ingest or become entangled in plastic, and humans may be exposed to harmful chemicals that leach into food and water. Although awareness of plastic pollution is increasing, recycling rates remain low. To reduce the long-term economic and environmental impacts of plastic waste, we must rethink how products are designed, used, and discarded. Figure 13.1 shows the different categories of plastic waste [18, 19].

13.1.2 The Plastic Waste Management: Need for Advanced Waste Management Solutions

The rising volume of plastic waste shows that traditional waste management methods such as landfilling, incineration, and limited recycling are no longer sufficient. These systems cannot handle the massive amounts of plastic being produced, leading to environmental damage, water contamination, and health risks. Landfills can pollute soil and groundwater, while incineration releases harmful chemicals and greenhouse gases (Figure 13.2). Although recycling is beneficial, it faces challenges such as contamination, inadequate infrastructure, and the high cost of processing certain plastics. As a result, much plastic waste remains untreated, harming wildlife, marine life, and overall ecosystem health. More effective waste management solutions are urgently needed to address these issues and reduce the environmental impact of plastic pollution.

The plastic waste problem can be addressed by adopting more effective disposal and recycling methods, such as biodegradable alternatives, chemical recycling, and AI-based sorting systems. Chemical recycling breaks plastics down into their basic components, making it easier to create new, high-quality products. Biodegradable and compostable plastics help reduce long-term pollution. Smart waste management technologies, such as automated sorting and block chain tracking, improve recycling efficiency and transparency [20–24].

Circular economy models further support reuse and recycling, benefiting both the environment and consumers. However, progress requires cooperation among individuals, businesses, and governments. Investments in new technologies, stronger regulations, and changes in consumer behavior are essential. Without better waste solutions, plastic pollution will continue to threaten human health, ecosystems, and global sustainability goals (SDGs).

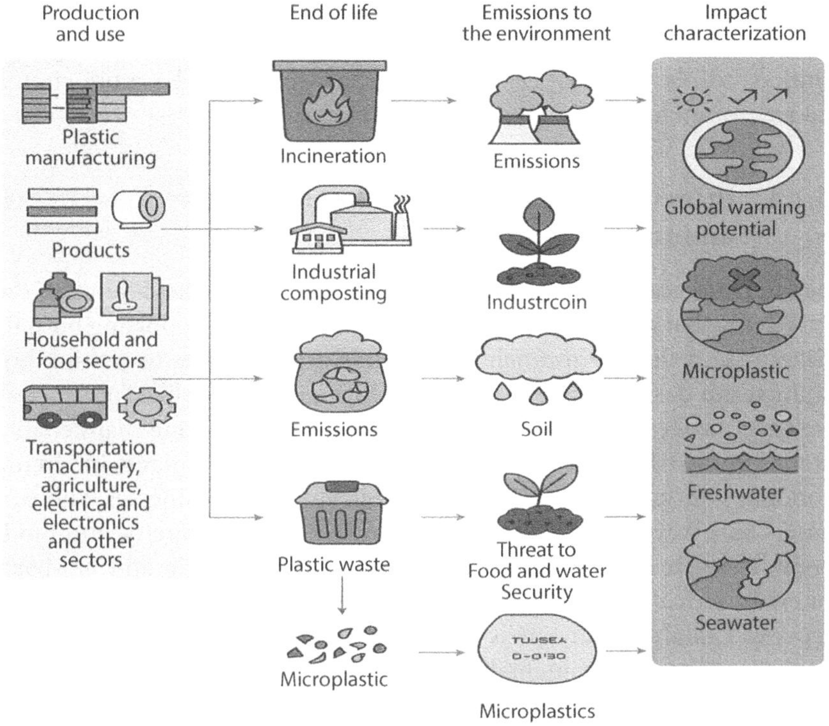

Figure 13.2 Schematic representation of plastic waste and its impact on environment.

13.1.3 Sustainable Solutions: Relevance in the Context of Circular Economy

Several techniques and tools can help manage plastic waste more effectively. These include advanced sorting and recycling technologies, chemical recycling methods that break plastics down into their molecular components for reuse, and biodegradable plastics that safely decompose in the environment [25]. Sustainable solutions also require new business models, such as deposit-return systems and plastic credit programs that make collecting and recycling plastic easier.

Collaboration among governments, businesses, and consumers is essential. Policymakers can support progress by regulating single-use plastics and holding manufacturers responsible for the waste they generate. Many companies are adopting circular design principles, developing recycling systems, and exploring alternative materials. Public awareness and proper disposal habits are also crucial. By combining these approaches, long-term solutions to plastic waste can be achieved, helping build a more circular and environmentally friendly economy [19, 26].

13.1.4 Principles of Circular Economy: Concept and Importance

The circular economy model goes beyond reducing waste; it aims to create systems that are restorative and regenerative while supporting both the economy and the environment. It requires organizations to rethink how products are designed, produced, and delivered. Models like "product-as-a-service," where companies retain ownership and provide maintenance, extend product life and reduce resource use [27, 28]. Interest in circular economy solutions has grown among policymakers, businesses, and consumers as a way to address climate change, resource scarcity, and biodiversity loss. This shift involves developing biodegradable and bio-based materials, advancing recycling technologies, and improving material recovery. Collaboration across industries can create new business opportunities, reduce environmental impacts, and support social equity.

Achieving a circular economy requires changes in behavior, business practices, and policy. Education and awareness campaigns help individuals and companies adopt circular habits, while governments must implement supportive regulations, such as extended producer responsibility (EPR) to encourage sustainable design [29, 30]. Ultimately, a circular economy reduces resource use, lowers greenhouse gas emissions, minimizes landfill waste, and supports long-term environmental and economic sustainability.

13.1.5 Linear Economy and Circular Economy

The comparison of linear economy and circular economy is shown in Figure 13.3 and Table 13.1 in detail.

13.1.6 Opportunities in Adopting Circular Economy Practices

The circular economy encourages people, businesses, and governments to use resources more efficiently by focusing on reuse, recycling, and renewal instead of the traditional "take, make, dispose" model. This approach reduces waste, saves resources, and lowers emissions. Practices such as eco-design, durable products, and improved recycling create jobs, cut

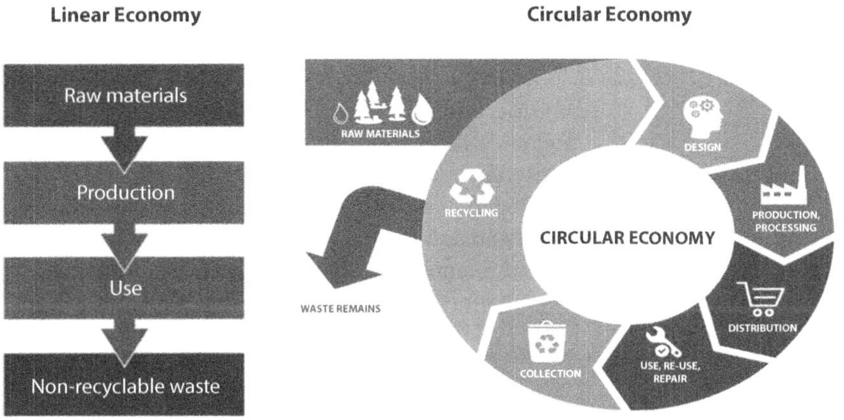

Figure 13.3 Linear economy (left) and circular economy (right) [30].

costs, support clean energy, and strengthen innovation. Companies that adopt circular strategies gain a competitive advantage as consumers and regulators increasingly prefer sustainable options [33].

A circular economy benefits both the environment and the economy. By minimizing waste and using resources wisely, organizations can increase revenue and open new markets in areas like reverse logistics, material recovery, and product-as-a-service models. It also drives innovation, leading to new technologies and solutions for environmental challenges. Communities benefit as well, through fairer resource use, greater participation in recycling and reuse programs, and access to more environmentally friendly products [34, 35]. The success of circular economy initiatives depends on strong collaboration among governments, businesses, and consumers. Supportive policies—such as extended producer responsibility and tax incentives for recycled materials—help accelerate progress. Education and awareness campaigns are also essential for building a culture of sustainability. As resource scarcity and climate change intensify, the circular economy offers a resilient and sustainable path forward for the environment, the economy, and society.

13.1.7 Benefits of Circularity in Plastic Waste Management

The circular economy offers more than environmental benefits as it provides a strategic advantage for businesses. By adopting circular practices, companies can drive innovation, meet regulatory requirements, and satisfy growing consumer demand for sustainable products (Figure 13.4).

Table 13.1 Comparison of linear economy and circular economy [31, 32].

Aspect	Linear economy	Circular economy
Aim	A linear economy follows a short-term wasteful approach, leading to resource depletion and environmental degradation	A circular economy focuses on sustainability, longevity, and regeneration, reducing environmental harm while creating economic value
Model	Take–make–dispose (one-time use)	Closed-loop system (reuse, recycle, regenerate)
Waste Management	Generates waste (landfills, incineration)	Minimizes waste (designs out waste, promotes recycling)
Resource Use	Depletes finite resources (extract-use-discard)	Maximizes resource efficiency (renewable/ regenerative materials)
Product Design	Short-lived, single-use, planned obsolescence	Durable, repairable, modular, and recyclable
Energy Consumption	Relies on fossil fuels, high energy demand	Prioritizes renewable energy and energy efficiency
Economic Impact	High costs of waste disposal and raw material extraction	Cost savings, new business models (leasing, sharing)
Environmental impact	Contributes to pollution, cause of climate change, and biodiversity loss	Reduces pollution, decrease carbon emissions, and ecosystem harm
Consumer role	Promotes disposable, buy-and-throw culture	Encourages responsible use (repair, reuse, recycle)
Example	Single-use plastics, fast fashion, disposable product	Refurbished electronics, biodegradable packaging, and industrial symbiosis

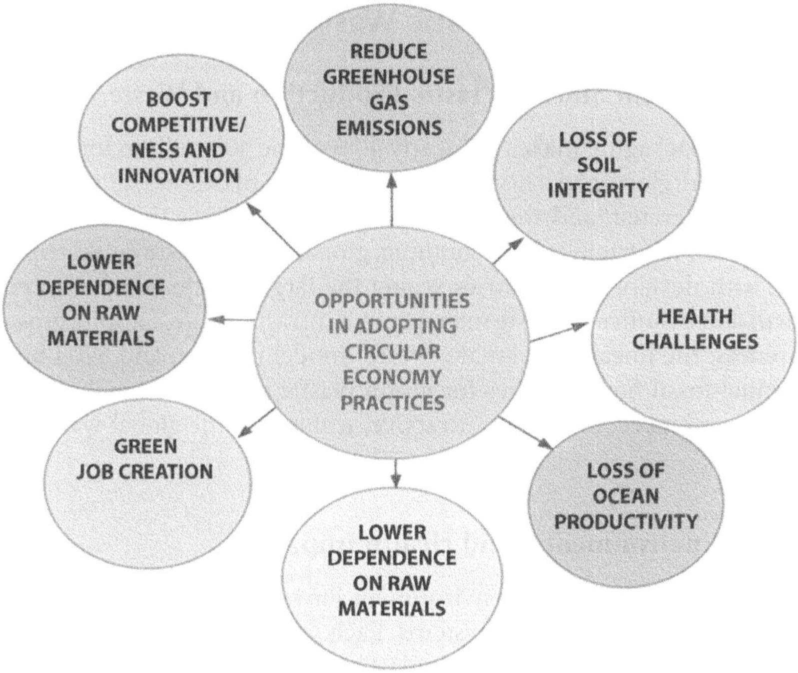

Figure 13.4 List on benefits in adopting circular economy practices.

Moving away from the traditional "take–make–waste" model toward reuse, recycling, and regeneration creates new opportunities for businesses, governments, and society. Circular approaches reduce material and waste management costs while opening new revenue streams through models such as product-as-a-service, remanufacturing, and industrial symbiosis [36–39].

This shift also strengthens business resilience by reducing dependence on scarce resources and improving supply chain stability. Successful examples such as Unilever's zero-waste factories and Tesla's battery recycling programs demonstrate the circular economy's potential across industries. However, challenges like high initial investment and limited infrastructure require collaboration through public–private partnerships and consumer education. As global awareness grows, the circular economy is becoming both an environmental necessity and a competitive advantage. Companies that adopt this model early can turn waste into value, support sustainable development, and build long-term economic and environmental resilience [40–42].

13.2 Challenges in Plastic Waste Management

13.2.1 Current State of Plastic Production and Waste

Single-use packaging makes up nearly 40% of the 400 million tons of plastic produced globally each year. Yet only 9% of plastic waste is recycled, 19% is incinerated, and the rest ends up in landfills or the environment (Table 13.2). If current trends continue, global plastic waste could triple by 2060, with developing countries seeing the largest increases due to rapid growth and insufficient waste management infrastructure. Despite growing awareness, the linear "take–make-waste" model persists, and $204 billion in annual fossil fuel subsidies for virgin plastic production make circular alternatives less economically attractive, worsening the global waste crisis [43].

13.2.2 Environmental and Health Impacts

Plastic pollution affects the environment far beyond visible waste, reaching even the most remote ecosystems. Each year, 11 million metric tons of plastic enter the oceans, and microplastics have been detected in the food chain and human bodies, raising concerns about long-term health effects [44]. In developing countries with poor waste management, burning plastic releases toxic chemicals like dioxins, which can cause cancer and respiratory problems, especially for communities near informal recycling sites. Chemicals in plastics, such as phthalates, can disrupt hormones, metabolism, and reproductive health and have been found in blood, breast milk, and the placenta. This highlights the urgent need for global strategies to reduce plastic production; improve waste management; and develop durable, safe alternatives to protect both current and future generations [45].

13.2.3 Technical and Design Limitations

Recycling flexible packaging faces significant technical challenges, especially with multi-material laminates, which are difficult to separate into individual components and hinder effective recycling [46, 47]. Mechanical recycling also degrades polymer quality, limiting how often materials can be reused and reducing overall recycling efficiency. These issues are further worsened by poor product design, as many items lack single-material construction or easily separable components that would simplify recycling.

Table 13.2 Current state of plastic production and waste generation in India and on global level [40–43].

	Indian scenario	Global scenario
Plastic production:	- Annual plastic production: about 17 million tons. - Expected to reach 22 million tons by 2025 - Major producers: Maharashtra, Gujarat, and Tamil Nadu.	- About 300 million tons of plastic waste are made every year. - Only 9% of plastic waste is recycled around the world. - 79% of plastic waste ends up in landfills or the natural environment.
Plastic waste:	- Amount of plastic waste made each year: 9.46 million tons. - Only 60% of plastic waste is recycled; the other 40% is not collected and usually ends up in landfills or bodies of water.	- About 300 million tons of plastic waste are made every year. - Only 9% of plastic waste is recycled around the world. - 79% of plastic waste ends up in landfills or in the atmosphere. (Sector-wise percentage of plastic consumption is shown in Figure 13.5)
Recent initiatives:	- Rules for Managing Plastic Waste, 2016 (changed in 2018). - No more single-use plastics (starting in July 2022). - Rules for Extended Producer Responsibility (EPR).	- The UN Environment Assembly passed a resolution to stop plastic pollution. - Many countries and regions have banned single-use plastics. - There is more focus on circular economy approaches.
Key challenges:	- Single-use plastics: It makes a lot of waste materials. - Inadequate facilities for managing waste. - Low rates of reuse and recycling. - Less awareness and typical mindset to change attitude toward the environment. - Pollution of the environment, especially plastic pollution in the ocean.	
Future outlook:	- More focus on biodegradable plastics and sustainable options. - More money going into better recycling and waste management technologies. - More need for stronger policies and cooperation between countries.	

13.2.4 Economic and Policy Barriers

The main reason plastics are widely used is their low cost, 30–40% cheaper than recycled plastics, because fossil fuel subsidies and environmental costs are not included in the price. Price fluctuations in oil further destabilize the recycling market. Only 30 countries have implemented Extended Producer Responsibility (EPR) laws, and enforcement is often weak, leaving 86% of plastic packaging unaccountable. Policy gaps are significant: while the EU has advanced circular economy programs, many Asian and African countries lack basic waste management systems. Global plastic waste trade can worsen environmental problems; for example, after China banned imports in 2018, waste exports to Southeast Asia rose by 35% [48].

13.2.5 Behavioral and Social Challenges

A deeply ingrained "throwaway culture" drives excessive plastic use. For example, an average plastic bag is used for just 12 minutes but takes 500 years to decompose. Surveys show only 14% of people actively avoid single-use plastics, and 25–40% of recycling streams are contaminated due to misunderstandings of recycling symbols. Informal waste pickers, who handle 60% of the world's plastic recycling, face social stigma and hazardous working conditions, reducing system efficiency. Additionally, green washing by the industry, 68% of plastic producers' sustainability claims are false according to regulatory reviews, slows meaningful cultural and behavioral change.

13.3 Road-Map for Transitioning to a Circular Economy

13.3.1 Sustainable Product Design and Materials Innovation

A circular economy is based on the idea that we must rethink how we design and make things. The goal of sustainable product design is to have as little impact on the environment as possible by using fewer materials, getting rid of harmful chemicals, and making sure that products last longer. People from different fields need to work together to set eco-design standards so that products are useful and good for the surroundings [49, 50].

13.3.1.1 Sustainable Plastic Design

Creating environmentally friendly plastics involves more than just selecting safe materials. Designers are focusing on lightweight, recyclable, and plant-based or biodegradable plastics, as well as innovative designs that use less material without compromising strength; for example, containers with ribs or honeycomb structures. Applying circular economy principles is also key: modular products that are easy to repair or upgrade, and using recycled plastics, help extend product life and reduce waste [51–53]. Eco-friendly design not only benefits the environment but also lowers material costs and waste for companies, making sustainable plastics increasingly practical and efficient.

13.3.1.2 Design for Recyclability

Product design should ensure modularity and allow items to be easily disassembled for recycling. This includes using a single type of polymer, avoiding mixed materials, and clearly labeling components to guide both consumers and recyclers. Features like removable adhesives and modular construction make it easier to recover materials. These strategies are essential for achieving high recycling rates and closing the loop in material flows.

13.3.1.3 Alternative Materials and Bioplastics

The development of novel materials, such as bioplastics, offers a way to reduce reliance on fossil fuel–based products. To be effective, these materials must be compatible with existing waste management systems and avoid issues like microplastic pollution. Innovations like mycelium-based packaging and seaweed-derived polymers provide more sustainable options, but scaling these solutions and making them affordable remains a major challenge [53].

13.3.2 Waste Collection and Segregation Systems

A circular economy requires that resources be reused and recycled rather than discarded, which depends on effective waste collection and sorting systems. System design should consider factors like population density, infrastructure, and local waste generation. Urban areas may use automated collection, strategically placed recycling bins, and scheduled pickups, while rural areas can rely on community collection points, mobile units, and education on proper sorting [54].

Advanced sorting facilities with AI and robotics improve recovery efficiency, but community participation is equally vital. Incentives, awareness campaigns, and educational programs help foster a culture of responsible waste management, encouraging proper separation and overall waste reduction [32].

13.3.3 Policy and Regulatory Frameworks for Plastic Waste Management

Effective policies and regulations are essential for transforming plastic waste management, ensuring accountability, and promoting sustainable behaviors. Around the world, governments are implementing laws, economic incentives, and stakeholder collaborations to shift from linear disposal models to circular systems. These frameworks not only reduce waste but also encourage innovative material use and recovery, supporting long-term sustainability [55, 56].

13.3.4 Business Models for Circularity

More and more companies are using models that focus on making products that last, using resources wisely, keeping materials in a closed loop, and making sure that making money doesn't hurt the environment.

13.3.4.1 Product-as-a-Service Models

Product-as-a-service (PaaS) is a new way of thinking about ownership that moves it from consumers to producers. Producers lease things like electronics and packaging and are responsible for keeping them in good shape, fixing them, and getting them back at the end of their lives.

13.3.4.2 Closed-Loop Supply Chains

Closed-loop supply chains reduce the need for new resources by incorporating recyclables into production. By using recycled materials and offering take-back programs, businesses can make supply chains more resilient, efficient, and environmentally friendly. These loops are further strengthened through cross-industry collaboration, such as trading materials between sectors. Key factors for transitioning to a closed-loop circular economy are illustrated in Figure 13.5.

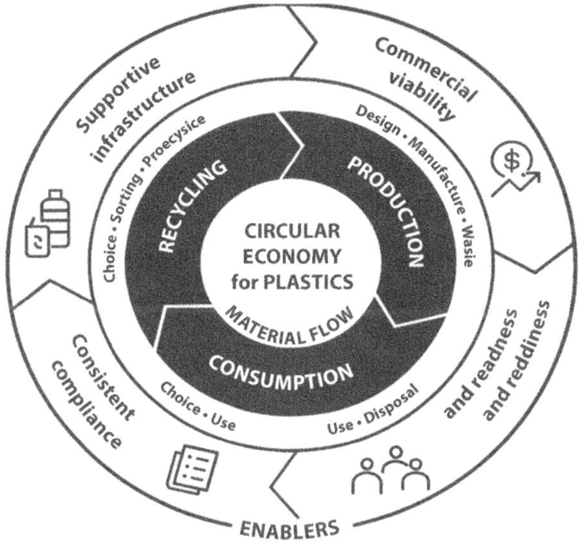

Figure 13.5 Important factors for transitioning to a circular economy.

13.4 Important Strategies for Plastic Waste Management in a Circular Economy

13.4.1 Eco-Friendly Design and Recyclability

Circular plastic management is rooted in eco-design, which prioritizes safety, simplicity, and recyclability. This approach uses mono-materials, such as single-polymer packaging, and eliminates harmful additives like phthalates. Examples include modular electronics with snap-fit parts that simplify material recovery. Standardized labeling, such as resin codes, further supports recycling and reduces contamination. By combining these strategies, businesses can prevent plastics from ending up as waste [57, 58].

13.4.2 Development and Use of Bioplastics

Bio-based plastics, such as polylactic acid (PLA) and polyhydroxyalkanoates (PHA), are plant-derived and offer a renewable alternative to fossil fuel-based plastics [59–61]. Their effectiveness depends on compatibility with end-of-life processes: compostable bioplastics require industrial

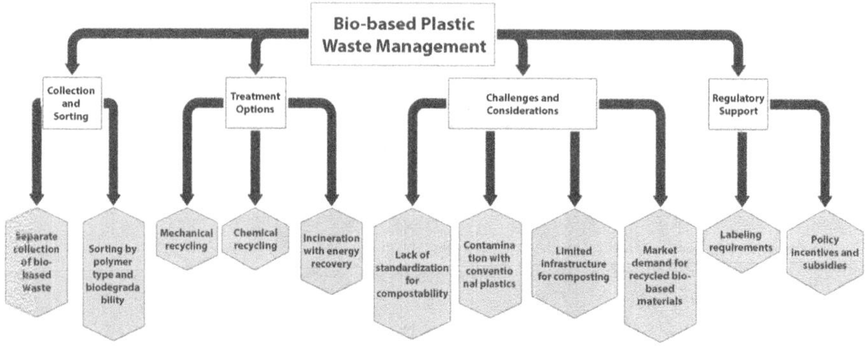

Figure 13.6 Bioplastic waste management.

composting facilities, while "drop-in" bioplastics, chemically identical to conventional plastics, can be integrated into existing recycling systems. Expanding their use requires addressing challenges like land competition and microplastic formation through effective policies and innovative solutions (Figure 13.6).

13.4.3 Waste Collection and Sorting Systems

Achieving circularity requires efficient collection and sorting systems (Figure 13.7) to prevent plastics from entering ecosystems. Strategies such as door-to-door collection, separate recycling bins, and deposit-return schemes (DRSs) increase participation. In developing countries, informal

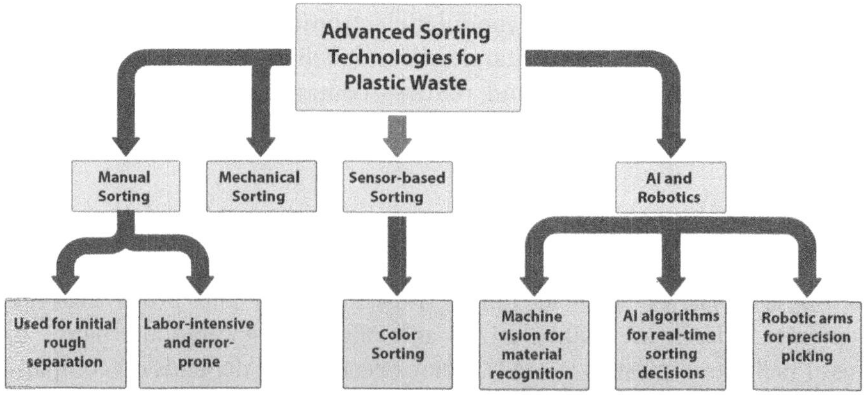

Figure 13.7 Different types of advanced sorting technologies.

sectors like waste picker cooperatives significantly boost recovery rates. Clear reporting and community education also help maintain cleanliness, ensuring recyclers receive higher-quality materials [62–67].

13.4.4 Enhancing Waste Collection Infrastructure

Investing in infrastructure improvements such as sensor-equipped containers that signal when they are full optimizes collection routes and reduces costs. Setting up decentralized micro-collection centers in collaboration with local recycling groups improves accessibility in resource-limited areas. Policies like Extended Producer Responsibility (EPR) can support these efforts by requiring producers to cover the costs of waste management [68–71].

13.4.5 Recycling Innovations

Recent advances in recycling technologies are helping address plastic waste and environmental challenges. Chemical recycling methods, such as pyrolysis and depolymerization, break plastics down into their original monomers, enabling the production of high-quality recycled materials. Enzymatic recycling uses engineered enzymes to degrade plastics like PET efficiently, with lower energy use and biodegradability. Plastic waste can also be converted into valuable materials like graphene. These technologies allow recycling of plastics that were previously unrecyclable, closing loops in complex waste streams. AI-powered sorting systems improve recovery efficiency, while block chain enables tracking of recycling supply chains. Innovations like upcycling, turning low-value plastics into durable construction materials or textiles, further support a circular economy. Combined with policy support and corporate initiatives, these advances are essential for reducing plastic pollution and achieving sustainability goals [72]. A systematic approach to the new plastics economy is illustrated in Figure 13.8.

13.4.6 Product-as-a-Service Models

Platform-as-a-service models, managed by producers, encourage durable and recyclable products because businesses aim to retain the value of their assets. Services like car-sharing reduce the need for vehicles and associated plastics by promoting more efficient use. A circular economy requires a comprehensive strategy emphasizing sustainability, innovation, and systemic change to reduce plastic waste. Products are designed to be

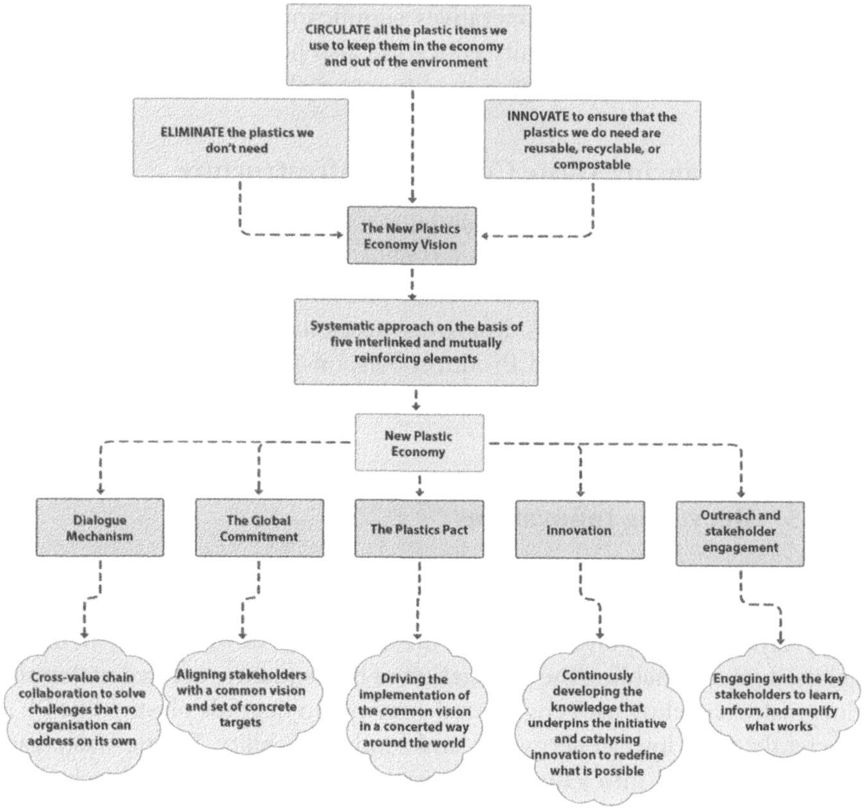

Figure 13.8 Flowchart of plastics economy vision.

eco-friendly, long-lasting, easy to disassemble, and recoverable, minimizing contamination and improving recycling rates [39].

13.5 Technological Innovations and Advancement in Plastic Waste Management

13.5.1 Role of Artificial Intelligence and Digital Tools

Artificial intelligence (AI) has greatly improved plastic waste management. Advanced sorting systems using computer vision and machine learning can accurately identify different polymers, while AI-powered robots in Material Recovery Facilities sort waste faster and more efficiently, reducing contamination. Digital tools, such as mobile apps with image recognition,

Table 13.3 Impacts of AI and robotics in revolutionizing waste management.

Technology	Impacts
Smart sorting systems	AI-powered vision systems can very accurately tell what kind of trash it is (plastic, metal, glass, paper, etc.) and sort it. Then, robotics picks them up and puts them in the right bins, which greatly increases the amount of recycling that happens.
Predictive analytics	AI uses information about trash collection routes, bins, and how often they are used to make collection schedules more efficient, use less fuel, and cut costs.
Robotic arms and drones	Robotic arms do tasks that take a lot of time and effort, like sorting or compacting. Drones keep an eye on landfills and illegal dumping, or they can help plan smart waste infrastructure.
Contamination detection	AI finds contaminated recyclables in real time, which helps keep recycling streams clean and lowers the cost of processing.
Smart bins	IoT-enabled bins with sensors inform collection trucks when the containers are full and AI strategies the best routes for quick pickups.
Circular economy optimization	AI keeps track of how materials move and finds ways to turn waste into raw materials, which helps the circular economy.

Figure 13.9 An Infographic view of AI and robotics in revolutionizing waste management.

guide consumers to recycle correctly. These technologies (Table 13.3 and Figure 13.9) enhance efficiency, reduce errors, and make waste management systems more cost-effective and scalable [73–75].

13.5.2 Digital Solutions and Data Analytics

Data analytics platforms are transforming plastic waste management by tracking waste generation, optimizing collection routes, and forecasting recycling needs. IoT-enabled smart bins monitor fill levels in real time, allowing collection schedules to be adjusted to reduce fuel use and emissions. Cloud-based dashboards consolidate data from cities and businesses to identify inefficiencies and track progress toward circular economy goals. By leveraging Big Data, stakeholders can make informed decisions that improve recycling rates, reduce landfill reliance, and align policies with actual waste patterns [76].

13.5.3 Innovations in Waste-to-Resource Technologies

Innovative waste-to-resource technologies are transforming plastic waste into useful materials. Chemical recycling methods like pyrolysis and depolymerization break plastics into their molecular components, enabling the production of new products. Enzymatic recycling uses engineered microbes to degrade plastics like PET with minimal energy. Examples of circular solutions include using recycled plastic filaments for 3D printing and creating building materials, such as plastic roads. These technologies extend the life of plastics and reduce the need for traditional recycling [77–80].

13.5.4 Blockchain and Traceability for Circular Supply Chains

Blockchain technology enables tamper-proof tracking of a product's lifecycle, from raw materials to disposal, increasing transparency in plastic supply chains. Smart tags and QR codes inform consumers and recyclers about recyclability and track material flows. Initiatives like IBM's Plastic Bank [81, 82] use blockchain to incentivize plastic collection in developing countries through digital tokens. This traceability reduces greenwashing, holds brands accountable for circular economy commitments, and builds stakeholder trust by verifying the recycled content in new products.

13.6 Policy and Governance for Circular Plastic Economy

13.6.1 Regulatory Mechanisms and Extended Producer Responsibility (EPR)

Governments worldwide are promoting a circular plastic economy, with Extended Producer Responsibility (EPR) as a key tool. EPR requires companies to manage the costs and logistics of collecting, recycling, or safely disposing of the plastic waste their products generate. Countries like Japan, India, and EU member states have implemented EPR programs, which push brands to use more recycled materials, improve recyclability, and adopt eco-design. These policies shift waste management costs from taxpayers to producers and encourage innovation, with penalties for non-compliance and rewards for sustainable solutions [41, 83].

13.6.2 International Agreements and Standards

Global cooperation is critical to tackling transboundary plastic pollution. Agreements like the UN Global Plastics Treaty, the EU Circular Economy Action Plan, and ASEAN's regional initiatives harmonize policies, close regulatory gaps, and promote technology transfer to developing countries. Standards such as the Ellen MacArthur Foundation's Circular Economy Protocols and ISO 18604 (packaging recyclability) provide guidance for businesses and governments to align practices globally.

13.6.3 Financial Incentives and Subsidies for Circular Economy Practices

Financial mechanisms—including tax breaks, grants, and low-interest loans—support businesses in building recycling infrastructure or adopting bio-based alternatives. By aligning fiscal policies with circular economy goals, governments can accelerate the transition, stabilize the economy, create green jobs, and drive systemic change [84–86].

13.7 Role of Stakeholders in Driving Circularity

13.7.1 Government Policies and Leadership

National and local governments drive the circular plastic economy through green public procurement, investment in waste management infrastructure, and funding research on new recycling technologies. By aligning economic incentives with sustainability goals and fostering multi-stakeholder partnerships, governments ensure fair progress across sectors and communities.

13.7.2 Corporate Initiatives and Industry Collaboration

Many businesses are adopting circular economy principles by redesigning products, minimizing waste, and implementing closed-loop supply chains. Companies invest in eco-friendly packaging, take-back programs, and recycled materials. Initiatives like the New Plastics Economy project demonstrate industry collaboration to standardize recyclable packaging and share best practices. Circular business models, such as product-as-a-service and eco-design, reduce environmental impact, open new revenue streams, and enhance brand appeal to eco-conscious consumers.

13.7.3 Consumer Awareness and Behavior Change

Consumers play a crucial role in the circular economy. Awareness of plastic pollution and access to recycling options encourage eco-friendly purchasing and proper waste sorting. Educational campaigns, clear labeling, and incentive programs like deposit-return schemes guide sustainable behavior. As more consumers favor sustainable brands, market pressure increases, prompting businesses and governments to prioritize circular solutions, linking production, consumption, and waste management.

13.7.4 Civil Society and NGO Contributions

NGOs and civil society groups promote circularity through advocacy, education, and practical projects. Environmental NGOs raise awareness, push for stricter regulations, and hold businesses accountable. Community groups manage local recycling programs, organize cleanups, and run upcycling workshops. Research institutions provide data and policy guidance, while social enterprises develop innovative solutions to plastic waste.

Together, these organizations drive bottom-up change that complements government and business efforts.

13.8 Barriers to Implementation: Transitioning to a Circular Economy-Based Plastic Waste

13.8.1 Technical Challenges and Solutions

Transitioning to a circular plastic economy faces technical challenges such as limited recycling infrastructure, difficulty in sorting mixed or layered plastics, and low-quality recycled materials. Many facilities cannot process contaminated plastics, and chemical recycling remains energy-intensive and costly at scale. Emerging solutions include AI-powered sorting, advanced pyrolysis, enzymatic polymer degradation, modular recycling units, and standardized material designs. Public–private partnerships and robust testing and certification systems can build trust in recycled plastics and accelerate innovation [87].

13.8.2 Economic and Market Barriers

Recycling often remains less cost-effective than producing virgin plastics due to fossil fuel subsidies and unaccounted environmental costs. Market instability and lack of guaranteed demand make investors hesitant, while small circular businesses struggle to compete with large linear companies. Solutions include full-cost accounting, guaranteed off take agreements, procurement policies favoring recycled materials, and financial tools like green bonds, recycling credits, and EPR fees. Harmonized global standards and material markets can further boost demand for circular plastics.

13.8.3 Behavioral and Social Resistance

Consumer and industry habits such as reliance on single-use products and linear production models impede the adoption of circular practices. Overcoming resistance requires educational campaigns, clear recycling instructions, and visible leadership from influencers and community leaders. Incentives, social norm marketing, deposit–return schemes, and single-use plastic bans can encourage behavioral change. Circular solutions must also be convenient, affordable, and as effective as traditional options to achieve widespread adoption.

13.9 Case Studies, Best Practices, and Comparative Analysis

13.9.1 Successful Circular Economy Models in Plastic Management

Circular economy models for plastic management aim to create a closed-loop system by reducing waste, reusing materials, and recycling plastics. Examples include deposit-return schemes in Germany and Norway, where small deposits on plastic bottles encourage returns, achieving recycling rates above 90% [88]. Extended Producer Responsibility (EPR) systems in Japan and South Korea make manufacturers responsible for the full life cycle of their products, promoting eco-friendly design and advanced recycling technologies [89, 90]. These models reduce plastic pollution, conserve resources, and create new business opportunities, supporting a more sustainable future. Successful projects such as the Netherlands' "Plastic Road," which uses recycled plastic for modular road parts, and Terra Cycle's Loop platform, offering reusable packaging for everyday items, demonstrate that large-scale circular systems are feasible. By combining innovative materials, business models, and consumer engagement, these approaches establish self-sustaining circular plastic ecosystems [91–96] (Table 13.4).

13.9.2 Regional and Industry-Specific Examples

Regional strategies highlight the importance of context-specific solutions for plastic waste. Japan's PET bottle recycling system achieves an 85% recovery rate through strict sorting, deposit schemes, and retailer participation. Norway's "Fishing for Litter" program engages fishermen in collecting ocean plastics, while Adidas partners with Parley for Oceans to convert ocean waste into sportswear. In construction, Kenya's plastic bottle schools and India's plastic roads transform waste into building materials. These examples demonstrate that circular solutions can be adapted to local waste streams, industrial capacities, and cultural practices, effectively addressing specific environmental challenges [97, 98].

Table 13.4 List of successful circular economy models in plastic management with examples.

Country	Model/ initiative	Key features	Impact/ outcome	Real-time example
Germany	Dual system (Grüner Punkt)	- Producers pay fees for plastic waste collection and recycling. - Strict EPR (Extended Producer Responsibility) laws	65% plastic packaging recycled (2023); high-quality recycled plastics market	Green Dot logo on packaging ensures compliance; used by brands like Henkel, ALDI
Japan	PET bottle-to-bottle recycling	- Advanced cleaning and sorting tech for food-grade recycled PET. - Consumer deposit system	85% PET bottle collection rate (2022); 60% reused in new bottles	Suntory uses 100% recycled PET for bottled water brands
Netherlands	PlasticRoad	- Roads made from recycled plastics (modular, lightweight, durable) - Public–private partnerships	50+ plastic roads built (2024); reduces carbon footprint vs. asphalt	Zwolle city installed the world's first PlasticRoad in 2018
India	Project REPLAN (by Khadi and Village Industries Commission)	- Mixes recycled plastic with cotton to make textiles - Empowers rural women	25+ tons of plastic waste upcycled annually; eco-friendly bags/ cloths sold globally	Fabindia and TerraCycle collaborate to scale production

(Continued)

Table 13.4 List of successful circular economy models in plastic management with examples. (*Continued*)

Country	Model/ initiative	Key features	Impact/ outcome	Real-time example
Sweden	TMR (Swedish Plastic Recycling)	- Chemical recycling plants convert non-recyclable plastics to fuel/ oil - Zero-waste cities initiative	99% household waste recycled; fuels public transport	Gothenburg's waste-to-energy buses run on recycled plastic fuel
Rwanda	Ban on Single-Use Plastics + Circular Hubs	- Strict plastic ban since 2008 - Local hubs collect and recycle plastics into construction materials	90% reduction in plastic waste; recycled bricks used in schools.	ECOPlast turns plastic waste into affordable roofing tiles
USA	Loop by TerraCycle	- Reusable packaging system for brands (e.g., Unilever, P&G) - Consumers return containers for refilling	100+ brands participate; 75% less plastic waste vs. single-use	Loop's reusable Haagen-Dazs containers sold in Walmart/ Kroger
Norway	**Deposit Return Scheme (Infinitum)**	- Consumers pay a deposit for bottles, refunded upon return - 97% recycling rate for bottles	1.2B bottles recycled annually; near-zero ocean leakage	Infinitum system recovers 95% of PET bottles within 6 months

13.9.3 Examples of Circular Economy in Action

Corporate and municipal initiatives are actively putting circularity into practice. Unilever's "Less, Better, No Plastic" program has eliminated 100,000 tons of packaging by using lighter and alternative materials. San Francisco's composting and recycling regulations retain 80% of waste from landfills, illustrating effective municipal circularity. Technological innovations like Carbios' enzymatic recycling break down PET plastics at the molecular level for repeated use, while startups like Notpla create edible seaweed-based packaging. These examples demonstrate that circular economy solutions can work at multiple scales, from molecular-level innovations to city-wide systems, showing that waste-free models are achievable with current technologies [99].

13.9.4 Challenges in Scaling Up Technologies

Scaling circular solutions remains difficult despite promising pilot projects. Chemical recycling adoption is slow due to high facility costs (over $200 million) and energy demands. Bioplastics face performance limitations and a lack of composting infrastructure. Many developing countries also lack basic collection systems, making advanced recycling impractical. This creates a paradox: small-scale solutions cannot reach economies of scale without investment, yet investors wait for proof of scalability. Effective scaling requires coordinated development of infrastructure, policies, and market mechanisms. The EU strategy exemplifies this approach, combining R&D funding with mandatory recycled content targets to ensure consistent demand.

13.9.5 Lessons Learned from Global Initiatives

Global experiences highlight key insights for circular plastic systems. In the US, recycling rates have stagnated, showing that voluntary measures alone are insufficient. In contrast, the EU's Single-Use Plastics Directive and Extended Producer Responsibility policies demonstrate the impact of strong regulatory frameworks. Rwanda shows that strict bans succeed with public compliance, and Scandinavian deposit–return systems highlight the effectiveness of economic incentives. Key lessons include tailoring solutions to local contexts, providing policy certainty to attract investment, emphasizing consumer education alongside infrastructure development, and fostering cross-value-chain collaboration. Successful initiatives integrate circularity into broader economic strategies, treating waste as a source of jobs, innovation, and environmental solutions.

13.10 Future Directions for Plastic Waste Management

The future of plastic waste management relies on a combination of advanced technologies, policy reforms, and circular economy strategies. Chemical and enzymatic recycling can recover most plastic materials, while biodegradable and compostable plastics offer renewable alternatives that degrade under specific conditions. AI and machine learning enhance sorting, processing, and traceability, improving efficiency across supply chains. Stronger international policies, including stricter single-use plastic regulations and expanded Extended Producer Responsibility (EPR) programs, shift accountability to manufacturers and encourage eco-design. Circular business models, such as product-as-a-service and refill systems, reduce dependency on single-use plastics and promote reuse. Coordinated efforts

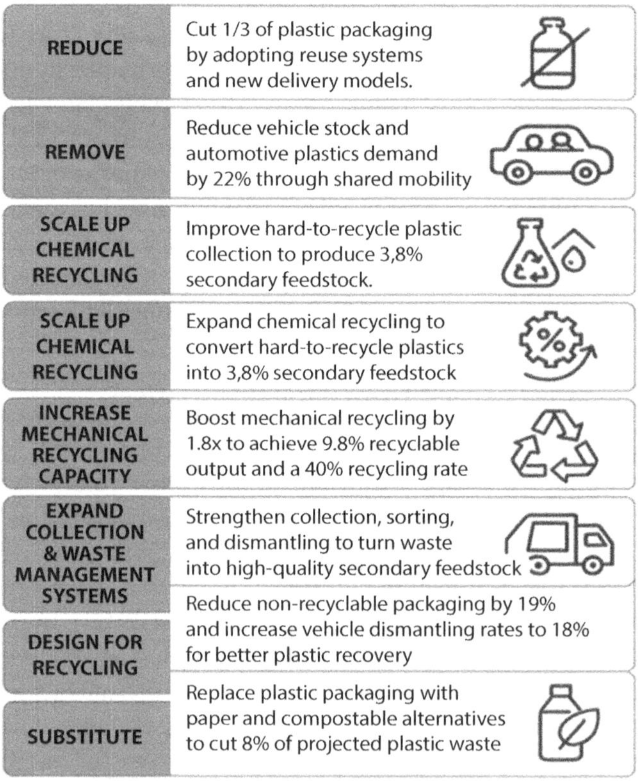

Figure 13.10 List of steps to achieve plastic waste reduction and circular economy goals by 2050.

among governments, businesses, and consumers, along with investments in bio-based materials and recycling infrastructure, are essential. By combining technological, regulatory, and behavioral strategies, a closed-loop plastic economy can transform waste into valuable resources while safeguarding the environment and supporting sustainable economic growth. Key steps toward achieving plastic waste reduction and circular economy targets by 2050 are summarized in Figure 13.10 [100–102].

13.11 Conclusion and Recommendations

Transitioning to a circular plastic economy is both an urgent environmental necessity and a major economic opportunity. Success depends on combining advanced recycling technologies, eco-design, Extended Producer Responsibility (EPR) schemes, and active consumer participation, supported by strong collaboration among stakeholders. Regional case studies show that large-scale circularity is achievable, but challenges such as market fragmentation, inadequate infrastructure, and resistance to behavioral change must be addressed. A circular economy not only reduces plastic pollution but also creates new business models, jobs, and resource security, demonstrating that sustainability and economic growth can go hand in hand. To accelerate this transition, policies should enforce strong EPR laws, promote research and development of circular materials, and establish global recyclability standards. Governments should set mandatory recycled content targets and support businesses investing in circular solutions, while companies should prioritize design-for-recycling, closed-loop supply chains, and transparency tools like blockchain. Policymakers, businesses, researchers, and consumers must work together to adopt circular products and reduce waste. The ultimate goal is a plastic economy where materials are continually reused rather than discarded, ensuring a healthier planet and a thriving green economy through visionary leadership, innovation, and collective commitment.

AI disclosure statement: The author confirms that no artificial intelligence (AI)-generated content or AI-assisted writing tools were used in the preparation of this chapter (including summaries and the abstract). All text, analysis, and conclusions are the result of the author's original work.

References

1. Li, X., *et al.*, Production Potential of Greenhouse Gases Affected by Microplastics at Freshwater and Saltwater Ecosystems. *Atmosphere*, 13, 1796, 2022.
2. Hilal, N., Widiyanto, T., Mulyasari, T., Lagiono, L., The Effect of Ultraviolet Exposure on the Quantity of Microplastics in the Air of Buildings Made from Plastic Waste. *Gema Lingkungan Kesehatan*, 22, 103, 2024.
3. Samadikun, B.P., Ramadan, B.S., Dewi, C.T., Rezagama, A., Andarani, P., Greenhouse Gas Estimation of Plastic Waste Reverse Logistic Networks System in Semarang City. *IOP Conf. Ser.: Earth Environ. Sci*, 448, 012030, Mar. 2020.
4. Miditana, S.R., Tirukkovalluri, S.R., Raju, I.M., Alim, S.A., Photocatalytic degradation of Orange-II by surfactant assisted Mn/Mg co-doped TiO_2 nanoparticles under visible light irradiation. *Curr. Chem. Lett.*, 13, 265–276, 2024, https://doi.org/10.5267/j.ccl.2023.6.003.
5. Mandych, O., Babko, N., Pasemko, G., Girzheva, O., Taran, O., Circular economy model: generating benefits for sustainable development. *Econ. Anal.*, 33, 1, 252, 2023.
6. Roleders, V.V., Circular economy model in industrial ecosystems. *Actual Probl. Reg. Econ. Dev.*, 1, 19, 243, 2023.
7. Ha, N.T.T., Factors Affecting the Transition to a Circular Economy Model in Banana Cultivation. *Trans. Chin. Soc. Agric. Mach.*, 55, 11, 2024.
8. Sysoieva, I., Derun, T., Oriekhova, T., Roleders, V., Mazur, V., Global experience of applying the circular economy model in light industry. *Manag. Theory Stud. Rural Bus. Infrastruct. Dev.*, 45, 1, 46, 2023.
9. Goel, A. and Singh, A.K., Measurement and Structural Assessment of Circular Economy Model. *J. Bus. Manag. Inf. Syst.*, 9, 1, 20, 2022.
10. Rinanda, R., *et al.*, Plastic Waste Management: A Bibliometric Analysis (1992–2022). *Sustainability*, 15, 24, 16840, 2023.
11. Villegas Pilay, Y.E., Cruz Ronquillo, S.I., Bravo Carrasco, A.R., Circular economy model for reuse of plastic in eco-efficient production of building materials. *J. Bus. Entrep. Stud.*, 7, 4, 9, 2023.
12. Ly, B., Competitive advantage and internationalization of a circular economy model in apparel multinationals. *Cogent Bus. Manag.*, 8, 1, 23311975, 2021.
13. Todorova, E., Yunussova, G., Formazyuk, X., Kostadinova-Slaveva, A., Kazkenova, G., Jetpysbai, G., Advancing Plastic Waste Management for a Circular Economy: Comparative Insights from Plovdiv (Bulgaria) and Kostanay (Kazakhstan). *Processes*, 13, 888, 2025, https://doi.org/10.3390/pr13030888.
14. Wamba, S.F., Mosconi, E., Fotso, M., Chai, J., Assessing the potential of plastic waste management in the circular economy: a longitudinal case study in an emerging economy. *Ann. Oper. Res.*, 8, 7, 2023.

15. Rapati, C., Plastic Waste Management to Support The Circular Economy in The Pulp and Paper Industry. *Business Review and Case Studies,* Graduate Program of Management and Business, Bogor Agricultural University, doi: 10.17358/BRCS.4.1.1.
16. Thanh, N.T., *et al.*, Assessment and Sustainable Management Strategies for Plastic Waste in Can Tho City, Vietnam: A Circular Economy Approach. *Water,* 16, 7, 951, 2024.
17. Bernardo, C., Environmental and economic life cycle analysis of plastic waste management options. A review. *AIP Conference Proceedings,* 2016, doi: 10.1063/1.4965581.
18. Zhao, C., Du, H., Liu, M., Gong, Y., The Evolutionary Trend and Impact of Global Plastic Waste Trade Network. *Sustainability,* 13, 7, 3662, 2021.
19. Shahnawaz, M., Sangale, M., Ade, A., Plastic Waste Disposal and Reuse of Plastic Waste, 2019, 10.1007/978-981-13-7492-0_3.
20. Ararat H, J.A., Restrepo-Morales, J.A., Jones Rafael, L.-A., Mastering the Circular Economy: Integrating Sustainable Business Practices and Advanced Waste Management Solutions, in: *Global Conference on Business and Social Sciences Proceeding,* vol. 16, 1, pp. 79, 2024.
21. Kwemoi, K.C., Engineering Sustainable Solutions for Waste Management. *Res. Invent. J. Eng. Phys. Sci.*, 4, 1, 36, 2025.
22. Singh, S. and Srivastava, G., Cleaner cities through smart waste solutions: A sustainable approach to urban waste management. *Int. J. Sci. Res. Arch.*, 15, 3, 325, 2025.
23. Kantola, J. and Abila, B., Proposed solutions in municipal solid-waste management. *Int. J. Environ. Waste Manage.*, 19, 4, 297, 2017.
24. Kocić, D., Tot, N., Beljin, M., *Development of sustainable waste management systems using advanced sensor technologies,* vol. 1, p. 257, 2024.
25. Balu, R., Dutta, N.K., Roy Choudhury, N., Plastic Waste Upcycling: A Sustainable Solution for Waste Management, Product Development, and Circular Economy. *Polymers,* 14, 22, 4788, 2022.
26. Mhamane, S.V., Shinde, R.D., Khemalapure, A.B., Review Paper on Sustainable Pothole Repair Using Plastic Waste in Bituminous Mixes. *Int. J. Res. Appl. Sci. Eng. Technol.*, 13, 4, 4982, 2025.
27. Velenturf, A.P.M. and Purnell, P., Principles for a sustainable circular economy. *Sustainable Prod. Consumption,* 27, 1437, 2021.
28. Haase, M., Bernegger, H., Meslec, M., Models of Circular Economy Principles, in: *4th International Conference "Coordinating Engineering for Sustainability and Resilience" & Midterm Conference of Circular B "Implementation of Circular Economy in the Built Environment"*, Ungureanu, V., Bragança, L., Baniotopoulos, C., Abdalla, K.M. (eds.), CESARE 2024. Lecture Notes in Civil Engineering, vol. 489, Springer, Cham, 2024, https://doi.org/10.1007/978-3-031-57800-7_42.

29. Gopalakrishna Pillai, S., Arasli, F., Haldorai, K., Rahman, I., Unlocking sustainable performance through circular economy principles. *J. Hosp. Tour. Insights*, 8, 5, 0680, 2024.
30. Prasetyo, Y.E., *et al.*, Plastic Waste Management: Opportunities and Challenges in Implementing Circular Economy in Gorontalo, in: *Proceeding of the International Conference on Multidisciplinary Research for Sustainable Innovation*, vol. 1, p. 65, 2024.
31. Raj, M., Ghosh, A., Ilame, T., Kumar, S., Plastic waste management strategies: planning through sustainable lens and way forward towards circular economy. *Discover Sustainability*, 6, 1, 01103, 2025.
32. Anjali, A., Reforming Plastic Waste Management: A Legal Perspective on Extended Producer Responsibility and Circular Economy Policies. *Legal Res. Dev.*, 9, 3, 09, 2025.
33. Ashatkar, A., Revolutionizing Plastic Waste Management in the Packaging Industry: An In-depth Exploration of AI-Driven Circular Economy Strategies. *Int. J. Res. Appl. Sci. Eng. Technol.*, 12, 3, 370, 2024.
34. Balamuragan, R., Rani, N.B.S., Reddy, M.V.N.S.S., Karasu, T., C, K., A Report on Plastic Waste Problem and Management. *Int. J. Innov. Res. Eng. Manag.*, 8, 1, 20, 2021.
35. Almadhi, A., Abdelhadi, A., Alyamani, R., Moving from Linear to Circular Economy in Saudi Arabia: Life-Cycle Assessment on Plastic Waste Management. *Sustainability*, 15, 13, 10450, 2023.
36. Hossain, R., *et al.*, Plastic Waste Management in India: Challenges, Opportunities, and Roadmap for Circular Economy. *Sustainability*, 14, 8, 4425, 2022.
37. Kiaušienė, I., Hladkova, V., Makūnaitė, G., Application of circular economy principles in the tourism sector. *Manag. Theory Stud. Rural Bus. Infrastruct. Dev.*, 46, 1, 31, 2024.
38. Manas, D., Recycle Plastic Waste Brick. *Int. J. Res. Appl. Sci. Eng. Technol.*, 10, 1, 1271, 2022.
39. Chow, C.-F., So, W., Cheung, Y.T.-Y., Yeung, S.K., Plastic Waste Problem and Education for Plastic Waste Management, 2017, 10.1007/978-981-10-3344-5_8.
40. Patil, S. and Professor, A., Integration of circular economy principles in green construction practices. *Eur. Chem. Bull.*, 2, 0366, 2023.
41. Helm, L.T., Mcgivern, A., Murphy, E.L., Borrelle, S.B., Impacts of plastic waste management strategies. *Environ. Rev.*, 31, 1, 45, 2022.
42. Yeboaa, C., Tetteh, E.K., Chollom, M.N., Rathilal, S., Sustainable Solutions for Plastic Waste Mitigation in Sub-Saharan Africa: Challenges and Future Perspectives Review. *Polymers*, 17, 11, 2025.
43. Nandabalan, Y. and Singh, S., *Plastic Waste: Environmental Hazards, Its Biodegradation, and Challenges*, 2020, 10.1007/978-981-13-1891-7_6.

44. Adekanmbi, A., Obaigbena, A., Abatan, A., Izuka, U., Ninduwezuor-Ehiobu, N., Ani, E., Assessing the environmental and health impacts of plastic production and recycling. *World J. Biol. Pharm. Health Sci.*, 17, 2, 232, 2024.
45. Bilal, M., Environmental and Health Impacts of Solid Waste Management Amidst Climate Change. *Knowl.-Based Eng. Sci.*, 6, 1, 1–19, 2025.
46. Adanu, K., The growing global plastic waste problem - lessons for environmental economics policy design and choice. *Int. J. Green Econ.*, 14, 2, 121, 2020.
47. Miditana, S.R., Dokka, W.T., Babu, A.R., Ampolu, S., Agarwal, N., Simma, N., Cyclodextrins for UV protection in food, beverages, and agricultural products: A comprehensive review. *North Afr. J. Food Nutr. Res.*, 9, 20, 219–235, 2025.
48. Brooks, A.L., Wang, S., Jambeck, J.R., The Chinese import ban and its impact on global plastic waste trade. *Sci. Adv.*, 4, 6, 1126, 2018.
49. Harmi Tjahjanti, P., Bamban Jakaria, R., Fahruddin, A., Akbar, A., Fernanda, R.E., Febriyan Ikhsanudin, A., Investigation Study of Semi-Bio Briquettes and Synthesis Briquettes From Used Polymer Waste Plastic Bottles. *J. Improsci*, 2, 5, 279, 2025.
50. Selvaraj, S., *et al.*, COVID-19 Biomedical Plastics Wastes—Challenges and Strategies for Curbing the Environmental Disaster. *Sustainability*, 14, 11, 6466, 2022.
51. Maragkaki, A., Sampathianakis, I., Katrini, K., Michalodimitraki, E., Gryparis, C., Raptis, V., Power, A., Lolos, T., Tsobanidis, C., Harmandaris, V., Velonia, K., Manios, T., Bio-waste to Bio-plastic (B2B): Production of Compostable Bio-Plastics from Food Waste. *Proceedings*, 30, 1, 47, 2019, https://doi.org/10.3390/proceedings2019030047.
52. Kanimozhi, B., Kumar, A., Kumar, A., Shinde, A.T., Formation of Bio-Fuel from Waste Plastic Scrap. *Appl. Mech. Mater.*, 766, 551, 2015.
53. Stasiškienė, Ž., *et al.*, Challenges and Strategies for Bio-Based and Biodegradable Plastic Waste Management in Europe. *Sustainability*, 14, 24, 16476, 2022.
54. Shakuto, S., Tan, Q.H., Reynolds, D., Rahadini, I.A., Yeoh, B.S.A., Pang, N., Household Plastic Waste Management and Gender Dynamics in Circular Economies. *Sociol. Compass*, 18, 12, 70023, 2024.
55. Talbott, T.C., Extended Producer Responsibility: opportunities and challenges for waste pickers, Edward Elgar, p. 13, 2022.
56. Patra, B.C. and Mishra, D.N., *Extended Producer Responsibility (EPR) and E-Waste Management*, vol. 10, p. 136, IGI Global, United States of America, 2024.
57. Milios, L., Yu, Y., Esmailzadeh Davani, A., Sustainability Impact Assessment of Increased Plastic Recycling and Future Pathways of Plastic Waste Management in Sweden. *Recycling*, 3, 3, 33, 2018.
58. Lu, Y.-Q., Liu, P.-L., Fu, Q.-R., Ding, X.-M., *Plastic Product Evaluation Based on Mold Conceptual Design*, Jan. 10, p. 80733, 2005.

59. Shah, S., Matkawala, F., Garg, S., Nighojkar, S., Nighojkar, A., Kumar, A., Emerging Trend of Bio-plastics and Its Impact on Society. *Biotechnol. J. Int.*, 1, 10, 2020.
60. Liwarska-Bizukojc, E., Effect of (bio) plastics on soil environment: A review. *Sci. Total Environ.*, 795, 148889, 2021.
61. Kumar, S.N., Jain, R., Anand, K., Haridhas, A.K., Utilization of Agro Waste for the Fabrication of Bio Composites and Bio plastics—Towards a Sustainable Green Circular Economy, 2022, 10.1007/978-3-030-75235-4_7.
62. Feil, A., Jansen, M., Pretz, T., Thoden Van Velzen, E.U., Separate collection of plastic waste, better than technical sorting from municipal solid waste? *Waste Manag. Res.: the Journal for a Sustainable Circular Economy*, 35, 2, 172, 2016.
63. Agarwal, N., Solanki, V., *et al.*, Bacterial Laccases as Biocatalysts for the Remediation of Environmental Toxic Pollutants: A Green and Eco-Friendly Approach—A Review. *Water*, 14, 4068, 2022, 10.3390/w14244068.
64. Bernat, K., Post-Consumer Plastic Waste Management: From Collection and Sortation to Mechanical Recycling. *Energies*, 16, 8, 3504, 2023.
65. Agarwal, N., Srivastava, A., Solanki, V., Singh, N., Emerging trends in the use of microbial isolates for the treatment of dye containing wastewater, 2024, 10.1016/B978-0-443-13561-3.00010-7.
66. Bernardo, M., Forehead, H., De Carvalho Vallin, I., Gonçalves-Dias, S.L.F., Management of Household Plastic Waste in Wollongong, Australia: The Role of Selective Waste Collection Systems. *Sustainability*, 15, 2, 1726, 2023.
67. Procházka, R., *et al.*, Collection of Plastic Packaging of Various Types: Sorting of Fractions of Plastic Waste Using Both Automated and Manual Modes. *IEEE Access*, 12, 44244, 2024.
68. Aikowe, L.D. and Mazancová, J., Plastic Waste Sorting Intentions among University Students. *Sustainability*, 13, 14, 7526, 2021.
69. Junjuri, R. and Gundawar, M.K., Low-Cost Sorting of Plastic Waste. *Opt. Photonics News*, 31, 12, 61, 2020.
70. Aiguobarueghian, I., Ogunbiyi, E., Adanma, U., Solomon, N., Reviewing the effectiveness of plastic waste management in the USA. *World J. Adv. Res. Rev.*, 22, 2, 1720, 2024.
71. Sörme, L., Jensen, S., Augustsson, A., Voxberg, E., Rosenlund, J., Coloured Plastic Bags for Kerbside Collection of Waste from Households—To Improve Waste Recycling. *Recycling*, 4, 2, 20, 2019.
72. Mojaver, P. and Khalilarya, S., Air Plastic Waste Gasification, 2023. 10.1007/978-3-031-31160-4_4.
73. Bobulski, J. and Kubanek, M., Deep Learning for Plastic Waste Classification System. *Appl. Comput. Intell. Soft Comput.*, 2021, 1–7, 2021.
74. Erkinay Ozdemir, M., Ali, Z., Subeshan, B., Asmatulu, E., Applying machine learning approach in recycling. *J. Mater. Cycles Waste Manage.*, 23, 3, 855, 2021.

75. Khaled, A.S.D., Sharma, D., Yashwanth, T., Reddy, V., Doewes, R., Naved, M., Evaluating the Role of Robotics, Machine Learning and Artificial Intelligence in the Field of Performance Management, 2023, 10.1007/978-981-19-0108-9_30.
76. Lubongo, C., Alexandridis, P., Bin Daej, M.A.A., Recent Developments in Technology for Sorting Plastic for Recycling: The Emergence of Artificial Intelligence and the Rise of the Robots. *Recycling*, 9, 4, 59, 2024.
77. Beigizadeh, R., Hanjani, S.E., Delgoshaei, A., Ali, A., Ariffin, M.K.A., A comprehensive model for determining technological innovation level in supply chains using green investment, eco-friendly design and customer collaborations factors. *Rairo-oper. Res.*, 56, 4, 2775, 2022.
78. Fayshal, M.A., Current practices of plastic waste management, environmental impacts, and potential alternatives for reducing pollution and improving management. *Heliyon*, 10, 23, 40838, 2024.
79. Wichai-Utcha, N. and Chavalparit, O., 3Rs Policy and plastic waste management in Thailand. *J. Mater. Cycles Waste Manage.*, 21, 10, 2018.
80. Apevienyeku, H.K., Ampofo, J., Amankwaa Amoako, J., Recycling potential of medical plastic waste. *Discov. Environ.*, 2, 1, 2024.
81. Vatin, N.I., Mohan, C., Singla, N., Yellanki, V.S., Singh Negi, G., Sustainability Measures: An Experimental Analysis of AI and Big Data Insights in Industry 5.0. *BIO Web Conf.*, 86, 01072, 2024.
82. Nwokediegwu, Z. and Ugwuanyi, E., Implementing ai-driven waste management systems in underserved communities in the usa. *Eng. Sci. Technol. J.*, 5, 3, 794, 2024.
83. Castro-Amoedo, R., Granacher, J., Kantor, I., Dahmen, A., Barbosa-Povoa, A., Maréchal, F., On the role of system integration in plastic waste management. *Resour. Conserv. Recycl.*, 201, 107295, 2023.
84. Al Nuwairan, M., Sabir, Z., Aldhafeeri, A., Asif Zahoor Raja, M., An advance artificial neural network scheme to examine the waste plastic management in the ocean. *AIP Adv.*, 12, 4, 045211, 2022.
85. Chea, J.D., Ruiz-Mercado, G.J., Benavides, P.T., Conway, M., Yenkie, K.M., Designing Better Plastic Management Processes Through a Systems Approach. 3, 691–697, 2024.
86. El-Rayes, N., Chang, A., Shi, J., Plastic Management and Sustainability: A Data-Driven Study. *Sustainability*, 15, 7181, 2023, https://doi.org/10.3390/su15097181.
87. Khan, R., Mahmood, H., Siddiqi, M., Asghar, M., Framework for Plastic Waste Management. *Int. J. Circ. Econ. Waste Manag.*, 2, 1, 1–21, Jun. 2022, doi: 10.4018/ijcewm.302204.
88. Narayan, S., Jha, A., Choudhary, R., Misra, S., Ranjan, A., Himanshu, Choudhury, M., Suman, P., Raj, C., Birje, T., Unveiling the Diversity of Waste: A Deep Dive into Sources and Types of Generated Waste, from Municipal Solid Waste to Electronic Waste, 2026, 10.1007/978-3-032-05302-2_2.

89. Agustina Hidayat, Y., Arya Zamal, M., Kiranamahsa, S., A study of plastic waste management effectiveness in Indonesia industries. *AIMS Energy*, 7, 3, 350, 2019.
90. Wen, Z., Chen, M., Dinga, C.D., Xie, Y., China 2019s plastic import ban increases prospects of environmental impact mitigation of plastic waste trade flow worldwide. *Nat. Commun.*, 12, 1, 2021.
91. Putri, A.R., Takaoka, M., Fujimori, T., Plastic waste management in Jakarta, Indonesia: evaluation of material flow and recycling scheme. *J. Mater. Cycles Waste Manage.*, 20, 4, 2140, 2018.
92. Judijanto, L. and Firdayetti, F., Bibliometric Analysis of Research Dynamics of Circular Economy in Plastic Management. *West Sci. Soc. Humanit. Stud.*, 3, 161, 2025.
93. Imawati, Y., Ganis Sukoharsono, E., Fefta Wijaya, A., K., Toward the Future: Circular Economy Models for Sustainable Waste Management. *Trans. Chin. Soc. Agric. Mach.*, 55, 4, 2024.
94. Dhawan, S.K., Waste Plastic Management – A Step towards Circular Economy. *Vantage: J. Themat. Anal.*, 4, 2, 21, 2023.
95. Dang, V.H., Gam, P.T., Xuan Son, N.T., Vietnam's Regulations to Prevent Pollution from Plastic Waste: A Review Based on the Circular Economy Approach. *J. Environ. Law*, 33, 137, 2020.
96. Ünlü, M.G., Midor, K., Dinç, V., Płaza, G., Plastic waste management in the smart cities: Poland and Turkey as a case study. *Sci. Pap. Silesian Univ. Technol. Organ. Manag. Ser.*, 166, 829, 2023.
97. Gill, Y.Q., Khurshid, M., Abid, U., Ijaz, M.W., Review of hospital plastic waste management strategies for Pakistan. *Environ. Sci. Pollut. Res. Int.*, 29, 7, 9408–9421, Dec. 2021.
98. Hishan, S., Changing Landscape of Plastic Waste Management in India, 2023, 10.1108/S2043-052320230000019007.
99. Zahrah, Y., Liu, X., Yu, J., How Indonesia's Cities Are Grappling with Plastic Waste: An Integrated Approach towards Sustainable Plastic Waste Management. *Sustainability*, 16, 10, 3921, 2024.
100. Husen, A., Seta, A.K., Rizkyprima, A., Safitri, D., Evaluation of Plastic Waste Management in Rawajati Waste Storage Depots, South Jakarta. *Edumaspul: J. Pendidik.*, 7, 2, 5506, 2023.
101. Ige, R.O., Oladiji, F., Abodunrin, O.L., Ilori, O.R., Community-Based Approach for Promoting Sustainable Plastic Waste Management in Oyo State. *LAUTECH J. Civ. Environ. Stud.*, 13, 0160, 2024.
102. Dermol, Š. and Dermol, V., Exploring Online Education for Circular Economy: Content Analysis and Didactical Approaches in Plastic Waste Management Courses. *Int. J. Manag. Knowl. Learn.*, 13, 209, 2024.

Index

Wiley The manufacturer's authorized representative according to the EU General Product Safety Regulation is Wiley-VCH GmbH, Boschstr. 12, 69469 Weinheim, Germany, e-mail: Product_Safety@wiley.com.

Printed and bound by CPI Group (UK) Ltd, Croydon, CR0 4YY

27/10/2026

14991178-0003